# Modelling and Simulation

SIXTH BATH INTERNATIONAL FLUID POWER WORKSHOP

**FLUID POWER SERIES**

*Series Editor:* **Professor C. R. Burrows**
*University of Bath, England*

1. Design, Modelling and Control of Pumps
   FIRST BATH INTERNATIONAL FLUID POWER WORKSHOP
   *Edited by* **C. R. Burrows** *and* **N. D. Vaughan**

2. Fluid Power Components and Systems
   SECOND BATH INTERNATIONAL FLUID POWER WORKSHOP
   *Edited by* **C. R. Burrows** *and* **K. A. Edge**

3. Computers in Fluid Power
   THIRD BATH INTERNATIONAL FLUID POWER WORKSHOP
   *Edited by* **C. R. Burrows** *and* **K. A. Edge**

4. Fluid Power Systems, Modelling and Control
   FOURTH BATH INTERNATIONAL FLUID POWER WORKSHOP
   *Edited by* **C. R. Burrows** *and* **K. A. Edge**

5. Circuit, Component and System Design
   FIFTH BATH INTERNATIONAL FLUID POWER WORKSHOP
   *Edited by* **C. R. Burrows** *and* **K. A. Edge**

6. Modelling and Simulation
   SIXTH BATH INTERNATIONAL FLUID POWER WORKSHOP
   *Edited by* **C. R. Burrows** *and* **K. A. Edge**

# Modelling and Simulation

SIXTH BATH INTERNATIONAL
FLUID POWER WORKSHOP

Held at the
University of Bath, England
23rd-24th September 1993

*Edited by* **C. R. Burrows** *and*
**K. A. Edge**
*School of Mechanical Engineering*
*University of Bath, England*

**RESEARCH STUDIES PRESS LTD.**
Taunton, Somerset, England

**JOHN WILEY & SONS INC.**
New York · Chichester · Toronto · Brisbane · Singapore

RESEARCH STUDIES PRESS LTD.
24 Belvedere Road, Taunton, Somerset, England TA1 1HD



**Marketing and Distribution:**

*Australia and New Zealand:*
Jacaranda Wiley Ltd.
GPO Box 859, Brisbane, Queensland 4001, Australia

*Canada:*
JOHN WILEY & SONS CANADA LIMITED
22 Worcester Road, Rexdale, Ontario, Canada

*Europe, Africa, Middle East and Japan:*
JOHN WILEY & SONS LIMITED
Baffins Lane, Chichester, West Sussex, England

*North and South America:*
JOHN WILEY & SONS INC.
605 Third Avenue, New York, NY 10158, USA

*South East Asia:*
JOHN WILEY & SONS (SEA) PTE LTD.
37 Jalan Pemimpin 05-04
Block B Union Industrial Building, Singapore 2057

**Library of Congress Cataloging-in-Publication Data**

Bath International Fluid Power Workshop (6th : 1993 : University of Bath)
Modelling and simulation : Sixth Bath International Fluid Power Workshop : held at the University of Bath, England, 23rd-24th September 1993 / edited by C.R. Burrows and K.A. Edge.
p. cm. — (Fluid power series)
Includes bibliographical references and index.
ISBN 0-86380-162-5 (Research Studies Press). — ISBN 0-471-95350-4 (John Wiley & Sons)
1. Fluid power technology—Mathematical models—Congresses.
2. Fluid power technology—Computer simulation—Congresses.
I. Burrows, C. R. (Clifford Robert) II. Edge, K. A. (Kevin A.)
III. Title. IV. Series.
TJ843.B37 1993
620'.1'06—dc20 94-29061
CIP

**British Library Cataloguing in Publication Data**

A catalogue record for this book
is available from the British Library.

ISSN 0963-0104
ISBN 0 86380 162 5 (Research Studies Press Ltd.)
ISBN 0 471 95350 4 (John Wiley & Sons Inc.)

Printed in Great Britain by SRP Ltd., Exeter

# Preface

Bath University Fluid Power Centre was, once again, delighted to host the Sixth in the annual series of Bath International Fluid Power Workshops. Authors from nine countries took part in the event and we are pleased to present their contributions in this volume.

For her considerable assistance in the compilation of this book we gratefully acknowledge the help of Mrs Janet Harvey.

Prof C R Burrows, Director
Prof K A Edge, Deputy Director
Fluid Power Centre
University of Bath

# Contents

# Contributors

Andersen, T.O.
Control Engineering Institute, Technical University of Denmark, DK-2800 LYNGBY, Denmark

Araki, K., Dr
Department of Mechanical Engineering, Faculty of Engineering, Saitama University, 255 Shimo-Okubo, URAWA 338, Japan

Backé, W., Prof Dipl.-Ing.
Institut für Hydraulische und Pneumatische Antriebe und Steuerungen, RWTH Aachen, Steinbachstrasse 53, 5100 AACHEN, Germany

Bouhal, A.
Laboratoire d'Automatique Industrielle, Institut National des Sciences Appliquées de Lyon, Bat 303, 20 Avenue Albert Einstein, 69621 VILLEURBANNE CEDEX, France

Burrows, C.R., Prof
Fluid Power Centre, University of Bath, Claverton Down, BATH, BA2 7AY, UK

Cheng, R.M.H.
Centre for Industrial Control, Mechanical Engineering Department, Concordia University, Montreal, Quebec, H3G 1M8, Canada

Conrad, F., Prof
Control Engineering Institute, Technical University of Denmark, Building 424, DK-2800 LYNGBY, Denmark

Daley, S., Dr
European Gas Turbines Ltd., Engineering Research Centre, Cambridge Road, WHETSTONE, Leics, LE8 3LH, UK

Edge, K.A. Prof
Fluid Power Centre, University of Bath, Claverton Down, BATH, BA2 7AY, UK

Eschmann, R., Dipl.-Ing.
Institut für Hydraulische und Pneumatische Antriebe und Steuerungen, RWTH Aachen, Steinbachstrasse 53, 5100 AACHEN, Germany

Fagan, M.J., Dr
Department of Engineering Design & Manufacture, University of Hull, HULL, HU6 7RX, UK

Gamble, J. Dr
Vickers Systems Division, Trinova Limited, PO Box 4, New Lane, HAVANT, Hants, PO9 2NB, UK

Grossschmidt, G., Prof
Department of Machine Design, Tallinn Technical University, Ehitajate tee 5, EE-0026 TALLINN, Estonia

Gunnarsson, S.
Department of Mechanical Engineering, Linköping University, S-58183 LINKÖPING, Sweden

Hansen, L.H..
Control Engineering Institute, Technical University of Denmark, Building 424, DK-2800 LYNGBY, Denmark

Hansen, P.E.
Control Engineering Institute, Technical University of Denmark, DK-2800 LYNGBY, Denmark

Harrison, D.
Flotron Limited, 445 Perth Avenue Trading Estate, SLOUGH, Berks, SL1 4TS, UK

Harrison, R.
Department of Manufacturing Engineering, Loughborough University of Technology, LOUGHBOROUGH, Leics, LE11 3TU, UK

Ikeo, S., Prof
Faculty of Science and Technology, Sophia University, 7-1 Kioicho, Chiyoda-Ku, 102 TOKYO, Japan

Kitagawa, A.
Department of Control Engineering, Tokyo Institute of Technology, 12-1O-okayama 2-chome Meguro-ku, TOKYO 152, Japan

Krus, P., Dr
Department of Mechanical Engineering, Linköping University, S-58183 LINKÖPING, Sweden

Larsen, H.S.
Control Engineering Institute, Technical University of Denmark, Building 424, DK-2800 LYNGBY, Denmark

LeQuoc, S., Prof
École de Technologie Supérieure, Université du Québec, 4750 avenue Henri-Julien, MONTRÉAL, Québec, H2T 2C8, Canada

Liu, X.
Technical Research Laboratory, Hitachi Construction Machinery Co. Ltd., 650 Kandatsu, TUCHIURA 300, Japan

Lo, J.K.W.
Fluid Power Centre, University of Bath, Claverton Down, BATH, BA2 7AY, UK

McConnachie, J.
Department of Engineering Design & Manufacture, University of Hull, HULL, HU6 7RX, UK

Maeda, T., Prof
Department of Mechanical Engineering, Seikei University, 3-3-1 Kichijoji Kitamachi, Musashino-shi, TOKYO 180, Japan

Moore, P.R., Dr
Department of Manufacturing Engineering, Loughborough University of Technology, LOUGHBOROUGH, Leics, LE11 3TU, UK

Ohnishi, T.
Department of Mechanical Engineering, Faculty of Engineering, Saitama University, 255 Shimo-Okubo, URAWA 338, Japan

Paoluzzi, R., Dr
Cemoter, Via Canal Bianco 28, 44044 CASSANA (FE), Italy

Pahapill, J.
SJV IE Software, Institute of Cybernetics, Estonian Academy of Sciences, Akadeemia tee 21, EE-0026 TALLINN, Estonia

Pu, J.
Department of Manufacturing Engineering, Loughborough University of Technology, LOUGHBOROUGH, Leics, LE11 3TU, UK

Puglia, M.
Cemoter, Via Canal Bianco 28, 44044 CASSANA (FE), Italy

Richard, E.
Laboratoire d'Automatique Industrielle, Institut National des Sciences Appliquées de Lyon, Bat 303, 20 Avenue Albert Einstein, 69621 VILLEURBANNE CEDEX, France

Sakurai, Y.
Faculty of Science and Technology, Sophia University, 7-1 Kioicho, Chiyoda-Ku, 102 TOKYO, Japan

Sanada, K.
Department of Control Engineering, Tokyo Institute of Technology, O-okayama 2-chome Meguro-ku, TOKYO 152, Japan

Scavarda, S., Prof
Laboratoire d'Automatique Industrielle, Institut National des Sciences Appliquées de Lyon, Bat 303, 20 Avenue Albert Einstein, 69621 VILLEURBANNE CEDEX, France

Sethson, K.M.R.
Visiting Scholar, Fluid Power Centre, University of Bath, Claverton Down, BATH, BA2 7AY, UK. Now at Department of Mechanical Engineering, Linköping University, S-58183 LINKÖPING, Sweden

Takahashi, K.
Faculty of Science and Technology, Sophia University, 7-1 Kioicho, Chiyoda-Ku, 102 TOKYO, Japan

Tilley, D.G. Dr
Fluid Power Centre, University of Bath, Claverton Down, BATH, BA2 7AY, UK

Tomlinson, S.P., Dr
Fluid Power Centre, University of Bath, Claverton Down, BATH, BA2 7AY, UK

Vaughan, N.D., Dr
Fluid Power Centre, University of Bath, Claverton Down, BATH, BA2 7AY, UK

Walters, R.B.
Flotron Limited, 445 Perth Avenue Trading Estate, SLOUGH, Berks, SL1 4TS, UK

Wang, H.
Department of Paper Science, UMIST, PO Box 88, MANCHESTER, M60 1QD, UK

Weston, R.H.
Department of Manufacturing Engineering, Loughborough University of Technology, LOUGHBOROUGH, Leics, LE11 3TU, UK

Xiong, Y.F.
École de Technologie Supérieure, Université du Québec, 4750 avenue Henri-Julien, MONTRÉAL, Québec, H2T 2C8, Canada

Zarotti, L.G. Dr
Cemoter, Via Canal Bianco 28, 44044 CASSANA (FE), Italy

Zhang, H.
Faculty of Science and Technology, Sophia University, 7-1 Kioicho, Chiyoda-Ku, 102 TOKYO, Japan

Zhang, M.
Control Engineering Institute, Technical University of Denmark, Building 424, DK-2800 LYNGBY, Denmark

Zhou, J.J.
Control Engineering Institute, Technical University of Denmark, DK-2800 LYNGBY, Denmark

# 1. Introduction

K A Edge *and* C R Burrows

The Sixth Bath International Fluid Power Workshop was held on 23rd and 24th September 1993 at the University of Bath Fluid Power Centre. This book contains 19 Sections corresponding to the papers presented at the Workshop with a focus on current international research and development in the areas of Modelling and Simulation. Each Section is concluded with a Written Discussion featuring issues raised by delegates at the Workshop and the authors' responses. This opening Section provided an overview of the contributions.

The first paper, by Sethson and Vaughan, reports on the modelling of the electro-magnetic behaviour of fast-switching hydraulic valves, with particular reference to the non-linear characteristics. The authors stress the need for improved models, particularly for applications where fast switching rates are anticipated; such models should allow better control techniques to be established at the design stage. For the experimental studies a Hall-effect sensor was employed to determine magnetic flux density variations during valve motion. Although the authors found force measurements to be of greater value when appraising their model they suggest that the Hall-effect sensor has the potential for reliable parameter identification, provided it is carefully positioned.

The problem of valve instability due to fluid momentum effects is well-known and can be a significant problem for valve designers. However, Maeda, in Section 3, shows that the phenomenon can be utilised beneficially in a novel design of 'turbo-motor'. The motor, which is a rotodynamic rather than positive displacement machine, is configured such that fluid momentum effects are used to enhance the speed capabilities. A maximum speed of 40000 rev/min was achieved in the experimental test programme. Maeda presents a detailed model of the machine and assesses both simulated and measured performance.

In Section 4, Fagan and McConnachie from the University of Hull, UK, discuss the design of piston pumps for water hydraulic systems. This is timely in view of growing concerns for the environment, and, clearly, systems capable of running on raw water should have a major impact in a number of important application areas. The authors present modelling and analysis techniques, covering geometric layout, portplate timing, force balance and contact of concentric components. It is argued that simple stress calculations are ineffective and that finite element analysis is required for accurate analysis. The latter has also proved to be highly beneficial as an aid in the assessment of component failures.

Research workers at the Earthmoving Machinery and Off-road Vehicles Institute in Italy (CEMOTER) are well-known for their research on the assessment of control schemes for hydrostatic pumps and transmissions. Paoluzzi, Puglia and Zarotti examine, in Section 5, triple pump controls (pressure and flow control combined with power matching control) with particular reference to design. The authors focus on open-circuit variable capacity pumps for mobile applications and present the results of simulation studies for a specific system.

In previous proceedings of the Workshop, the editors have drawn attention to the limited number of papers reporting on work and development in the field of pneumatics. The situation appears to be improving and for the Sixth Workshop, six of the nineteen papers dealt with various aspects of pneumatic systems and their control. Sections 6, 7 and 8 are all concerned with pneumatic system simulation. Backé and Eschmann, in Section 6, present details of a simulation package developed at the Institute for Fluid Power Transmission at the Aachen University of Technology. The authors stress the importance of accurate information on cylinder friction characteristics and discuss the test procedures employed in their study. Provided such information is available, it is shown that accurate prediction of both open-loop and closed-loop system performance can be achieved.

In Japan, there appears to be a certain preference for the use of a bond graph approach to simulation. Researchers from Sophia University are extending an existing bond-graph software package developed in Japan (BGSP) to allow the simulation of "multi-port" fields. Heat transfer and fluid compressibility effects are included in the bond graph elements, and the authors demonstrate the effectiveness of the approach through comparison with experimental data obtained from a valve-controlled cylinder.

Tomlinson, Lo and Tilley from the University of Bath address an entirely different area of "pneumatics" - that of the human respiratory system. The authors are involved in the simulation of underwater breathing equipment and as part of this research study have developed a series of models for the principal elements of the respiratory system including the cardiovascular system and metabolic processes. The models of the various elements are presented in the form of analogues of fluid power components. In a study of a diver using a semi-closed breathing system, it is shown that simulation can be used to identify safe and unsafe operating conditions. It is pleasing to see from this paper how far fluid power system simulation methodologies can be extended.

Sections 9 and 10 are devoted to two significantly different approaches to the simulation of hydraulic systems. In Section 9 Grossschmidt and Pahapill discuss the use of knowledge-based programming in the development of a simulation package. In this generalised approach, elements of a circuit are modelled by "multi-port" elements and can be linked together to create a form of block diagram. The authors claim that the knowledge-processing approach means that it is unnecessary to provide an explicit algorithm for the solution of a problem as the simulation system produces this automatically. Models may contain non-linearities and can be either lumped parameter or distributed parameter. The accuracy of simulations of a variety of distributed parameter systems is illustrated by comparison with the results of experimental studies.

In Section 10, Walters and Harrison concentrate on the particular domain of electro-hydraulic control systems. The principal features of their simulation package are outlined and this is followed by a case study of a closed loop position control system. The alternative solutions are compared, both in the frequency domain and time domain. This 'domain-specific' approach to simulation can be contrasted to the use, in a number of other papers, of more general simulation packages. Walters and Harrison argue the need for a user-friendly tool for application engineers which is also capable of meeting the requirements of the experienced designer.

Three papers on various aspects of adaptive control are presented in Sections 11, 12 and 13. In the first of these Gamble, from Vickers Systems Division of Trinova Ltd, describes an adaptive scheme for a pilot-operated solenoid valve. It is shown that the valve can be modelled as a variable-gain integrator, and the use of a self-tuning adaptive control can lead to consistent dynamic performance in the presence of variations in supply pressure. Significant improvements in step response characteristics were achieved compared with a conventional valve. This contribution was judged by delegates attending the Workshop to be worthy of the Best Paper award.

Sanada and Kitigawa examine, in their paper (Section 12) the problems of unmodelled dynamics in adaptive control schemes. In a model reference adaptive control scheme for an electro-hydraulic servo-motor, the authors adopt a linear reduced-order model and go on to show that under certain circumstances the system can become unstable. A fuzzy-reasoning procedure which imposes adjustable limits on the gain of an adaptive feedback loop, is demonstrated to be a highly effective means of stabilisation.

An electro-hydraulically driven robot is used by Andersen and co-workers (Section 13) as a means of illustrating the problems of controlling interacting servo systems. Unknown external disturbances and variations in system parameters suggest the use of adaptive control, and the authors derive a novel scheme based on hyperstability theory. In their approach only position feedback is required and no exact knowledge of system parameters is necessary. The effectiveness of the controller, compared with two other schemes, is clearly demonstrated through simulation and an experimental study.

Various aspects of pneumatic systems are examined in Sections 14, 15 and 16 including the problem of control. Pu and his co-workers from the University of Loughborough present an experimental study of gain-scheduling in the control of pneumatic servos. State feedback is employed, with the gains of the feedback loops adjusted solely as a function of the load position. Although a number of simplifying assumptions are adopted, the potential benefits are clearly identified.

In Section 15, a test technique for establishing the frequency response characteristics of pneumatic servo systems is described. A pseudo-random binary test signal, know as an M-Sequence, is used to evaluate the frequency response of a valve-controlled cylinder. Bias errors incurred in the identification due to the presence of the valve non-linear characteristics are examined and compensated for in the analysis.

Problems of non-linear behaviour in pneumatic systems are also of concern to Bouhal, Richard and Scavarda. In Section 16, they draw attention to the limitations of linear adaptive controllers when used for systems containing appreciable non-linearities, and present an alternative non-linear adaptive scheme. The relative merits of linear and non-linear controllers are examined with particular reference to a pressure regulator.

Papers on the control of the force developed by hydraulic actuators are relatively uncommon so it is pleasing to include the contribution by Conrad and co-workers in Section 17. They show that a digital control scheme based on a combination of force feedback and velocity feedforward offers significant improvement in performance compared to a traditional PI controller.

Fluid power systems are frequently employed in applications where failures can lead to a substantial loss of productivity or can be hazardous to human life. The ability to perform effective fault diagnosis should, therefore, be a high priority in many installations. Daley and Wang contribute to developments in this area in their paper on the use of neural networks for automated fault diagnosis. In their simulation study, the authors show the effectiveness of their approach for the particular case of changes in leakage and efficiency occurring in a hydraulic motor.

The problems of controlling non-linear systems, addressed in Section 16 by Bouhal and his co-workers, are also of concern to the authors of the paper presented in Section 19. In a simulation study of an electrohydraulic servo, LeQuoc and his colleagues propose the inclusion of a "preload" in order to eliminate the servovalve deadband. A numerical procedure is described which allows both the linear and non-linear characteristics of the plant to be identified.

In the final Section of the book, Gunnarsson and Krus investigate the merits of LQG control applied to a large hydraulically-driven crane. This is a particularly interesting application because of the flexibility of the mechanised structure. It is shown that with appropriate selection of control parameters well-damped behaviour can be achieved. An extension of the scheme to include integral action in the regulator allows disturbance-generated steady state errors to be eliminated.

# 2. A Model of the Electromagnetic Characteristics of Fast Switching On/Off Valves

K M R Sethson *and* N D Vaughan

**Abstract**

A scheme of modelling the electromagnetic part of fast switching on/off valves is described. It is important to establish a good dynamic model for this family of valves in order to study potential improvements in control methods. The use of a special test system in conjunction with a Hall-element device is described. The structure of the magnetic circuit model and evaluation of critical terms are illustrated. Simulation results are compared with experiments for current, magnetic flux, and force.

## 1 Introduction

The solenoid on/off valve is a component with considerable advantages in the sense that it is cheap, robust and compact. However, the on/off characteristic would appear to be a limitation for more sophisticated applications. It would be highly desirable to use the valve where modulation of flow or pressure is required. Earlier work in which the valve was switched to produce a pilot pressure signal [1] has indicated that such a characteristic is possible to achieve. It also indicated that high switching speeds for the valve were beneficial for this type of demodulation, although other methods are possible [2]. Simulation methods provide the vehicle to examine the potential of higher speed operation and investigate suitable methods of valve and system control. This is the ultimate aim of the project to which the work described here contributes. However, for satisfactory simulation it is necessary to establish a good dynamic model of an on/off valve. Such a model must be simple if it is to be useful in time domain simulations of a complete system.

The frequencies of interest are in the range 300-400 Hz. At such high frequencies the model needs to handle accurately the dynamics of the mechanical components within the valve as well as the fluid flow characteristics. However, one of the most difficult aspects of modelling any electrohydraulic valve is the electro-magnetic conversion process [3]. Partially because it contains concepts of less familiarity to mechanical engineers but principally because it is highly non-linear. In particular the electromagnetic hysteresis which can significantly affect the dynamic response appears almost as a time lag.

Another major problem in producing a model is the lack of any analytical technique which can faithfully predict the performance of the magnetic circuit. This is of course possible using sophisticated Finite Element techniques but these are not appropriate here, even if they have been used in the valve design. Construction of a suitable model therefore requires experimental data as a basis. This paper examines methods to obtain data including measurements using a Hall-element on top of the valve.

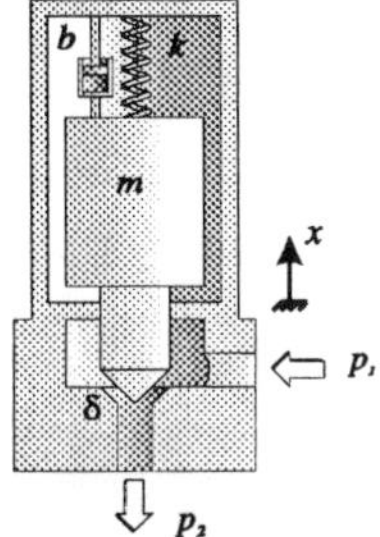


Figure 1: The principle of the On/Off valve.

The valve used in this work is a very straight forward design whose principle can be seen in Fig. 1. The complete valve model includes well established relations in the fluid power area based principally on orifice flow. The valve flow area is determined by the position of the poppet and valve core. This assembly is described simply as a two degree of freedom system responding to a force balance. The forces acting include a pre-compressed spring, a velocity dependent damping term with a displacement dependent coefficient, flow forces, gravity force and the force $F_c$ acting from the coil, see Fig 2.

This paper concentrates on the aspect of establishing a good model of the electro-magnetic force developed by the solenoid assembly. The aim has been to evaluate the possibilities to build a model with as little knowledge as possible of the inside of the valve.

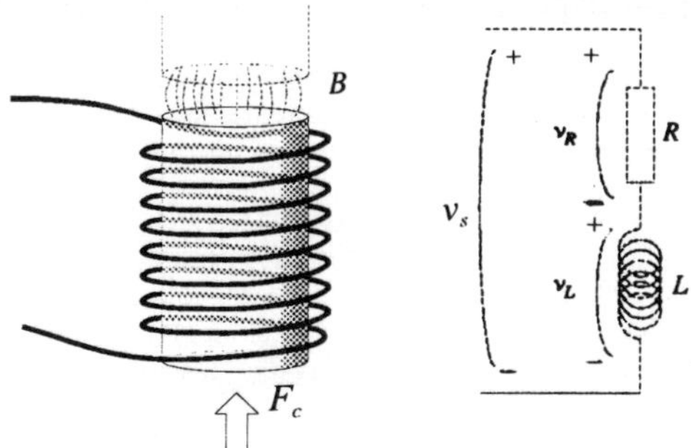


Figure 2: The principle of the electromagnetic part.

## 2 Nomenclature

| | |
|---|---|
| $x$ | Displacement of core. |
| $\dot{x}$ | Velocity of core. |
| B | Magnetic flux density. |
| H | Magnetic field strength. |
| $\mu_0$ | Permeability of free space. |
| $\lambda$ | Magnetic flux linkage. |
| $F_c$ | Magnetic force. |
| $i$ | Current through the coil. |
| $v_s$ | Supply voltage. |
| $v_L$ | Voltage across inductor. |
| $v_R$ | Voltage across resistor. |
| $A$ | Area of magnetic flux path. |
| $l_0$ | Maximum air-gap. |
| N | Number of turns. |
| L | Inductance. |
| R | Resistance. |
| $K_{vv}$ | Velocity/voltage gain. |
| $K_B$ | Gain of secondary flux. |
| $K_{ce}$ | Effective flow coefficient. |
| $B_{cl}$ | Magnetic model parameter. |
| $H_{cl}$ | Magnetic model parameter. |
| $\mu_s$ | Magnetic model parameter. |
| $\mu_{cl}$ | Magnetic model parameter. |
| $H_c$ | Magnetic model parameter. |
| $\mu_c$ | Magnetic model parameter. |
| $A_1$ | Magnetic model parameter. |
| $A_2$ | Magnetic model parameter. |
| $A_3$ | Magnetic model parameter. |
| $A_4$ | Magnetic model parameter. |

## 3 The Structure of the Electromagnetic Model

The model of the electro-magnetic circuit effectively describes the transformation of an input voltage signal into an output force and hence represents a number of effects. It comprises some well established equations combined with less common methods to represent the hysteresis effects as the valve core moves. Some well known phenomena in the magnetic path such as the skin-effect and eddy-currents [4] [8], have been neglected here.

There have been a number of workers in the field of hysteresis modelling many based on the work of Chua & Bass [5]. The technique described here is also a development from Chua & Bass and has been derived from the theory of Hodgdon and co-workers [6] [7]. Some modifications on the work of Hodgdon to find certain parameters have been investigated by the authors. Hodgdon's work defines an equation structure to represent the magnetic hysteresis in ferro-magnetically soft materials. This allows the time dependent behaviour of the magnetic field to be described by a series of differential equations. This represents the physical situation in a solenoid for a fixed position of the core. The effects do not account for changes taking place with movement of the core material and this has required the modifications to the basic equations as described below. The addition of position dependency has led to a set of equations highly appropriate for use in time domain simulation. The basic relationships used in the model are derived from standard theory.

The model is based on a simple circuit representation of the solenoid as a resistor in series with an inductor as shown in Fig. 2. The voltage to current relationship is described by:

$$v_s = iR + \frac{d}{dt}(Li) = iR + \frac{d\lambda}{dt} \tag{1}$$

Application of Faraday's law means that the voltage across the inductor may also be written as a function of the flux linkage giving:

$$\frac{d\lambda}{dt} = v_s - Ri = v_l \tag{2}$$

This arrangement using flux linkage ($\lambda$) is more convenient, allowing the hysteresis curve characteristics to be lumped into one relationship as discussed below. This may be considered as including non linear inductance effects. Flux linkage can also be related to flux density through the standard relationships:

$$\lambda = N\phi \tag{3}$$

$$\phi = BA \tag{4}$$

Hence the derivative of flux linkage with time:

$$\frac{d\lambda}{dt} = \frac{d}{dt}(BNA) = N(\frac{dB}{dt}A + \frac{dA}{dt}B) \tag{5}$$

Where both the working area and flux density are time dependent. The working area is shown in Fig. 2 as the section of end face area plus a component due to the collar of the core diameter and air gap in length. This may described in the form:

$$A(x) = A_{10} - A_{11}x \tag{6}$$

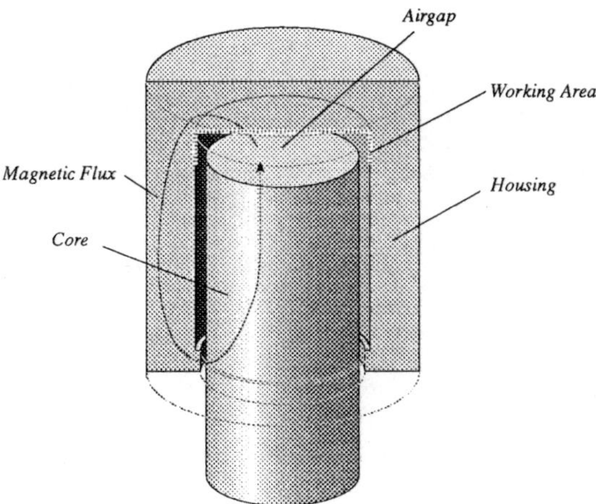


Figure 3: The magnetic path as a moving core inside a tube.

It should be noted that this is an approximation to the actual working area which in practice will be a more complicated function which might need to be non-linear if the core diameter is small in comparison with the air gap. However, this term is easily approximated in terms of the solenoid component dimensions and any subsequent inaccuracy compensated in the hysteresis characteristic. The time derivative of area can thus be related to solenoid core velocity:

$$\frac{dx}{dt} = -\frac{1}{A_{11}}\frac{dA}{dt} = K_{vv}\frac{dA}{dt} \tag{7}$$

The non-linearities in the magnetic path can be modelled by a relationship between the time derivatives of field strength ($H$) and flux density ($B$) [7]. This can be formulated into a set of equations which handle all the non-linearities associated with the magnetic path including saturation, hysteresis and rate dependent field changes. Collectively these are referred to as the hysteresis model. This requires both $B$ and $H$ as time derivatives and Eq. 5 can be re-arranged in the form:

$$\frac{dB}{dt} = \frac{1}{A_0(x)}\left(\frac{1}{N^2}\frac{d\lambda}{dt} - B_0\frac{dA}{dt}\right) \tag{8}$$

Where $A_0$ and $B_0$ denote area and magnetic flux density for a certain operating point. Normally $B_0$ is equal to $B$. The field derivative can be obtained from standard theory using the form:

$$H = \frac{Ni}{l_m} \tag{9}$$

Where $l_m$ is a characteristic length of the magnetic path. This will vary with core position and a linear relationship has been used:

$$l_m(x) = l_{m10} - l_{m11}x \tag{10}$$

An approximation to this value can again be obtained from consideration of the dimensions of the solenoid but it has also been used as a tuning parameter to give a match between experimental and simulation results. The exact number of turns $N$ is unknown and this can also be used as a tuning factor within a sensible range. Differentiating and re-arranging Eq. 9 gives:

$$\frac{di}{dt} = \frac{1}{N}\left(H_0\frac{dl_m}{dt} + l_{m0}\frac{dH}{dt}\right) \tag{11}$$

Where $H_0$ and $l_{m0}$ values are defined for the current operating point. The conventional relationship between field strength and flux density is modified as a time varying magnetic permeability as proposed by Hodgdon [7]. This hysteresis model is described by a set of equations 12 to 15.

$$\dot{H} = \alpha|\dot{B}|[f(B) - H] + \dot{B}g(B) \tag{12}$$

$$f(B) = \begin{cases} A_1\tan(A_2B_{bp}) + (B - B_{bp})/\mu_{cl} & \text{if } B > B_{bp} \\ A_1\tan(A_2B) & \text{if } |B| \leq B_{bp} \\ -A_1\tan(A_2B_{bp}) + (B + B_{bp})/\mu_{cl} & \text{if } B < -B_{bp} \end{cases} \tag{13}$$

$$f'(B) = \begin{cases} 1/\mu_{cl} & \text{if } B > B_{bp} \\ A_1A_2(1 + \tan^2(A_2B) & \text{if } |B| \leq B_{bp} \\ 1/\mu_{cl} & \text{if } B < -B_{bp} \end{cases} \tag{14}$$

$$g(B,\dot{B}) = \begin{cases} f'(B)(1 - A_3e^{-\frac{A_4|B|}{B_{cl}-|B|}}) & \text{if } |B| < B_{cl} \\ f'(B) & \text{otherwise} \end{cases} \tag{15}$$

The functions $f$ and $g$ in Eq. 12 describe the hysteresis effects. The saturation limits are included as Eq. 13, and the rate effects included in time constant form in Eq. 14. As can be seen non-linearities in the time constant are also included in this formulation. It is helpful to relate these equations to the basic magnetic hysteresis loop characteristic as can be seen in Figures 4 to 7. Hodgdon suggests methods to estimate the required parameters from such a hysteresis curve. This procedure was used with standard hysteresis curve data [11] as a base for the simulation. Further optimisation was possible within the HOPSAN simulation package and this was also used to give adjustments. It would also be possible to introduce a position dependency into these magnetic properties to describe changes with core position. This was not found necessary in this work.

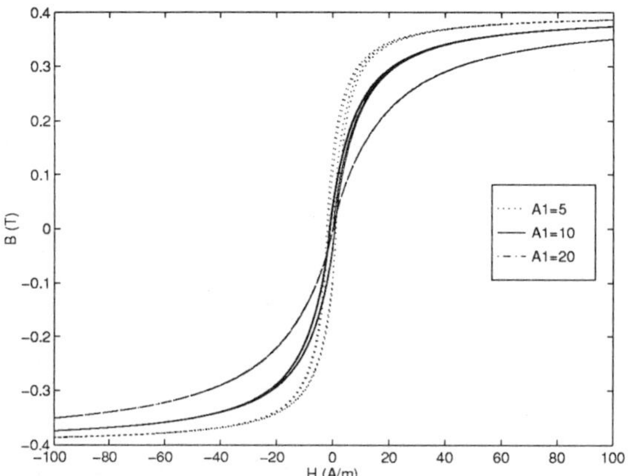


Figure 4: The hysteresis loop for different values of A1

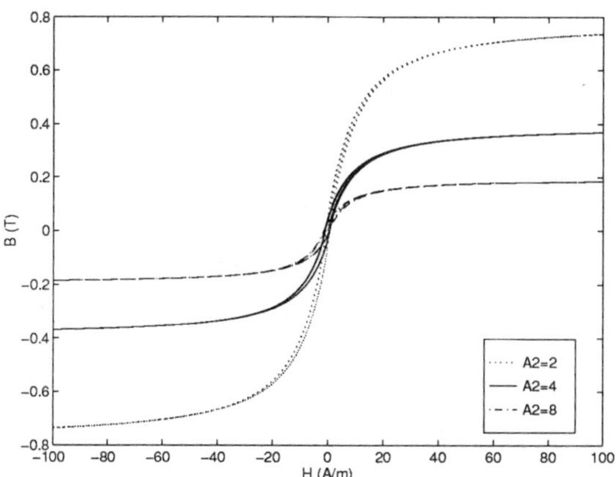


Figure 5: The hysteresis loop for different values of A2

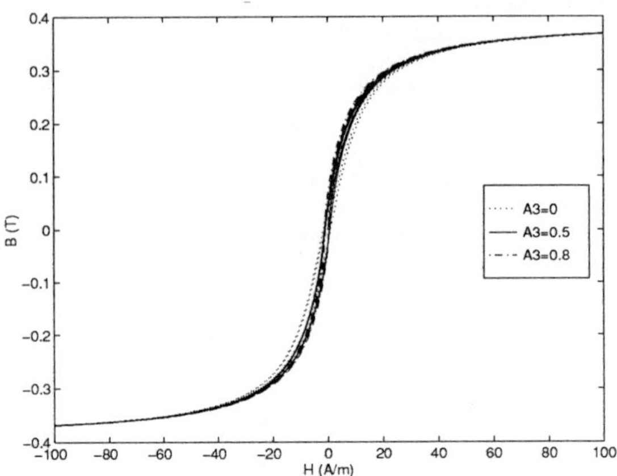


Figure 6: The hysteresis loop for different values of A3

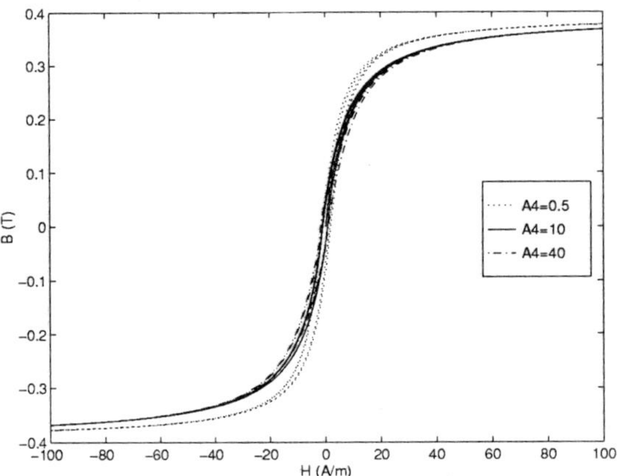


Figure 7: The hysteresis loop for different values of A4

The final stage of the modelling is the relationship between the flux and the force. The magnetic path is shown in Fig. 3, with the coil omitted for clarity. The force, $(F_c)$, is produced by the attraction between the core and the housing and is described by Eq. 16.

$$F_c = \frac{B^2 A(x)}{2\mu_0} \tag{16}$$

As can be seen from Eq. 16, the force is proportional to the square of the magnetic flux density which gives opportunities to achieve high forces by increasing the current. This provides opportunities for performance improvement with control strategies. The power consumption increases as well but by sensible control it can be held at a low average level.

The overall structure of the model is shown in the block diagram of Fig. 8. This shows the other force components and the moving mass of the system comprising of solenoid core and valve poppet. The model gives an output of flow $q$, which is determined from the valve position $x$ using standard relationships but represented here simply by $K_{ce}$. The highly non-linear nature of this model can be clearly seen.

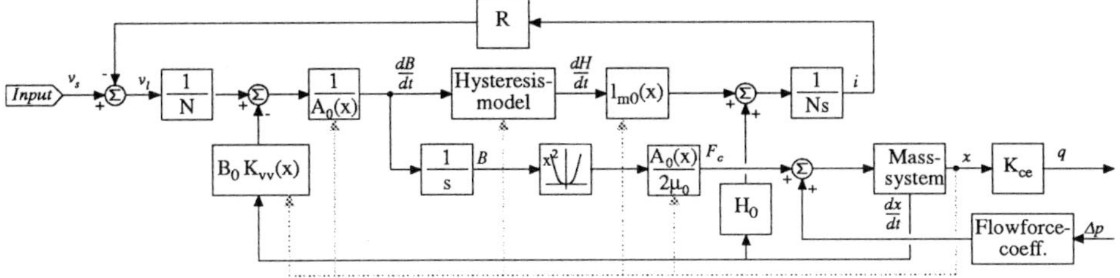


Figure 8: Block diagram of the overall structure of the model.

# 4 The Test System

To be able to verify simulations and theoretical work a test stand has been built. It incorporates a number of features for measuring signals in a dynamic sense, see Fig. 9. A Hall-effect sensor, discussed in [9], has been used for measuring the magnetic flux density on the outside of the valve. This signal can be of help in modelling the valve since it gives useful information on the magnetic path, which shortens the optimization time. However, there is a potential problem with this since it is the working flux in the air gap which is responsible for producing the force whereas the flux measurement is outside the valve. As a check on the results produced for the flux it was considered necessary also to measure the force produced by the solenoid. Two kinds of tests can be made with this arrangement:

1 Measuring the force with the core fixed at a certain displacement gave a good opportunity to identify the variations in magnetic path and inductance as a function of displacement. This has been achieved here by positioning the core with a stepping motor. Special notice has been taken to minimize the friction in these measurements and therefore the only metallic contact area is in between the poppet and the seat.

2 Removing the force transducer it is possible to measure the displacement through a window using optical sensors. This arrangement allows the valve poppet position to be measured dynamically in a non-intrusive way.

For either of these tests it is possible to have the system pressurised (with more or less loss of oil). This then allows the possibility of simultaneously measuring the dynamic flow through the valve orifice using two high response pressure-transducers. No results of the second tests will be presented here.

# 5 Experimental and Simulation Results

A number of measurements have been made recording force, current, voltage and magnetic flux density. Several alternative input signal waveforms have been used for the data collection. The

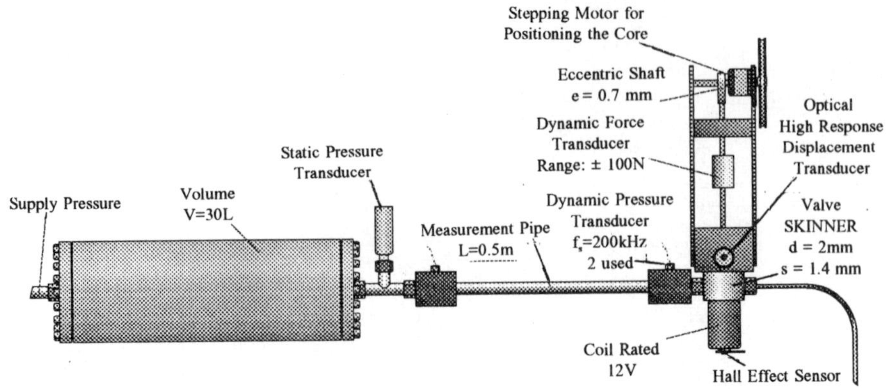


Figure 9: The test stand arrangement.

signal which was found to give the quickest identification of system parameters was the simultaneous application of a sinewave and a small amplitude square wave of the same frequency. The results presented show the output for just over one half cycle of this waveform.

In combination with the measurements there have been several simulations using the HOPSAN-package [10]. At this stage in the project most of these have been optimisation loops to tune the different parameters. The procedure for the optimisation has been:

1 Finding the basic hysteresis shape from measurements of current and magnetic flux density and then the procedure described by Hodgdon [7] enables initial selection of values for parameters $\mu_{cl}, B_{cl}, H_{cl}, \mu_s, \mu_c, H_c$ and $A_4$. Most of this was done by using material constants for iron [11].

2 Adapting the simulations to the measurements of current for different displacements enables the variation of magnetic path length $l_m(x)$, to be established. The number of turns $N$ has been set to a fixed number.

3 Adjusting simulations of force to the measured values. In this stage the effective area $A(x)$ was determined.

4 Repeating step 2 and 3 above to adjust $l_m(x)$ and $A(x)$.

Results comparing the result of simulations with experimental measurement are presented in Figs 10, 11 and 12. Fig 10 shows the current transient for the applied waveform. The agreement in magnitude obtained is good. The simulated rise time appears to be lagging relative to the measured trace but after this the two curves maintain the same phase relationship.

Figure 11 shows the magnetic flux density as measured externally by the Hall effect device and the simulated value in the air gap. Note that there are two scales which are 2 orders of magnitude different. The shape of the curves is however very similar. This shows the difficulty of relating external measurements to the relatively high flux densities in the air gap.

Fig 12 shows the equivalent results for the force output. As with current there is a good fit for magnitude with a slight lag evident from the simulated force. There is a superimposed spike on the measured results which was attributed to some movement within the valve due to a mechanical resonance. In other respects the simulation was considered to be a good representation of the measured values.

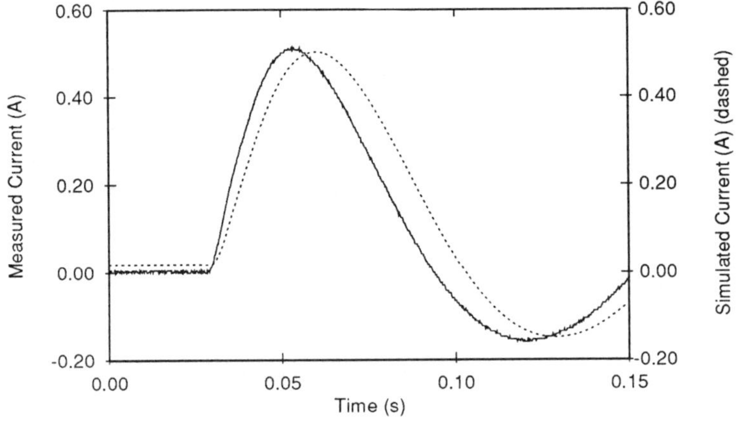


Figure 10: Measurement and simulation of current.

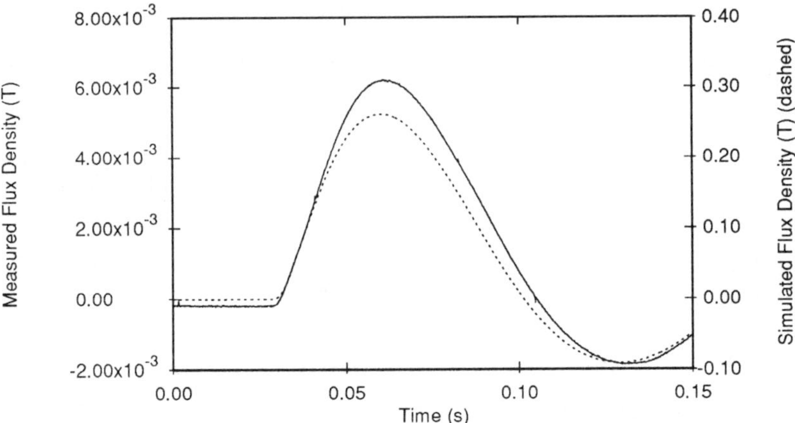


Figure 11: Magnetic flux as measured by the Hall-element and simulation of the magnetic flux density in the air gap. (Different scales).

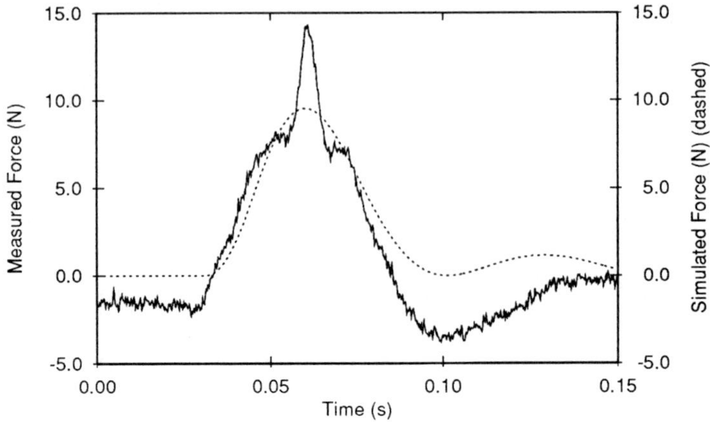


Figure 12: Force, measurement and simulation.

# 6 Conclusions

A model structure capable of describing the hysteresis and saturation effects in a moving core solenoid has been proposed. This has been shown able to represent the behaviour of the solenoid, in particular the current and force, with a suitable selection of parameters. The model is highly suitable for use in time domain simulation programmes.

However, the way of adjusting the parameters needs to be investigated in further work. One way of doing that is to use one Hall effect sensor closer to the airgap or several Hall effect devices spread out over the body of the valve. With a sensible placement of the Hall-element somewhere inside the valve/coil assembly, more accurate information of the magnetic state can be found.

Direct measurement of force is believed to give useful information in combination with the flux measurement externally to the valve. However, a better way of measuring the force is needed for future work.

The simulation results achieved are very dependent on the parameter set, $A1$, $A2$, $A3$, $A4$, $\mu_{cl}$ and $l_{m0}$. Most of the optimisations resulted in a very narow magnetic hysteresis loop, implying that the need for a hysteresis model in the lower frequency region is obsolete.

# 7 Acknowledgements

The authors of this paper would like to thank the reviewer for many constructive comments on the draft of this paper wich have contributed significantly to the final version. Magnus Sethson would also like to thank the board of Bengt Ingestrms Stipendiefond scholarship that has served as a financial basis during Mr Sethson's visit to Bath.

# 8 References

## References

[1] N D Vaughan, I M Whiting.
Use of pulse-width modulation pressure control for hydraulic valves
*Proc 2nd Bath Int Fluid Power Workshop.* Sept 1989.

[2] U Becker, E Schnieder.
Different Mode of Operation of Pulse-Modulation with the Drive of 2/2 Fast switching Valves
*The Third Scandinavian International Conference on Fluid Power.* May. 1993, Sweden

[3] N D Vaughan, J B Gamble.
The Modelling and Simulation of a Proportional Solenoid Valve
*ASME Winter Annual Mtg.* Nov. 1990, 90-WA/FPST-11

[4] Prest P, Vaughan N D
Drive circuits for pulse width modulated valves
*Proc. International conference on fluid power,*
*March 24-26 1987, Tampere Finland,* pp. 217-225.

[5] Chua L O, Bass S C
A generalized hysteresis model,
*IEEE Transactions on Circuit Theory,* Vol. CT-19, No. 1, pp. 36-48

[6] C D Boley, M L Hodgdon
Model and Simulation of Hysteresis in Magnetic Cores
*IEEE Transactions on Magnetics,* Vol. 25, No. 5,
Sept. 1989, pp. 3922-3924.

[7] B D Coleman, M L Hodgdon
A constitutive relation for rate independent hysteresis in ferromagnetically soft materials
*Int J Engg Sci,* Vol. 24, No. 6,
1986, pp. 897-919.

[8] Raul I Rabinovici, Ben Zion Kaplan
Effective Magnetization and Forces due to Eddy Currents,
*IEEE Transactions on Magnetics,* Vol. 28, No. 3,
May 1992, pp. 1863-1869.

[9] K M R Sethson
Identifying the Dynamics of Fast Response Solenoid Valves
*The Third Scandinavian International Conference on Fluid Power,*
Linkoping 25 - 26 May 1993, pp. 271-287.

[10] HOPSAN a Simulation Package, User's guide,
Division of Fluid Power Technology,
Department of Mechanical Engineering,
*Linkping University,*
LiTH-IKP-R-704, Sweden 1991

[11] Christos Christopoulos
An Introduction to Applied Electromagnetism,
*JOHN WILEY & SONS,* EEE ISBN0471927619, UK 1990

## WRITTEN DISCUSSION

## A model of the electromagnetic characteristics of fast switching on/off valves

**KMR Sethson (University of Linköping) & ND Vaughan (University of Bath)**

**Question:** **PJ Chapple**
**Fluid Power Centre, University of Bath**

How well did you know the relationship between flux and displacement ($K_B$) or did you obtain this only from the tests?

**Answer:**

$K_B$ was primarily used for modelling the measurement error of the flux density due to the position of the Hall-element. We are not measuring the flux-density in the airgap. $K_B$ is only obtained from measurements.

**Question:** **JB Gamble**
**Vickers Systems Division, Trinova Ltd, UK**

(i) The technique described refers to the 'major' hysteresis loop. Can the model handle operation on a 'minor' loop, or indeed with arbitrary voltage inputs?

(ii) Are there residual magnetism effects generating force for no current? (Fig 12).

**Answer:**

(i) Yes, it should. The reference sources used show that their correlations of minor loops based on parameters established with a major loop are good. There will be limitations with large voltage inputs.

(ii) These have not been observed experimentally, the force goes back to zero when the current is zero. An offset on the load cell output was present which appears in the result presented.

**Question:** **W Rampen**
University of Edinburgh

(i) Back Emf has an effect - is this included in your study?

(ii) Are eddy currents really negligible at these frequencies?

(iii) Have you considered applying FE techniques?

**Answer:**

(i) Yes, back Emf effects are included within Eqn. 1.

(ii) We had taken the view that the method used to model the more obvious magnetic hysteresis and saturation would also cope with other non-linear effects. We have not yet tested experimentally at different frequencies to ensure that the model developed has frequency dependency suitable to account for eddy current losses. This could be an interesting development for the future.

(iii) The use of Finite Element analysis would be a very helpful tool to use alongside the time domain model which was the objective of this work. In particular it could be very helpful in conjunction with the Hall effect measurements of flux density.

# 3. Development of a High Speed Hydraulic Turbo-Motor

T Maeda

Abstract

In this paper, the author presents a High Speed Hydraulic Turbo-motor utilizing the negative damping flow force which is the cause of the self-excited oscillations of a valve.

Traditional turbomachines utilize the flux of momentum of fluid passing out and into its control volume. On the contrary, the turbo-motor studied here is driven by the flux of momentum passing out and into the control volume plus the rate of change of momentum within the control volume of the turbo-motor.

The author has made the theoretical analysis on the equations of motion of the turbo-motor and some experiments. By the results of the simulations by Runge-Kutta Method, it is made clear that the rate of change of momentum within the control volume contributes to the rotational speed of the turbo-motor.

By the experiments, about 40,000 rpm was obtained under the conditions of 10 MPa and 60 °C. Theoretical computations show good qualitative agreement with the experimental results.

## 1 Introduction

Because of ever increasing demands for the improvement of performances of hydraulic apparatus and their compactness, the recent trend is that the operating pressure in the conduit is becoming higher and higher. Thus the self-excited oscillation of a valve has come to the fore as a great hindrance to its stable operation,[1]. If a valve is inherently unstable and the self-excited oscillations could not be avoided, then why not that we should utilize this unstable phenomenon more positively and make a hydraulic oscillator out of it. Thus the hydraulic oscillators used self-excited oscillation were reported by the author [2,3,4].

Conventional hydraulic motors are low speed and high torque using high pressure, and they utilize the flux of momentum of fluid passing out and into its control volume. On the contrary, the turbo-motor studied here is driven by the flux of momentum of fluid passing out and into its control volume plus the rate of change of momentum of fluid within the control volume of the turbo-motor. Due to the rate of change of momentum of fluid within the control volume, negative damping occurs. By utilizing this effect, we will be able to make a high speed turbo-motor. One example of the application is to use as a turbo-motor to drive super-charger for an engine.

## 2 Nomenclature

$A$ =opening area of the nozzle
$A_c$ =cross sectional area of the turbine chamber
$C_d$, $C_v$ =discharge and velocity coefficient of the supply port respectively
$c_2$ =coefficient of viscous damping proportional to $\dot{\theta}^2$
$c_f$ =coefficient of hydrodynamic viscous damping due to the time rate of change of momentum within the control volume
$c_0$ =coefficient of viscous damping force due to the viscous friction between a spool valve and its bore
$cs$, $cv$ =control surface and cotrol volume respectively
$F_p$, $F_\tau$ =pressure force and shearing force acting on the control surface respectively
$h$ =nozzle depth
$I$ =moment of inertia of the turbine rotor
$i$ =nondimensionalized adding moment of inertia of the fluid in the turbine chamber
$k_f$ =coefficient of hydrodynamic restoring force due to the flux of momentum passing out and into the control volume
$k_m$ =coefficient of restoring force due to mechanical spring
$l_1$, $l_2$ =geometrical damping length as indicated in figure 4
$l_1'$, $l_2'$ =damping length which has over runs of the fluid as indicated in figure 4
$m$ =mass of a spool valve
$p$ =pressure
$Q$ =flow rate
$R_m$ =mean radius of a turbine rotor chamber as indicated in figure 2
$R_0$ =radius of a turbine rotor as indicated in figure 2
$T$ =torque which acts on the turbine rotor
$t$ =time
$t_r$ =rise time
$V$ =velocity
$V_{rn}$ =relative velocity of fluid normal to the control volume
$V_{out}$, $V_{in}$ =entering and out-going velocity through the control surface
$x=\theta/\theta_a$ =nondimensional rotational angle
$x'=dx/d\tau$ =nondimensional rotational velocity
$\alpha_1$, $\alpha_2$ =nondimensionalized viscous damping coefficients as indicated in equ.25
$\theta$ =rotational angle of the turbine rotor as indicated in figure 2
$\theta_a$ =reference rotational angle of the turbine rotor (width of the nozzle)
$(\ )_\theta = \theta$ component
$\rho$, $\rho_t$ =density of working fluid and turbine rotor respectively
$\phi$ =nozzle angle as indicated in figure 2
$\tau=\omega_0 t$ =nondimensionalized time

## 3 Theoretical analysis

### 3.1 PRINCIPLE OF HYDRAULIC TURBO-MOTOR

As an example of the existence of the negative damping force, it is well-known fact that the instability of the spool valve shown in figure 1. The equation of motion of the spool valve is given as follows;

$$m\ddot{x} + (c_0 - c_f)\,\dot{x} + (k_m + k_f)\,x = 0 \qquad (1)$$

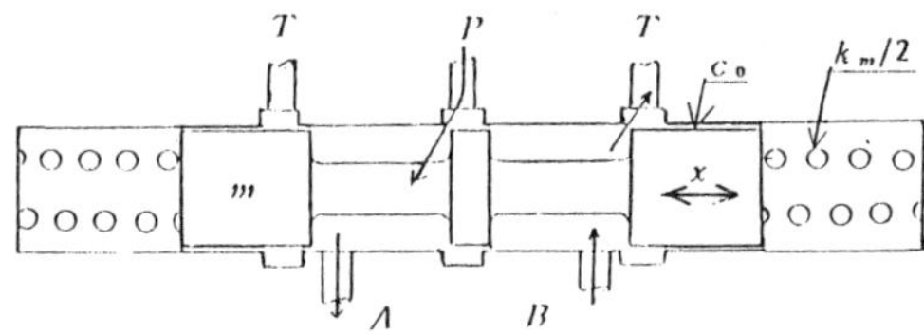


Figure 1. Instability of spool valve

For the region $(c_0 - c_f) < 0$ in eq.(1), the spool valve becomes unstable. As there is non-linearity of the flow force, there occurs self-excited oscillation of the spool valve. If coefficient of $\dot{x}$ is negative, the fluid force acting on the object works in the same moving direction as that of the object. Therefore, if we apply this kind of torque to a rotor, we can make a turbo-motor which rotates faster than the one which has not this kind of torque.

### 3.2 EQUATION OF MOTION OF TURBO-MOTOR UTILIZED NEGATIVE DAMPING TORQUE

#### 3.2.1 Turbo-motor with five blades and four nozzles

Now, we will consider an example of turbo-motor which has five triangle blades and four nozzles as shown in figure 2. The geometrical relationship of the turbo-motor are given as follows;

$$\theta_b = \theta_d = 4\theta_a \quad , \qquad \theta_a = 0.1\pi\,[\mathrm{rad}] = 18^\circ \qquad (2)$$

But if we want another set of blades and nozzles, we can make an arbitrary one.
Figure 3 shows the development of each stages of blades and nozzles. In every instance, one of the blades is in the mouth of the nozzle as indicated by $(m)$ in figure 2. So we will be able to classify the position relationship between nozzles and blades into two cases during the rotation as shown in figure 4. Those are:

Case 1 is in the process of full open of the nozzles.

Case 2 is in the process of passing the mouth of the nozzle. In this case, a part of the fluid goes backward and the rest of the fluid goes forward. At this moment, there occurs the rate of change of momentum of the fluid within the control volumes $l$ and $r$.

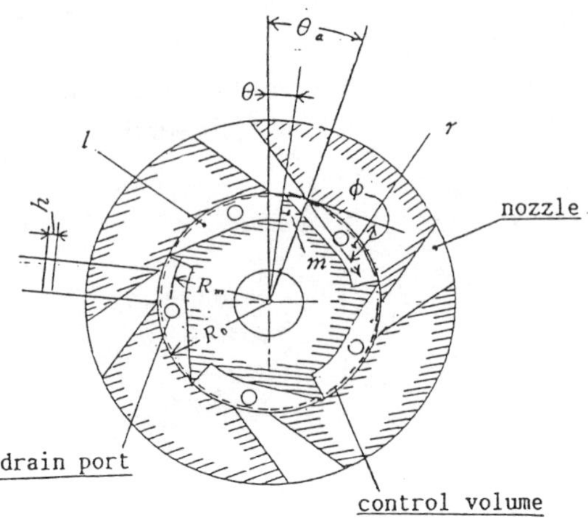


Figure 2. Turbo-motor with five blades

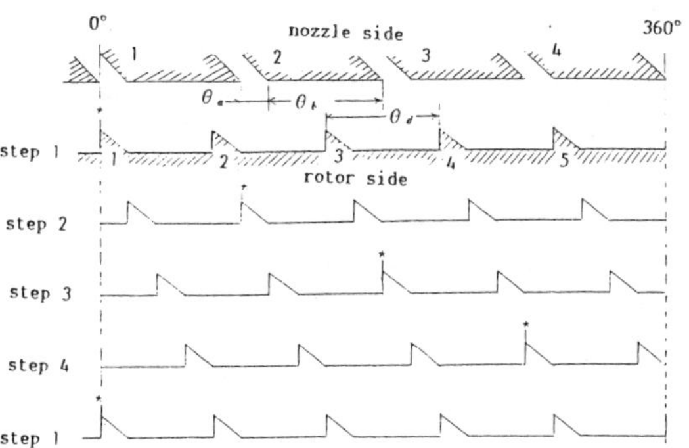


Figure 3 Development of nozzles and blades

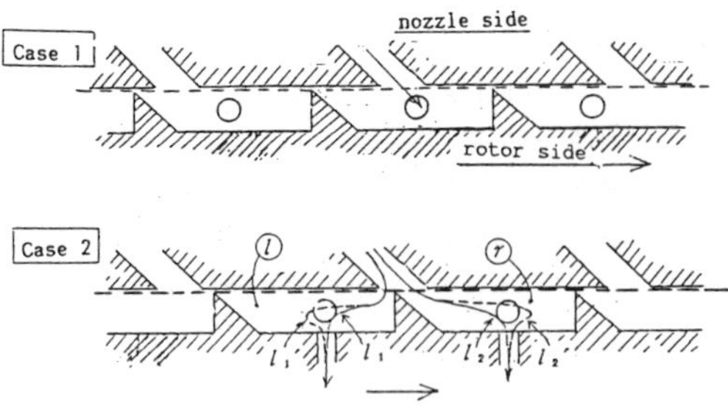


Figure 4 Relationship between nozzle and blade

As the turbine rotates at very high speed, we will assume the viscous friction proportional to $\dot{\theta}^2$, then the equation of motion of the turbine will be given by the following equation.

$$I\ddot{\theta}+c_2\dot{\theta}^2=T \qquad (3)$$

In order to obtain the torque $T$ by the fluid force acting on the turbine blades, we will apply the momentum theory [4] to the turbine blades which are surrounded by the control volume indicated in figure 4 with dotted line. Then the moment of momentum of the fluid T acting on the turbine blades for an inertial control volume takes the form as follows:

$$T=R_m\left(F_p+F_\tau-\iint_{cs}\rho \boldsymbol{V} V_{rn}\,ds-\frac{d}{dt}\iiint_{cv}\boldsymbol{V}\rho\,dv\right)_\theta \qquad (4)$$

If we take the control surface as shown in figure 2, then $(F_p)_\theta=0$ and $F_\tau$ is negligibly small compared to the other terms in equation (4). So, equ.(4) can be reduced to:

$$T=-R_m\left(\iint_{cs}\rho \boldsymbol{V} V_{rn}\,ds+\frac{d}{dt}\iiint_{cv}\boldsymbol{V}\rho\,dv\right)_\theta \qquad (5)$$

3.2.2 Fluid torque

Now, let us caluculate the torque $T$ in each cases.

(I) Process of full open of the nozzle ( case 1)

Since the area of the nozzle is constant, that is:

$$A=hR_0\theta_a=\text{constant} \qquad (6)$$

and considering

$$Q=C_dA\sqrt{2p/\rho} \qquad V=C_v\sqrt{2p/\rho} \qquad (7)$$

then the first term in equ.(5) may be written as follows:

$$\left(\iint_{cv}\rho \boldsymbol{V} V_{rn}\,ds\right)=\rho Q\,(V_{out}-V_{in})_\theta=\rho Q\,(R_0\dot{\theta}-V\cos\phi) \qquad (8)$$

The second term in equ.(5) becomes zero because of no change of velocity in the control volume of the turbine, that is:

$$\left(\frac{d}{dt}\iiint_{cv}V_r\rho\,dv\right)_\theta=0 \qquad (9)$$

Therefore, torque $T_1$ is given as follows:

$$\boldsymbol{E_{av}=\sum\left[G_s(j\omega)-G_m(j\omega)\right]\,/\,(\text{Data number})} \qquad (10)$$

(II) Process of passing the mouth of the nozzle (case 2)

In this moment, one nozzle is divided by a blade into upward side and downward side.

(A) Left hand side of the nozzle

As the area of the nozzle is

$$A=hR_0\theta \qquad (11)$$

then equ.(5) may be written as follows:

$$\left(\iint_{cs}\rho V V_{rn}\,ds\right)_\theta=\rho Q\,(R_0\dot{\theta}-V\cos\phi)=-2C_dC_vR_0hp\cos\phi\cdot\theta+C_dR_0^2h\sqrt{2\rho p}\,\theta\dot{\theta} \qquad (12)$$

and,

$$\left(\frac{d}{dt}\iiint_{cv} V_r \rho\, dv\right)_\theta$$

$$=\frac{d}{dt}\int_0^{l_l}\left(-\frac{Q}{A_c}+R_m\dot{\theta}\right)\rho A_c\, dy = -C_d R_0 h\, l_l\sqrt{2\rho p}\,\dot{\theta}+\rho A_c l_l R_m \ddot{\theta} \tag{13}$$

Therefore, torque $T_{2l}$ due to the left hand side of the nozzle is given as follows:

$$T_{2l}=2C_d C_v R_m R_0 h p \cos\phi\cdot\theta + C_d R_m R_0 h\, l_l\sqrt{2\rho p}\,\dot{\theta}$$
$$-C_d R_m^2 R_0 h\sqrt{2\rho p}\,\theta\dot{\theta}-\rho A_c l_l R_m^2 \ddot{\theta} \tag{14}$$

(B) Right hand side of the nozzle

As the area of the nozzle is

$$A=hR_0(\theta_a-\theta) \tag{15}$$

then

$$\left(\iint_{cs}\rho V V_{rn}\, ds\right)_\theta=-2C_d C_v R_0 h p(\theta_a-\theta)\cos\phi + C_d R_0^2 h\sqrt{2\rho p}(\theta_a-\theta)\dot{\theta} \tag{16}$$

and

$$\left(\frac{d}{dt}\iiint_{cv} V\rho\, dv\right)_\theta=\frac{d}{dt}\int_0^{l_r}\left(-\frac{Q}{A_c}+R_m\dot{\theta}\right)\rho A_c\, dy=-C_d R_0 h\, l_r\sqrt{2\rho p}\,\dot{\theta}+\rho A_c l_r R_m\ddot{\theta} \tag{17}$$

Therefore, torque $T_{2r}$ due to the right hand side of the nozzle is given as follows:

$$T_{2r}=2C_d C_v R_m R_0 h p(\theta_a-\theta)\cos\phi - C_d R_m R_0^2 h\sqrt{2\rho p}\,(\theta_a-\theta)\,\dot{\theta}$$
$$+C_d R_m R_0 l_r h\sqrt{2\rho p}\,\dot{\theta}-\rho A_c R_m^2 l_r\ddot{\theta} \tag{18}$$

Hence the torque $T_2$ in case 2 is given as follows.

$$T_2=T_{2l}+T_{2r}=2C_d C_0 R_m R_0 h p\theta_a\cos\phi$$
$$+C_d R_m R_0 h\sqrt{2\rho p}(l_l+l_r-R_0\theta_a)\dot{\theta}-\rho A_c R_m^2(l_l+l_r)\ddot{\theta} \tag{19}$$

(Ⅱ) Total fluid torque

Thus the total fluid torque $T$ acting on the turbine blades becomes as follows:

$$T=3T_1+T_2=8C_d C_v R_m R_0 h\theta_a p\cos\phi + C_d R_m R_0 h\sqrt{2\rho p}\,(l_l+l_r-4R_0\theta_a)\,\dot{\theta}-\rho A_c R_m^2\,(l_l+l_r)\,\ddot{\theta}$$
$$=8C_d C_v R_m R_0 h\theta_a p\cos\phi + C_d L_1 R_m R_0 h\sqrt{2\rho p}\,\dot{\theta}+\rho A_c R_m^2 L_2\ddot{\theta} \tag{20}$$

where

$$L_1=l_l+l_r-4R_0\theta_a\ ,\qquad L_2=l_l+l_r \tag{21}$$

3.2.3 Equation of motion of the turbo-motor

Equation of motion of the turbo-motor will be expressed using equation (3) and (20) as follows:

$$(I+\rho A_c L_2 R_m^2)\,\ddot{\theta}-C_d L_1 R_m R_0 h\sqrt{2\rho p})\dot{\theta}+c_2\dot{\theta}^2-8C_d C_v R_m R_0 h\theta_a p\cos\phi=0 \tag{22}$$

Now, let us nondimensionalize $\theta$ and $t$ as to

$$x=\theta/\theta_a, \quad \tau=\omega_0 t \tag{23}$$

Then, equation (22) becomes

$$(1+i)x''-\alpha_1 x'+\alpha_2 x'^2-8=0 \tag{24}$$

where

$$i=\frac{\rho A_c L_2 R_m^2}{I}, \quad \omega_0^2=\frac{C_d C_v R_m R_0 h p \cos\phi}{I}$$

$$\alpha_1=\frac{C_d L_1 R_m R_0 h\sqrt{2\rho p}}{I\omega_0}=\sqrt{\frac{2C_d R_m R_0 h\rho}{I C_v \cos\phi}}L_1, \quad \alpha_2=\frac{c_2\theta_a}{I} \tag{25}$$

Where, $i$ is an adding moment of inertia of the fluid in the control volume. In this case, $i \ll 1$. Therefore, equation (24) is reduced to

$$x''-\alpha_1 x'+\alpha_2 x'^2-8=0 \tag{26}$$

3.3 SIMULATION OF THE EQUATION OF MOTION OF THE TURBO-MOTOR

In order to simulate the equation (26), let us estimate the rough values of $\alpha_1$ and $\alpha_2$. As an example of the numerical calculation, consider the following values which are cited from the experimental apparatus.

$$R_m \fallingdotseq R_0=1.5[\text{cm}], \quad h=0.01[\text{cm}], \quad I=\pi b\rho\pi b\rho_i R_0^4/2=0.124[\text{kg}\cdot\text{cm}^2],$$
$$\phi=40^\circ, \quad C_d=0.4, \quad C_v=1.0, \quad \rho_i/\rho_0=9.0, \tag{27}$$

If the flow path in the turbine rotor is as that shown in figure 4 by solid lines ($l_1, l_2$), then from the geometrical relationship:

$$l_1+l_2=2\pi R_0/5 \tag{28}$$

This case means that the flow path is the same as the geometrical damping length. In this case, from equs. (2) and (21),

$$L_1=l_1+l_2-4R_0\theta_a=0 \tag{29}$$

Thus, from equ. (25) we get

$$\alpha_1=0.014L_1=0 \tag{30}$$

On the contrary, in case of the traditional turbine, because of no damping length,

$$L_2=l_1+l_2=0, \quad L_1=-4R_0\theta_a=-1.9 \text{ [cm]} \tag{31}$$

Thus,

$$\alpha_1=0.014L_1=-0.027 \tag{32}$$

But, if the flow path in the turbine rotor is as that shown in figure 4 by broken lines ($l_1'$, $l_2'$), and if ($l_1'+l_2'$) is 10 % greater than ($l_1+l_2$), then:

$$L_1=1.1(l_1+l_2)-4R_0\theta_a=0.19 \text{ [cm]} \tag{33}$$

Thus,

$$\alpha_1 = 0.014 L_1 = 2.66\times10^{-3} \tag{34}$$

Therefore, we choose the values of $\alpha_1$ which covers the above values:

$$\alpha_1 = -2.7\times10^{-2}\ ,\ 0\ ,\ 0.5\times10^{-2},\ 10^{-2} \tag{35}$$

Now, let us estimate the values of $\alpha_2$.
As the order of $\alpha_2 x'^2$ is roughly equal to $\alpha_1 x'$, so that:

$$\frac{\alpha_2 x'^2}{\alpha_1 x'} = \frac{\alpha_2 x'}{\alpha_1} \approx 1 \tag{36}$$

and if we take $x' = 50$ ( that corresponds to about 8,000 rpm ), then

$$\alpha_2 \approx \frac{\alpha_1}{x'} \approx \frac{10^{-2}}{50} = 2\times10^{-4} \tag{37}$$

So, we choose the values of $\alpha_2$ as to:

$$\alpha_2 = 10^{-3}\ ,\ 5\times10^{-4}\ ,\ 10^{-4} \tag{38}$$

The result of calculation by Runge-Kutta Method are shown in figures 5.
In the figures, nondimensional rotational speed $dx/d\tau$ is shown against nondimensional time $\tau$.
It is clear from the figures that:
(1) an increase of $\alpha_1$ results in a large increase of rotational speed. By examining equ.(25), an increase of $\alpha_1$ corresponds to an increase of $L_1$, $h$ and $C_d$ and to a decrease of $R_0$ and $\rho_l$. A curve of $\alpha_1 = -2.7\times10^{-2}$ shows the result of a case of the non-damping length. Comparing the curves, we can see that the existence of the damping length has a large influence on the rotational speed of the turbo-motor.
(2) the effect of $\alpha_2$ is very large and a decrease of $\alpha_2$ results in a large increase of the rotational speed. From equation (25), value of $\alpha_2$ mainly depends on $c_2$, which is the viscous damping coefficient proportional to $\dot{\theta}^2$.
(3) under the condition of $\alpha_2 = 5\times10^{-4}$ ( see figure 5-b ), the rise time $\tau_r$ of rotational speed is about 22.0. If we convert $\tau_r = 22.0$ to a real time, it becomes $t_r = \tau/\omega_0 \fallingdotseq 0.21$ second ( where $\omega_0 = 106$ rad/sec). Here, the rise time $\tau_r$ is defined as to 90 % of the final value of the rotational speed of the turbo-motor. It can be seen from the figure that the rise time is almostly the same if the value of $\alpha_2$ is the same.
Actual rotational speed of turbo-motor is, from equations (23) and (25),

$$\frac{d\theta}{dt} = \frac{\theta_a x'}{R_0}\sqrt{\frac{2 C_d C_v h p \cos\phi}{\pi b \rho_l}} \tag{39}$$

So, in order to increase the rotational speed, we should increase $\theta_a$, $h$ and $p$ and decrease $R_0$, $\phi$, $\rho_l$ and $b$.
If the rotational speed becomes very high, the drag torque becomes very large, so that the rotational speed tends to saturate.

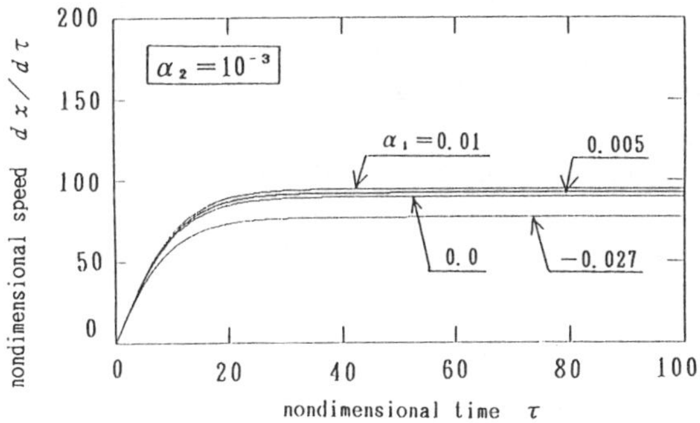


Figure 5-a  Simulation of pressure step response

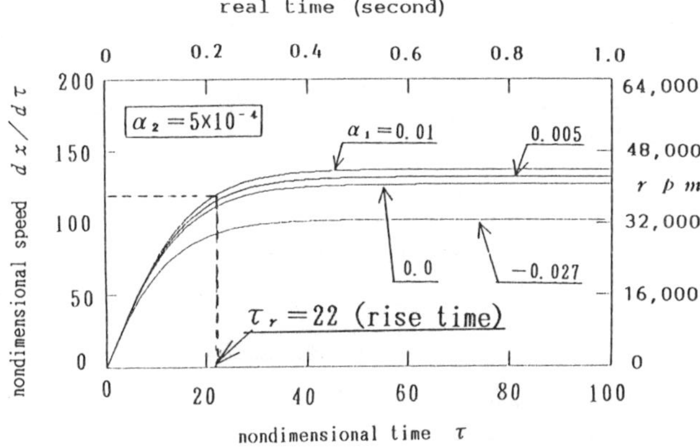


Figure 5-b  Simulation of pressure step response

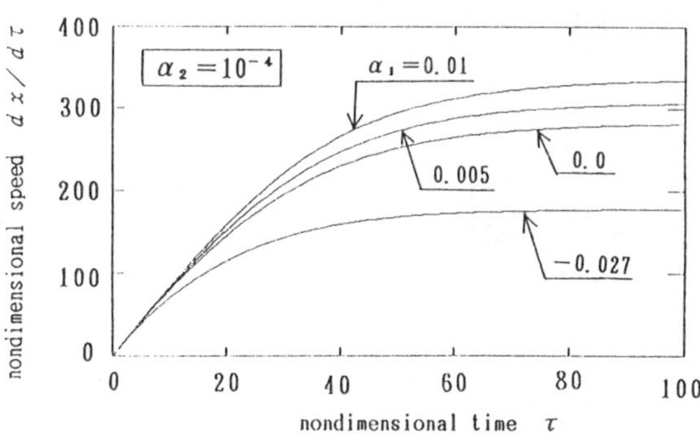


Figure 5-c  Simulation of pressure step response

## 4 Experiments and comparison with theoretical analysis

Figure 6 shows an oil circulating system which is used to test the turbo-motor. The working fluid used in the experiments is ESSO UNI POWER 32 and its kinematic viscosity is shown against the temperature in figure 7.

Figure 8 shows the rotational speed of the turbo-motor at some supply pressures. From the figure, the rotational speed increases with the increasing of the supply pressure and temperature of the working fluid. About 40,000 revolution per minute is obtained under the conditions of 10 MPa and 60 °C.

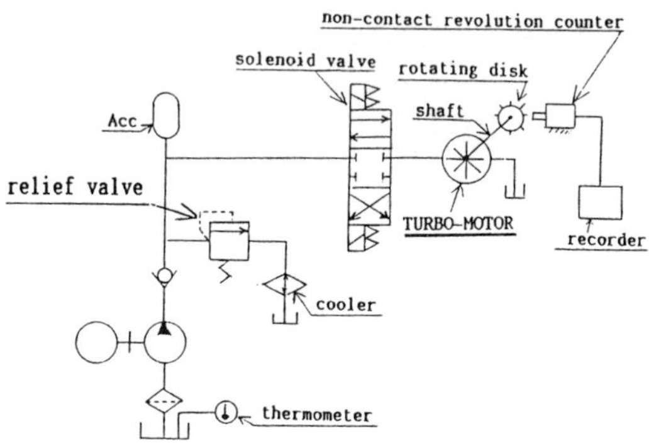


Figure 6 Oil circulating system

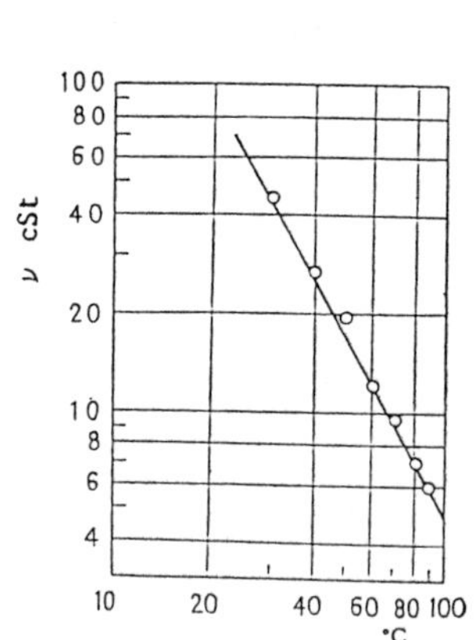


Figure 7 Kinematic coefficient of viscosity ν of working fluid

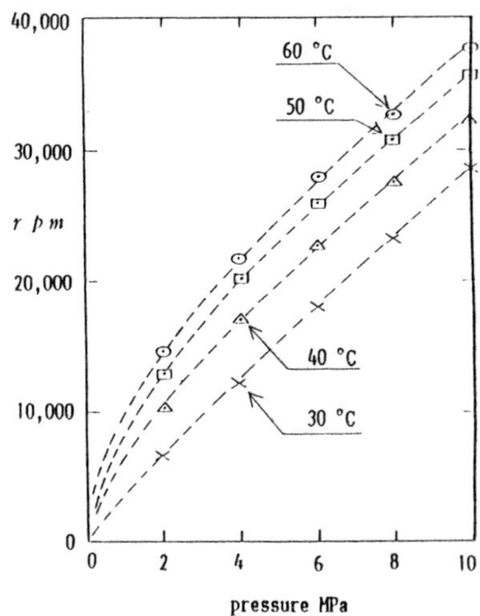


Figure 8 Experimental results of rotational speed

Experiments of pressure step input response of the turbo-motor are obtained as follows. After setting the pressure by a relief valve, pressure step inputs are applied by opening anelectro-hydraulic solenoid valve. At this moment, transients of the rotational speed of the turbo-motor are picked up digitally and analogically by a non-contact revolution counter and recorded with a recorder. Sampling frequencies is set at 1 kHz in order to make the curve smooth.
One of the experimental curves recorded is shown in figure 9.
The experimental results have a range of pressures as given in figure 10.

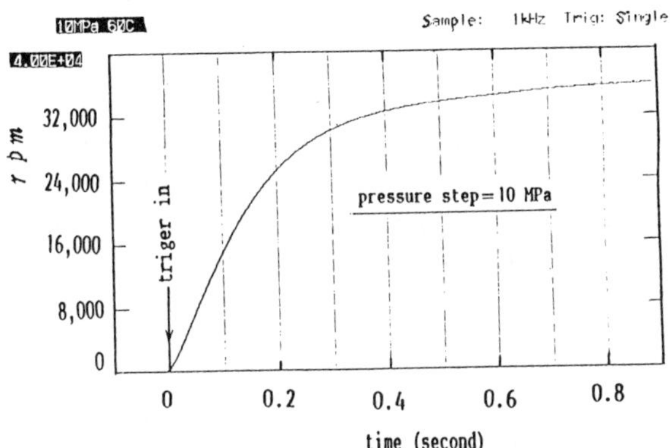


Figure 9 Experiment of pressure step input response

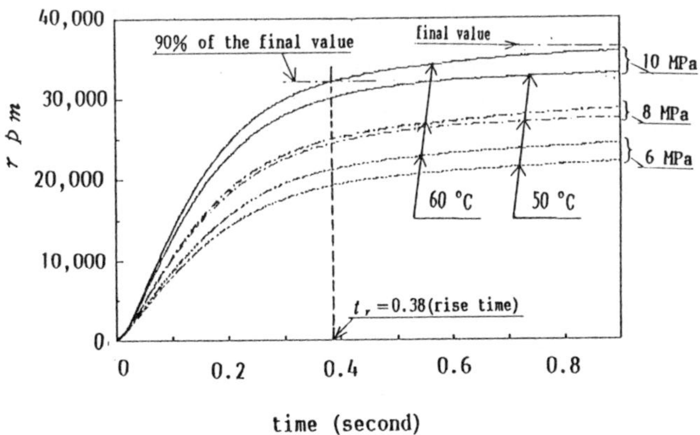


Figure 10 Experimental response curve by pressure step input

From figure 10, the rise time $t_r$ of the experimented turbo-motor is about 0.38 second. Comparing with figure 9 and figure 5-b, they are similar and coincide with each other with respect to the shape of the curves and the magnitude of revolutions. But there are some deviation of the rise time. This deviation is considered as to come from the following reasons. We have not considered the compressibility of the working fluid in the theoretical analysis. But in the experiments, as the large pressure step inputs are applied to the turbo-motor, there occurs the compression of the working fluid and the expansion of the rubber hose . So, there occurred the lag of the rotation of the turbo-motor.

Considering these conditions, we will be able to estimate $\alpha_2$ in the experiments as to $\alpha_2=5\times10^{-4}$. Then, from equation (25), $c_2=\alpha_2 I/\theta_a=(5\times10^{-4}\times0.124)/0.1\pi$ $=1.97\times10^{-4}$ [kg・cm].

## 5 Conclusion

A turbo-motor utilizing the negative damping force is presented and the theoretical analysis and simulation are made. The results of the simulation revealed the effectiveness of the negative damping effect. As the result of the theoretical analysis, the necessary guiding principles for the design of the hydraulic turbo-motor are obtained. From the experiments, about 40,000 rpm is obtained under the conditions of 10 MPa and 60 °C.

## References

[1] Funk J.E. " Poppet Valve Stability ", Trans. ASME, Ser. D, Vol.86, No.2, 1964, pp207~212.

[2] Maeda T. " Studies on the Dynamic Characteristics of a Poppet Valve ", Bull. JSME, Vol.13, No.56, 1970, pp281~289, pp290~297.

[3] Maeda T. " Studies on Hydraulic Oscillator ", Japan Soc. of Mech. Engrs. Vol.44, No.377, 1978, pp109~117.

[4] Maeda T., Konami S., " Studies on Hydraulic Oscillator (2nd Report) ", Bull. JSME, Vol.24, No.187, 1981, pp117~123.

**WRITTEN DISCUSSION**
**Development of high speed hydraulic turbo-motor**
**T Maeda (Seikei University, Japan)**

**Question:** **W Backé**
**IHP, RWTH, Aachen**

It is very interesting to hear about such a fast running motor. However, you did not show the dependence between speed and torque. This is important because it is the purpose of a motor to generate torque. Did you measure the torque-speed dependence?

**Answer:**

As you indicated, I think that the relationship between speed and torque is the most important characteristic of this turbo-motor.

To my regret, there are no commercial instruments to measure the torque under these high speeds. I am now considering a method to measure the torque by using an eddy current method.

**Question:** **CR Burrows**
**Fluid Power Centre, UK**

What type of bearings are used in the motor?

Are there any vibration problems at high speeds?

**Answer:**

In this motor, two precision ball bearings are used. There is no vibration at all at any speed. Lubrication of the bearings is the most important factor to allow operation at high speed.

**Question:** **PJ Chapple**
**Fluid Power Centre, UK**

What applications are there for the high speed turbo-motor?

**Answer:**

One example of an application is to use the turbo-motor to drive a super-charger for an engine. Other examples include drives for a grinding wheel and concrete cutter.

**Question:** **ND Vaughan**
**Fluid Power Centre, Bath, UK**

Do you have any idea of the efficiency of this type of motor under loaded conditions?

**Answer:**

There are many factors which affect the efficiency of this type of motor. Especially, the shape of nozzles and turbine blades. Bearings and lubrication of bearings are the most important factors.

**Question:** **GL Zarotti**
**CEMOTER, Italy**

What is the flow rate required by the turbo-motor in the region of 40,000 rev/min?

**Answer:**

The depth of the nozzle is 0.1 [mm] and the circumferential length is 4.7 [mm]. There are four nozzles in this case, so the total area of the nozzles is $A = 2\text{x}10^{-2}$ [cm$^2$]. So, the flow rate of this nozzle is expressed as follows:

$$Q = C_d A \sqrt{\left(\frac{2 \; X \; \Delta p}{\rho}\right)}$$

where,

$$\Delta P = 10 \; [MPa] \; , \rho = 8 \; x \; 10^{-5} \; [Ns^2/cm^4]$$

therefore,

$$Q = 2.5 \; [\text{L/min}]$$

The flow rate from the experiment is about 3.5 [L/min] under the conditions of 10 MPa and 60°C. It is considered that the difference in the flow rates comes from the measurement of the nozzle depth and the estimation of $C_d$.

**Question:** **RE Koski**
**Sun Hydraulics Corpn., USA**

For a motor with 5 blades working over a speed range from say 8000 rev/min to 40000 rev/min, the fundamental frequency varies from 666 Hz to 3333 Hz. This covers a frequency range where the human ear is most sensitive; how noisy is your motor?

**Answer:**

Thank you very much for your question on the noise from the turbo-motor. It is very gentle to the human ear until about 20,000 rev/min; it becomes noisy for speeds greater than 25,000 rev/min.

The figure below shows the spectrum of the noise pressure level. We can see from the figure that the noise levels at 3,760 Hz and 4,320 Hz are particularly strong, the noise pressure levels being 76.5 dB(A) and 68 dB(A) respectively. Here, the overall background noise is 80.5 [dB A].

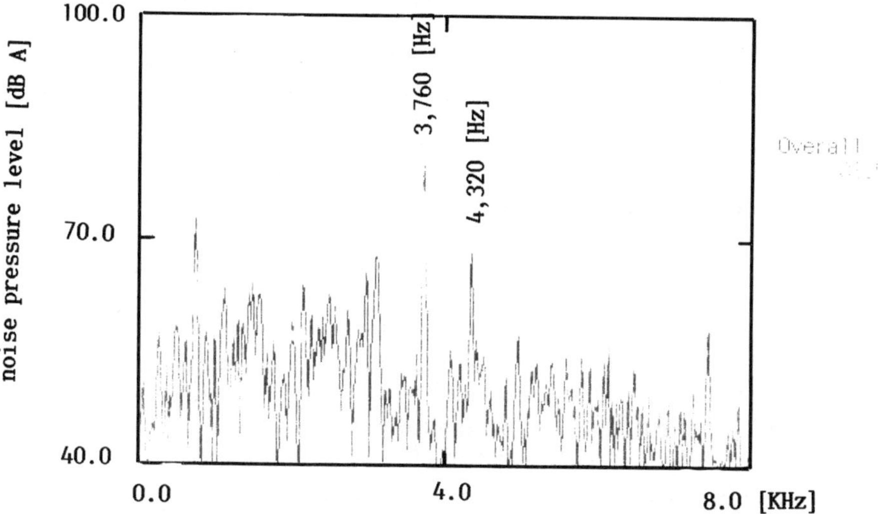


Spectrum of the noise pressure level with driving a turbo-motor
10 [MPa], 60 [°C], 36,000 [rpm] , distance from mic.: 1 [m]

# 4. Modelling and Analysis Techniques used in the Design of Axial Piston Pumps for Water Hydraulics

M J Fagan *and* J McConnachie

## ABSTRACT

The Water Hydraulics project at Hull is aimed at developing water powered (and lubricated) axial piston pumps and motors using advanced engineering materials, such as ceramics, with the ultimate goal of developing a sea-water pump. This paper will discuss the design, modelling and analysis undertaken during the project, illustrating each phase with examples of the outputs from the different models. It will conclude with a progress report on the project so far and the current position of the sea-water pump.

## 1 BACKGROUND

The University of Hull began the Water Hydraulics project in 1989 in collaboration with J H Fenner plc with funding provided by the DTI under the Support For Innovation (SFI) initiative. The purpose of the grant was to further develop the basic pump units already available using new materials and in particular engineering ceramics, and although this particular source of funding has now finished, research continues towards an electrically neutral sub-sea sea-water pump. It is envisaged that the sea-water pump will require minimal filtration and ultimately be suitable for a variety of 'dirty water' applications.

The pump used in the research was originally developed at the NEL (National Engineering Laboratory) and more recently at J H Fenner plc (Hull). It is a straight axis axial piston pump as illustrated in figure 1. The cylinder block is coupled to a shaft, and as it rotates the pistons reciprocate in the bores causing water to be drawn in and pushed out, producing a pumping action. At present, the standard (Fenner) units operate reliably on clean, filtered water at pressures of up to 210 bar.

There are three aspects to the work at the University of Hull, namely,

- material selection and screening
- pump/component design and analysis
- pump testing.

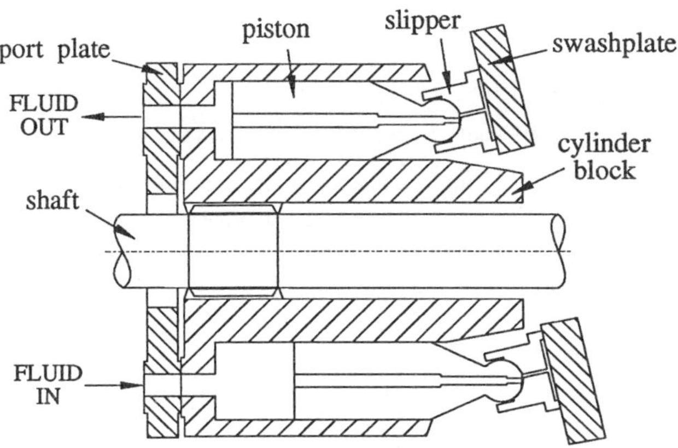


Figure 1 Schematic layout of the axial piston pump.

Material selection is of course vital to the successful development of the pumps, and much effort has been spent on identification of suitable materials. Since there is negligible lubrication of the running surfaces, the pump's operation relies on the correct material pairing. Potential material pairs are tested on a 'pad-on-plate' test rig, which is similar to a 'pin-on-disc' testing machine, but uses a (piston-loaded) slipper running on a rotating swashplate. This pad-on-plate test operates at representative loads and speeds which are much higher than those used in pin-on-disc tests. The cavitation erosion resistance of the materials is tested using a 20 kHz ultrasonic vibratory cavitation test.

The majority of the design and analysis work is computer based, centred around a number of computer models. In particular there are three aspects to the modelling,

- basic design and analysis of the pump system and its individual components
- modelling of the contact conditions
- diagnosis of component failure.

Each of these areas of modelling is now discussed with some typical results.

## 2 DESIGN AND ANALYSIS OF THE PUMP AND COMPONENTS

The design and analysis of the components is one of the most important areas of work in the project. Because of the complexity of the pump and the many interactions between the different components, it was decided to attempt to develop a computer model of the complete pump to allow a speedy evaluation of the effects of any design changes or material substitutions. It was also planned in the longer term, to use these models to undertake detailed research into the effects of varying individual features of the pump. The design and analysis has been undertaken in four stages, namely

- development of the basic geometry,
- timing analysis,
- force analysis,
- finite element analysis of individual components.

The overall dimensions of the components are derived in the first phase, and then from the port-plate geometry a detailed timing analysis is performed, to design the specific features of the port-plate and obtain the pressure variations during the pump's operation. With this information a three-dimensional force analysis of the complete pump is then undertaken, allowing the calculation of the loadings on all the components. A series of finite element models of the components then allows a detailed examination of the stress distributions as necessary. All of the analyses and models are completely parametric allowing the analysis of any sized pump or component with minimal effort.

## 2.1 Geometry designer

When attempting to calculate the basic dimensions of a pump to operate at a given flow and pressure, it soon becomes evident that many of the dimensions are closely interrelated. Furthermore, other secondary requirements such as minimum casing length, impose additional restrictions on the range of acceptable component sizes. Hence an involved selection and combination of dimensions is required. To aid in the derivation of the basic geometry, a geometry designer has been developed, which works by accessing a large number of simple geometrical relationships which link the components' dimensions. This has already been described in detail elsewhere (Radcliffe et al. 1993). The results of a session with the geometry designer is a large data file with all the pump's dimensions which are used in the subsequent timing and force analysis stages.

Figure 2 shows how the program has been used to explore the relationship between swash angle, pitch circle diameter of the piston bores and piston diameters in a series of pumps with the same flow. The graph shows that by increasing the swash-angle and thereby increasing the piston stroke, the piston diameter may be decreased to maintain the same flow. Furthermore, increasing the pitch circle diameter of the bores for a given swash-angle also increases the stroke and thus leads to a further possible decrease in piston diameter. The lower limits of the piston diameter and pitch circle diameter for each swash-angle are also shown; this arises from the conditions of zero material thickness between bores.

## 2.2 Timing analysis

The timing of an axial piston pump controls the transition of cylinder pressure between delivery and boost pressure and vice versa, and might typically last 20% of the pressure cycle of a pump. The design of the port-plate has a major effect on the pressure cycle of a cylinder, whilst other factors such as swash angle, pump speed and fluid properties are less

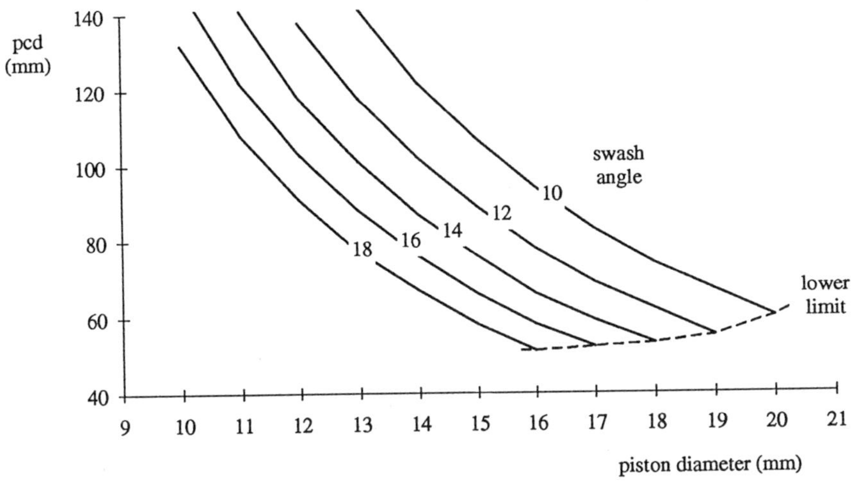


Figure 2 Graph showing the relationship between swash-angle, piston diameter and pitch circle diameter.
(Pump delivering 30 cc/rev at 1500 rpm).

significant. An incorrectly 'timed' pump can lead to large pressure overshoots on the delivery stroke and pressure undershoots, giving cavitation on the suction stroke. Thus the pump timing has a very large effect on the forces acting in the pump and so must be considered before the force analysis of the other components and in particular the balancing of the cylinder block can be examined.

Methods have been developed to model the port-plate and take account of different groove layouts and designs with variable leakage rates. The resulting differential equations are solved by a fourth order Runge-Kutta method, and implemented in a computer program which predicts the variation of pressure and flows with angular position (Radcliffe, 1992).

Figure 3 shows a graph of cylinder bore pressure versus rotation angle for two designs of port-plate. When the port-plate is designed correctly (dotted line) the pressure rises and falls to the required high and low pressure values without any undershoot or overshoot; but when the silencing grooves are omitted, significant pressure deviations occur (solid line).

## 2.3 Force analysis

The force analysis is central to the component modelling. It is a full three-dimensional equilibrium analysis of the pump's components using vector mechanics techniques. It works by dividing each revolution of the pump into a finite number of steps, and then solving all the equilibrium equations to find the reactions and moments at every time step. Inertia loads

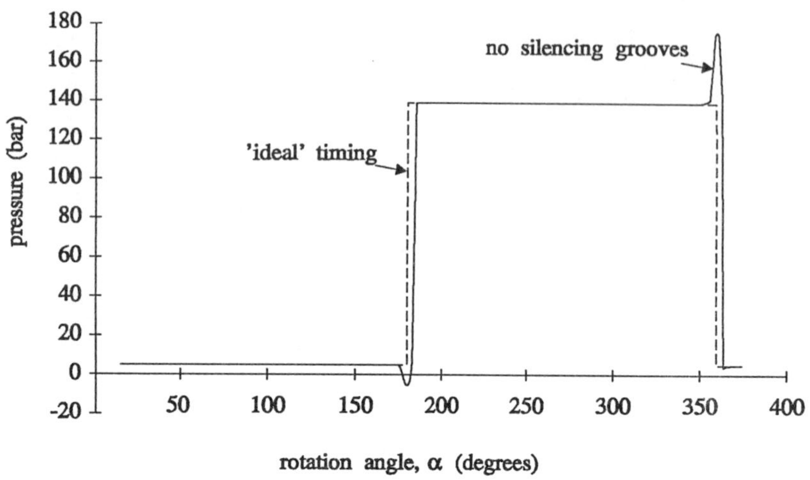


Figure 3 Variation of the bore pressure versus rotation angle for two designs of port-plate.
— — — ideally timed
———— silencing grooves omitted

are considered by calculating the accelerations of the components from their positional vectors. This work has been published in detail elsewhere (Radcliffe et al. 1992), but the complexity of the analysis is illustrated in figure 4, which shows the forces acting on the cylinder block. An explanation of the different forces in figure 4 is given in table 1.

| Vector | Force/reaction |
|---|---|
| $\underline{R}_{16f}$ | 9 × lower reaction forces from pistons |
| $\underline{R}_{17f}$ | 9 × upper reaction forces from pistons |
| $\underline{F}_{9h}$ | 6 × pressure forces on lands |
| $\underline{R}_{37}$ | upper snout bearing reaction |
| $\underline{R}_{38}$ | lower snout bearing reaction |
| $\underline{R}_{42}$ | solid contact reaction with port plate |
| $\underline{R}_{20f}$ | 9 × frictional shear from shaft |
| $\underline{R}_{46}$ | spring force |
| $\underline{F}_{3f}$ | 9 × pressures over end bore areas |
| $\underline{F}_{10f}$ | 9 × pressures in cylinder bores |
| $\underline{F}_{99}$ | inertia force |

Table 1. Loading and reaction forces on the cylinder block in the force analysis.

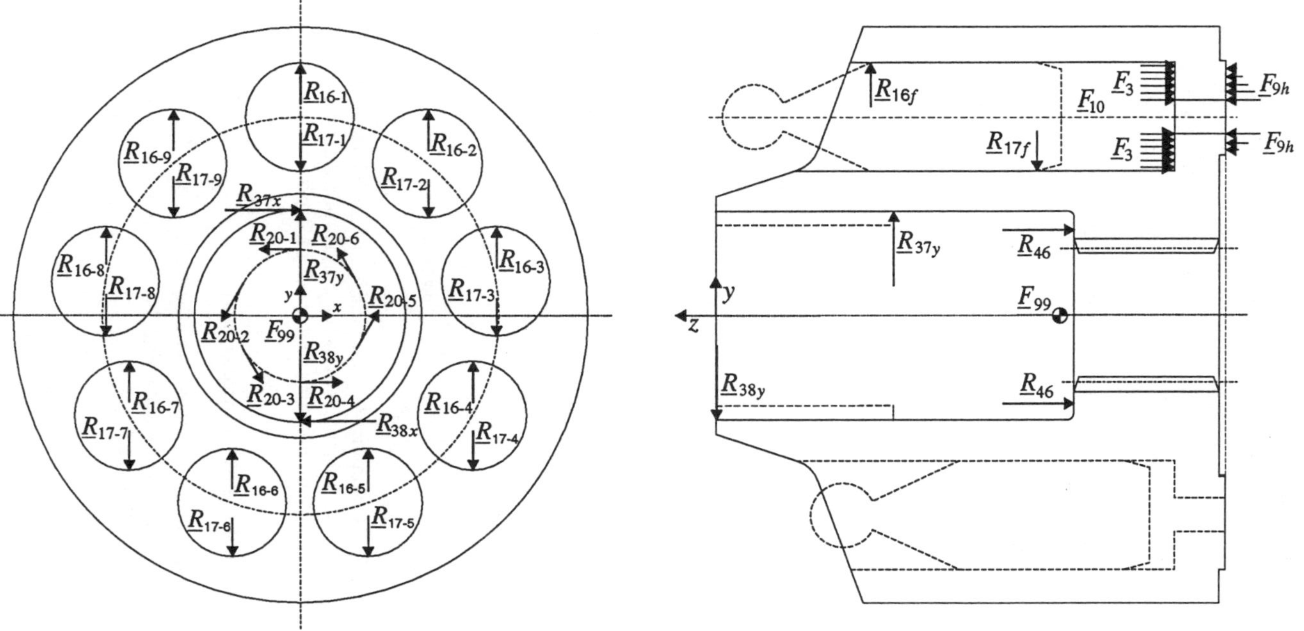


Figure 4 Loading and reaction forces on the cylinder block.

The force analysis program has been used to produce figure 5, which shows some of the loads and reaction forces in pumps with different sized pistons but the same flow. (The pumps have the same bore pitch circle diameter, but the swash angles and piston diameters are varied to achieve a constant flow rate). Figure 5 presents the piston diameters, snout bearing reactions, slipper force on the piston ball and the resultant piston bending stress. Note that for ease of presentation, all the results are normalized with respect to the values predicted with the standard configuration, that is with a swash angle of 13°.

Thus the snout bearing reactions increase significantly as the swash angle is increased (from 9° to 20°). However, the magnitude of the slipper reaction on the piston ball decreases, but note that the resultant stress actually increases because the diameter of the piston decreases with increasing swash angle.

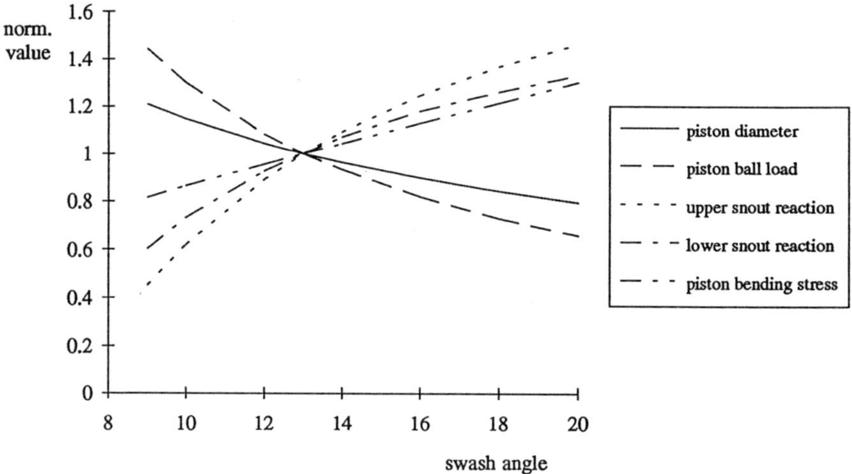


Figure 5 Graph showing the variation of some internal pump forces and stresses with swash angle and piston diameter. (Pump delivering 30 cc/rev at 1500 rpm).

## 2.4 Finite element analysis of components

The geometry of most of the components in the pump is complex, and as has already been described, the loading is certainly non-trivial, hence simple stress calculations (such as beam and thick cylinder theories) are quite inadequate for detailed design considerations. To accurately analyse the pump, finite element methods are required, and with suitably designed parametric models any sized component can be easily considered. A series of models has been developed for the majority of components in the pump. For example, figure 6 shows the models for three early designs of slippers. The type I slippers were machined from solid polymer and steel materials, while the polymer seat in the type II design was injection

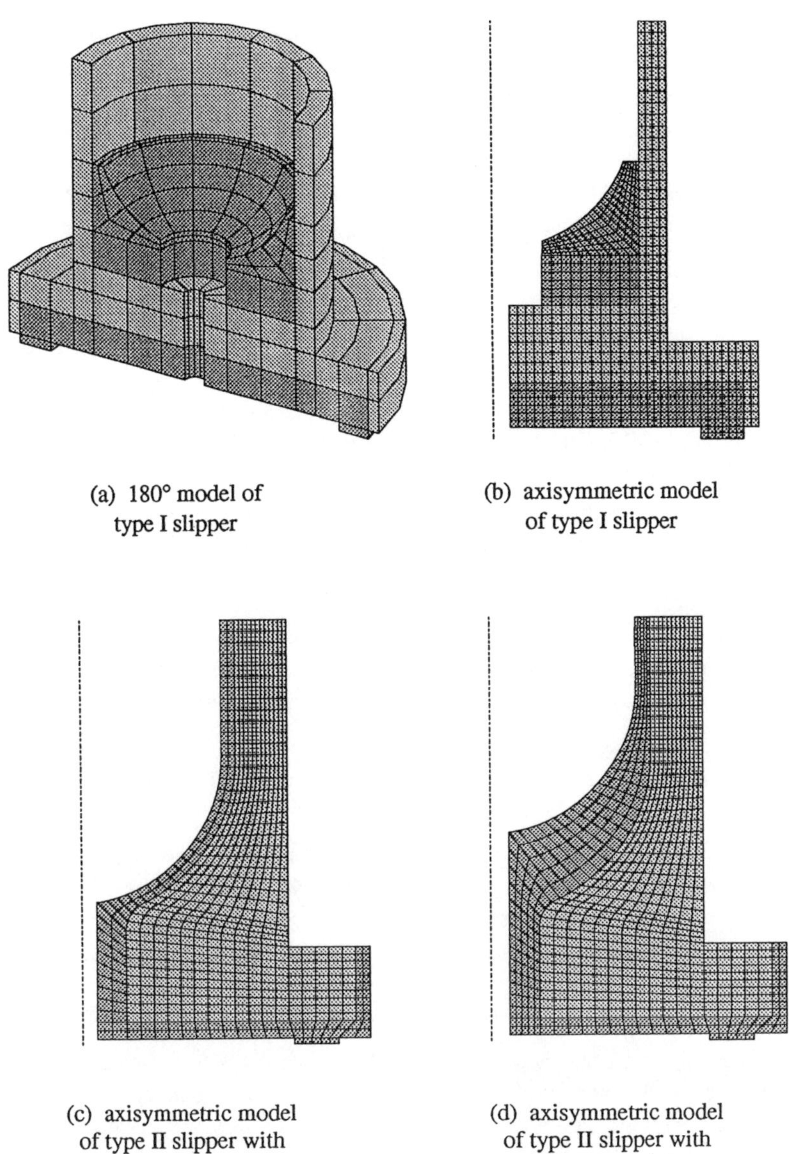

(a) 180° model of type I slipper

(b) axisymmetric model of type I slipper

(c) axisymmetric model of type II slipper with thin polymer seat

(d) axisymmetric model of type II slipper with thick polymer seat

Figure 6 Finite element models of different slipper designs.

moulded in to a steel core. Two versions of the latter design are shown with thick and thin seats. Typical results from these models are presented and discussed in the following section on component failures.

## 3 MODELLING OF THE CONTACT OF CONCENTRIC COMPONENTS

Achieving suitable contact conditions is crucial to the successful operation of the pump, and in particular the performance of the piston in the bores and the piston head in the slippers are of primary importance. In view of this, a fundamental study of layered concentric components has been undertaken as part of the overall project, which sheds some light on the correct selection of materials and running conditions.

A number of two- and three-dimensional models have been developed (McConnachie and Fagan, 1993a and 1993b). For example, the most simplistic model consists of a 90° segment of sleeved piston in a lined bore (figure 7a). A plane strain model of three cylinder bores then allows the loading pattern predicted in the first model to be combined with the pressure loads to give a realistic stressing regime around the cylinder bores (figure 7b). A three-dimensional cylindrical contact model accurately predicts the contact stresses for sleeved and lined components (figure 7c). Finally an accurate model of a section of the cylinder block then allows the stress levels around the porting to be calculated due to the contact and pressure loads (figure 7d). The contact models require a non-linear analysis, with the two parts of the model being separated by gap elements requiring an iterative solution.

Validation of these contact models has been achieved to some extent by comparison of the model results with the wear measurements obtained from pumps. The position and area over which wear occurred generally agreed with that predicted by the models.

An examination of the contact conditions occurring in the slipper models described above has also been undertaken, in particular to investigate why a number of slippers failed prematurely, as discussed in the following section.

## 4 FINITE ELEMENT ANALYSIS OF COMPONENT FAILURES

During the development of the pump units, there have naturally been a number of failures of the components. While such failures are disappointing, much information can be learned from a thorough examination of the failed components, and they are a valuable means of verifying the analytical and finite element models. The time, type and position of the failure, can be predicted by the models if the models are constructed and loaded correctly. In conjunction with the structural investigation of the failures, a detailed material investigation of the components also confirms the material's composition and preparation, and a detailed examination of the wear mechanisms gives useful information on the contact conditions.

Finite element models of the early slipper designs are shown in figure 6. They were developed to analyse the different designs and quantify the failures. The Von-Mises stress

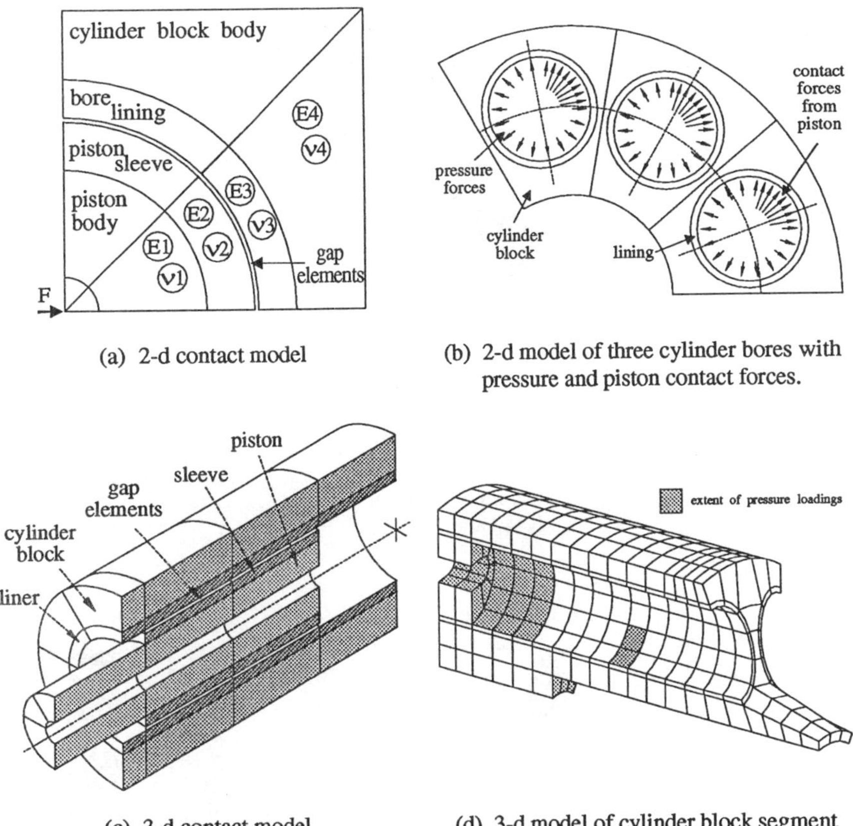


(a) 2-d contact model

(b) 2-d model of three cylinder bores with pressure and piston contact forces.

(c) 3-d contact model

(d) 3-d model of cylinder block segment

Figure 7 Finite element models to examine the contact conditions between the pistons and cylinder bores.

distribution in the type I slipper is shown in figure 8a. The stress concentrating effect of the corner in the steel is obvious, and indeed a number of slippers have failed from this point. Increasing the radius of the corner and the wall thickness have eliminated this problem.

Figures 8b and 8c show the Von-Mises stress distribution in the type II slipper designs. The thin seat leads to a higher stress in the steel, and the maximum stress in the polymer occurs at the location where the seat has been seen to crack and fragment. Failure of the polymer was not seen in the slipper with the thicker seat. Overall, the type II design leads to a significant reduction in the stresses occurring in the slipper, primarily due to the greater steel content, but also because of the improved load transfer path.

In addition to these failures, a number of cylinder blocks and pistons have failed during the

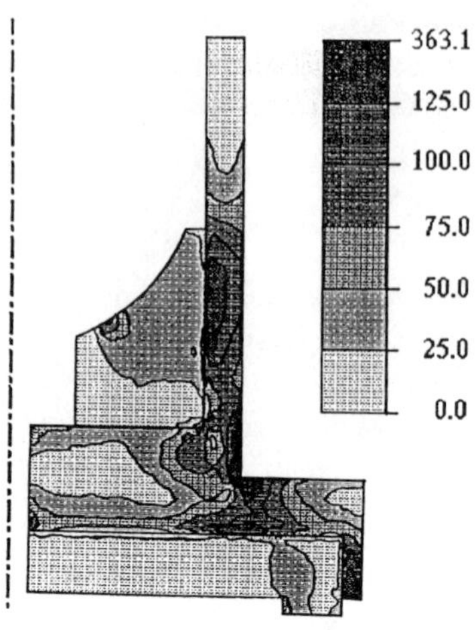


(a) axisymmetric model
type I slipper

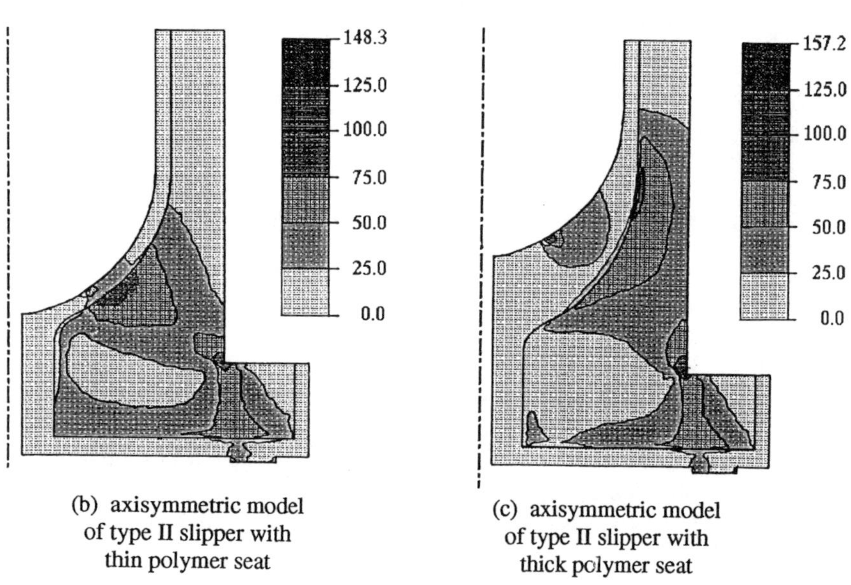


(b) axisymmetric model of type II slipper with thin polymer seat

(c) axisymmetric model of type II slipper with thick polymer seat

Figure 8 Von-Mises stress distributions in the different slipper designs.

project. Again finite element models of these components have been used to examine and explain the failures and to aid in the redesign of the components to prevent the failures from re-occurring (McConnachie and Fagan, 1993b).

## 5 CONCLUSIONS

Clearly the detailed design and analysis of axial piston pumps is a complex process. The computer models described in this paper serve a number of functions. They allow a particular pump specification to be developed and analysed in detail in a way that was previously not possible and allow a rapid evaluation of the new design, but they also provide the opportunity for the study of the components' interactions and general axial piston pump design.

Regarding the development of the sea-water pump, the combination of ceramic running against fibre-reinforced polymer has proven to be a successful low wear, low friction combination, (not only at the piston/cylinder block interface but also at other critical interfaces in the pump). A unit with ceramic pistons and swashplate has recently undergone sea-water trials with a filtration level of just 120μm. The pump was tested for more than $60 \times 10^6$ cycles (700 hours) at pressures up to 140 bar on tap water in the laboratory, and then for nearly $15 \times 10^6$ cycles at similar pressures on sea-water before failure of the cylinder block occurred. The reason for the failure appears to be simple fatigue of the stainless steel, which can be easily remedied.In any case, work is now underway to develop a monolithic ceramic cylinder block, which will eliminate the fatigue problems. It will also improve the wear characteristics of the backface of the cylinder block which was suffering as a result of the low filtration levels.

## ACKNOWLEDGEMENTS

This work was funded by the DTI - Support for Innovation (SFI) initiative in collaboration with J H Fenner plc.

## REFERENCES

**McConnachie J. and Fagan M. J.**, 1993a. "Design and analysis of the cylinder block of an axial piston pump". 3rd Scandinavian International Conference on Fluid Power, 25-26 May, 1993, Linkoping, Sweden.

**McConnachie J. and Fagan M. J.**, 1993b. "Computer modelling of layered conformal contact". Contact Mechanics 93. First International Conference, 13-15 July, 1993, Southampton.

**Radcliffe P. W.**, 1992. "Timing of an axial piston pump", Internal report, Fenners/DTI Water Hydraulics Project, University of Hull.

**Radcliffe P. W., Fagan M. J. and McConnachie J.**, 1992. "Computer modelling and analysis of axial piston pumps". In: *Systems Modelling and Control*, (Eds. C R Burrows and K A Edge), Research Studies Press.

**Radcliffe P. W., Fagan M. J. and McConnachie J.**, 1993. "Computer aided desing of axial piston pumps - a geometry designer". In: *Circuit, Component and Systems Design*, (Eds. C R Burrows and K A Edge), Research Studies Press.

**WRITTEN DISCUSSION**
**Modelling and analysis techniques used in the design of axial piston pumps for water hydraulics**
**MJ Fagan & J McConnachie (University of Hull, UK)**

**Question:** **W Backé**
**IPH, RWTH Aachen, Germany**

You told us that your pump was running with a filter-level of 120 μm. We have found that contamination with solid particles leads to a rapid reduction in volumetric efficiency and life. This is because of the very narrow sealing gaps in water-hydraulic pumps.

Did you find a dependence between life and contamination concentration?

**Answer:**

Yes we certainly find that reducing the level of filtration leads to increased wear particularly of the backface of the stainless steel cylinder block. This is why we are introducing a ceramic backface to the cylinder block, and designing a monolithic ceramic cylinder block. However, we haven't run enough tests to quantify the relationship between life and contamination concentration.

**Question:** **M Sethson**
**University of Linköping, Sweden**

How much more will a ceramic pump cost than a unit constructed from 'traditional' materials?

**Answer:**

At present a pump with ceramic pistons and swashplate and polymeric liners will cost about 2-3 times more than a traditional unit. For arduous operating conditions (for example, sub-sea with minimal filtration) we are investigating a pump with a ceramic cylinder block and pistons which might cost 5 times that of a 'traditional' unit. However, the traditional stainless steel-based unit would only have a very short life in these conditions. For those applications which would benefit from sub-sea water hydraulics, such as autonomous sub-sea control systems, this premium would be well worth paying. Furthermore, note that these prices are for prototype units, with economies of scale in the manufacture of the ceramic components, the price differential would be significantly less.

**Question:** **J Darling**
**Fluid Power Centre, Bath, UK**

Did you find cavitation to be a problem in your pump?

Were the failures associated with the pumping of water or were they structural failures which would have occurred regardless of the working fluid?

**Answer:**

Cavitation has not been a problem with our pumps, and we believe shouldn't be if the timing, porting and port-plate are carefully designed, and the unit is then operated at the design conditions. However, we have seen evidence of cavitation on the metallic components of other designs of pumps with stainless steel components; in particular, steel pistons and exposed steel in the port-plate have been damaged. Our ceramic pistons and polymer coated port-plates show no such cavitation damage, presumably because of the hardness of the ceramic and the resilience of the polymer.

The failures may have been related to the fact that we were pumping water, in that particles in the water combined with smaller operating clearances lead to increased 'frictional' forces and thus more severe loading conditions.

# 5. Triple Pump Controls with Speed Sensing Power Management

R Paoluzzi, M Puglia *and* G L Zarotti

*There is a time for departure even when there's no certain place to go.*

T. Williams, 1953

## 1 Introduction

So far, three is the maximum order of multiple automatic controls of open circuit variabile displacement pumps for mobile applications. Two members of these controls are common to all architectures, i.e. the *pressure compensator* or *limiter* (PC) and the *flow compensator* or *limiter* (FC), which bound the components of the pump output power. The third member deals with the pump mechanical power and its match with the (diesel) engine. Two implementations are currently found:

a) the *torque limiter* looks at the pump itself and bounds the input torque. It is sometimes extended to two pumps by means of the symmetric exchange of pressure signals (*summing* torque limiter)

b) the *speed limiter* (SL) looks at the engine and bounds its minimum speed by acting on the pump, whichever is the cause of the speed droop (pump itself or additional torque)[1]. In case of two pumps, one of them works as *master* unit (with full triple control), and the other (with double control only) shares the same SL signal.

In the simplest configuration (engine plus one or two controlled pumps) the torque and speed limiter have the common advantage of making a pump/engine power ratio greater than *one* feasible. On the other side they are different in several respects. According to the current qualitative comparison, the torque limiter is said to be faster in response to load changes, operate on a wider range of engine speed, have less power consumption, and require a simpler and modular hardware. Conversely, the speed sensing is said to use the full engine torque, adjust for altitude or engine deterioration effects, and allow for additional power demands. These pros (and implicit cons), however, miss the key difference, namely that the torque limiter implies *high priority* in use of engine power, whereas the speed limiter implies *low*

1. A similar principle is applied in the hydrostatic transmissions area [1].

priority. Such a priority attribute is probably one among the reasons why the torque limiter is much more popular in practical applications. In the same perspective, some problems are also to be seen which will be discussed in this paper.

## 2 Nomenclature

$A_i$ i-th spool active area, $mm^2$
$A_l$ control piston active area, $mm^2$
$A_s$ pumping piston active area, $mm^2$
$A_u$ bias piston active area, $mm^2$
$C_i$ i-th spool viscous damping coefficient, Ns/mm
$C_p$ swashplate viscous damping coefficient, Nms/deg
$C_0$ overall orifice flow coefficient, -
$d$ equivalent orifice diameter, mm
$G_i$ i-th spool area gain, $mm^2$/mm
$J$ swashplate inertia, g $m^2$
$K_i$ i-th spool spring stiffness, bar/mm
$K_p$ bias spring stiffness, bar/deg
$L_b$ swashplate c.o.g. distance from the swashplate tilting axis, mm
$L_c$ control & bias piston distance from the swashplate tilting axis, mm
$M_c$ average control/bias piston mass, g
$M_i$ i-th spool equivalent mass, g
$M_p$ swashplate mass, g
$M_s$ pumping piston mass, g
$N$ number of pumping pistons (9)
$Q$ orifice flow (generic), $dm^3$/min
$\Delta p$ differential pressure (generic), bar
$p_j$ j-th volume fluid pressure, bar
$p_n$ nominal pump pressure, bar
$p_{si}$ i-th valve setting, bar
$r$ feedback ratio (valve 3), -
$Re$ transition Reynolds number, -
$R_s$ pumping piston distance from centerline, mm
$t$ time, s
$T_a$ additional torque demand, Nm
$T_e$ engine output torque, Nm
$T_M$ full injection engine torque, Nm
$T_m$ null injection engine torque, Nm
$T_p$ portplate timing torque, Nm
$U_i$ i-th spool central overlap, mm
$V_j$ j-th fluid volume, $cm^3$
$X_{0i}$ i-th spool maximum travel, mm
$Z$ fuel rack actuator position, -
$\beta$ swashplate angle, deg
$\beta_0$ maximum swashplate angle, deg
$\nu$ fluid kinematic viscosity, cSt
$\rho$ fluid density, $kg/m^3$
$\omega$ pump & engine speed, rev/min
$\omega_n$ nominal pump speed, rev/min
$\omega_s$ speed limiter nominal set, rev/min

## 3 Triple Control Model

The simulation environment chosen for the axial piston pump with the triple control circuit is shown in Figure 1. The pump delivery goes to tank through two orifices in series (*flow orifice* and *load orifice*) and a relief valve in parallel (*load valve*). The pump swashplate is operated by a control piston which works against a bias piston and a bias spring. The bias piston senses the delivery pressure in volume $V_1$, and the control piston senses the pressure metered by three tandem valves in the control line (lumped in volume $V_{3,4,5}$). Valve 1 and valve 2 have normally closed 3-way spools and work as pressure compensator and flow compensator respectively. The third valve is part of the speed limiter subsystem shown in Figure 2. It includes a fixed displacement pump driven by the engine shaft, and a variable opening orifice which work together and develop the *speed sensing* signal. Valve 3, which has a normally closed

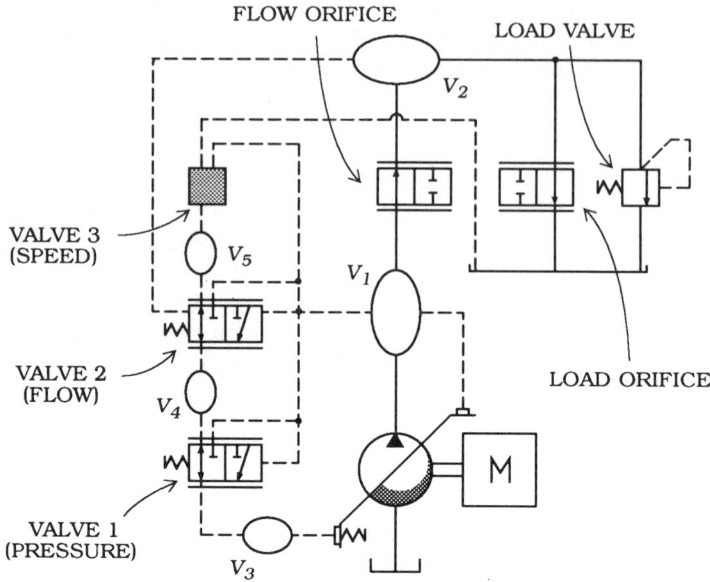


Figure 1 *Schematic of the circuit under study except the speed sensing subsystem*

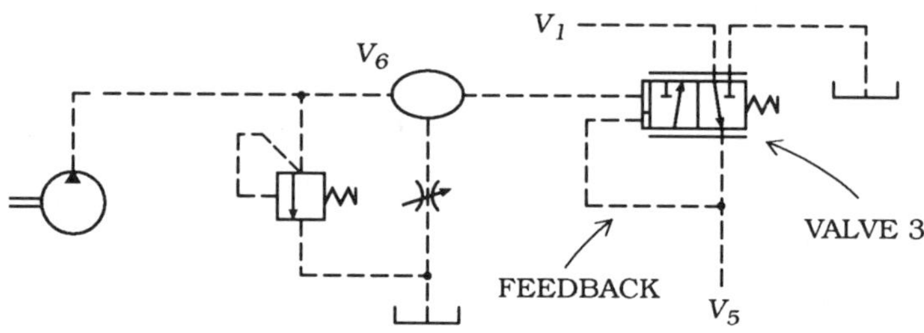


Figure 2 *Schematic of the speed sensing subsystem*

3-way spool, is operated by the speed sensing signal and a pressure feedback (related to the swashplate position) from control volume $V_5$. The spring setting in spool 3 is fixed, which allows of an energy saving relief to bypass the orifice.

The mathematical model of the system describes the interaction of several modules, mostly coincident with the pieces of hardware in Figure 1. In particular:

1) the *engine* module simulates the torque vs speed characteristic of a unit rated at 150 Nm @ 2200 rev/min, as a combination of its extreme limiting curves (maximum and null fuel injection curves) piloted by the position of the fuel rack actuator and affected by any torque demanded by auxiliaries, equipment, locomotion gear, etc.. (additional torque)

$$T_e(\omega) = Z \cdot T_M(\omega) + (1-Z) \cdot T_m(\omega) - T_a \qquad (1)$$

(the full injection curve has an 11% torque margin at 60% of the nominal speed). The rack actuator position is outputted by the *governor*, simply treated as a first order all speed controller that tries to maintain the engine speed set by the throttle position along the $T_M$ curve. The time constant of the controller is 0.1 s and the maximum overspeed (steady state) is about 160 rev/min. More sophisticated models (e.g. [2]) would be available, though the pump SL control has minor interactions with the engine governor

2) the *pump* module simulates flow and torque of a 100 $cm^3$/rev unit whose nominal pressure and speed rating are 300 bar and 2200 rev/min respectively (the ratio between the corner torque and the engine torque is 3.2). The description of flow and torque losses are taken from [3]. The torque balance of the swashplate is similar to the equation found in [4], namely

$$\frac{10^{-9}\pi}{180}\left[10^6 J + \frac{M_s R_s^2 N}{2\cos^4\beta} + 2M_c L_c^2 \cos^2\beta\right]\frac{d^2\beta}{dt} = \frac{10^{-11}\pi^2}{9}\left(\frac{M_s R_s^2 \omega^2 N}{2\cos^2\beta}\right)\tan\beta$$

$$+ 10^{-6} g M_p L_b \cos\beta - \frac{d\beta}{dt}C_p - \frac{10^{-9}\pi^2}{180^2}\left[\frac{M_s R_s^2 N}{\cos^4\beta}\tan\beta + M_c L_c^2 \sin 2\beta\right]\left(\frac{d\beta}{dt}\right)^2$$

$$+ 10^{-4}\left[p_1 A_u - p_3 A_l + K_p(\beta_0 - \beta) A_u\right] L_c \cos\beta + T_p \qquad (2)$$

with the exception of the average portplate timing torque. To this end the simulation results available in [5] are used in the following form

$$T_p = 10^{-4}\frac{p_n A_s R_s}{\cos^2\beta} F\left(\frac{p_1}{p_n}, \frac{\omega}{\omega_n}, \frac{\beta}{\beta_0}\right) \qquad (3)$$

which includes the dimensionless multilinear interpolating function $F$ based on the set of data collected in Table 1. Although these data do not come from laboratory testing, they are useful as warnings to potential problems in the control performance. Some design parameters of pump body and swashplate are collected in Table 2. Most of them are rescaled by similitude from a smaller unit, except $A_u$, $A_l$ and $K_p$ which had to be tuned separately [2]

3) each *3-way valve* module simulates the relevant flows vs differential pressures, and the balance of forces as well (at this stage the flow forces are disregarded). The opening of the metering orifices are linear functions of the spool position with fixed

2. For instance, the bias piston area has an inferior bound to counteract the spontaneous high pressure destroke caused by the portplate timing torque.

Table 1 *Dimensionless multiplier of the portplate timing torque*

| pressure, % | speed, % | angle, % | $F$ |
|---|---|---|---|
| 100 | 100 | 100 | -0.69 |
| 100 | 100 | 20 | -0.72 |
| 100 | 30 | 100 | -0.61 |
| 10 | 100 | 100 | +0.124 |
| 10 | 30 | 100 | +0.043 |
| 10 | 100 | 20 | -0.035 |
| 100 | 30 | 20 | -0.61 |
| 10 | 30 | 20 | -0.037 |

Table 2 *Design parameters of pump body and swashplate*

| | | | |
|---|---|---|---|
| $A_l$ = 500 mm$^2$ | $J$ = 20.7 gm$^2$ | $L_c$ = 88 mm | $M_s$ =157 g |
| $A_s$ = 418 mm$^2$ | $K_p$ = 1 bar/deg [a] | $M_c$ = 200 g | $R_s$ = 45.5 mm |
| $A_u$ = 156 mm$^2$ | $L_b$ = 26 mm | $M_p$ = 6125 g | $\beta_0$ = 16.5 deg |

a) Pressure is referred to the control piston active area

proportional gains. The motion of spools is constrained by two end stops (the same applies to the fuel rack actuator and swashplate)

4) all fixed and variable size *orifices* are simulated by flow vs pressure equations in two phases (linear and quadratic) whose transition is checked by the Reynolds number. This helps avoid computational "shooting" effects and handle the no flow condition as well. The equivalent form of the generic equation is the following

$$Q = 1.5\pi\sqrt{10}\ \mathrm{sgn}\,(\Delta p)\, C_0 d^2 \sqrt{\frac{2}{\rho}|\Delta p|}\ \min\left(1,\ 10^{5.5}\frac{C_0 d}{\nu Re}\sqrt{\frac{2}{\rho}|\Delta p|}\right) \qquad (4)$$

5) some fluid properties are constant throughout all simulation experiments, namely $\rho$ = 850 kg/m$^3$ and $\nu$ = 30 cSt. Conversely, the bulk modulus is pressure dependent (in the same manner as a liquid-gas mixture), and is asymptotic to 10.000 bar.

The system model has been developed, assembled and solved by means of EASY5x, a general purpose simulation package [6]. The numerical integrator is Stiff Gear or its modified version BCS Gear (both implicit, featuring variable order and step). Most discontinuities are modelled by the so called *switch states*, which work as some kind of slack variables.

## 4 Pressure and Flow Compensator

The first two steps of the system analysis are focussed on the pressure and flow compensators. The design parameters of the relevant valves have been derived through a refinement procedure which had the parametric simulation and the root locus inspection as basic tools, and the control *flow consumption* as a primary constraint.

Some parameters are collected in Table 3. As to flow consumption, two definitions

Table 3 *Design parameters of pressure and flow compensating valves*

| VALVE | $p_{si}$ | $A_i$ | $M_i$ | $G_i$ | $K_i$ | $C_i$ | $U_i$ | $X_{0i}$ |
|---|---|---|---|---|---|---|---|---|
| 1 | 250 | 71 | 77 | 3 | 15 | 0.1 | 0.35 | 3 |
| 2 | 10 | 71 | 60 | 1 | 4 | 0.1 | 0.40 | 3 |
| | bar | mm$^2$ | g | mm$^2$/mm | bar/mm[a] | Ns/mm | mm | mm |

a) Pressure is referred to the spool active area

are applicable:

1) the *high pressure* consumption is defined as the total flow subtracted from the pump delivery, and includes the flow due to the bias piston motion
2) the *low pressure* consumption is defined as the total flow discharged from the control valves to the tank or pump case, and it is not reversible.

In all steady state conditions, the two definitions coincide but they can lead to significantly different results in dynamic conditions.

## 4.1 Pressure Compensator Testing

The PC control is tested by leaving the flow orifice wide open (pressure differential about 1.2 bar), closing the load orifice at various rates (phase 1) and going back to the initial condition (phase 2) at the same rate. Both the FC and SL control are inhibited, and the load valve is inactive. Two starting pressure levels are considered at the pump delivery port, namely 20 and 80 bar, the latter being close to the maximum attainable without saturation of the governor ($Z = 1$), always set at 2100 rev/min.

Some performance indexes of phase 1 (pressure rise) are collected in Table 4, and

Table 4 *Pressure compensator performance (starting pressure $p_1$= 20/80 bar)*

| orifice clo/op | pressure overshoot | pressure undershoot | response time (SAE) | speed droop[a] | consumption (peak)[b] | pressure gradient |
|---|---|---|---|---|---|---|
| 50 | 118/76 | 73/71 | 60/32 | 3/56 | 64/57 | 19/13.6 |
| 100 | 51/34 | 53/41[c] | 18 | 30/75 | 41/38 | 11.4/8.4 |
| 250 | 20/13 | 7/5 | 18 | 130/186 | 28/20 | 5.2/3.6 |
| 500 | 9/6 | 3/1 | 18 | 275/394 | 16/11 | 2.6/1.7 |
| ms | bar | bar | ms | rev/min | dm$^3$/min | $10^3$ bar/s |

a) Evaluated from the governor set speed (2100 rev/min)
b) High pressure flow consumption (steady state value = 1.4 dm$^3$/min)
c) The maximum undershoot occurs in the second cycle

they generally show regular trends as the orifice closure and opening time increases. The fastest transient has almost no effect on the engine speed, and its behaviour in terms of overshoot and response time is also the combined effect of the limited opening of valve 1 (chosen in view of flow consumption) and the relatively long saturation time ($p_1 = p_3$). During phase 2 (pressure decay) the initial absolute pressure gradients are about 80% of those given in Table 4, and a pressure recovery follows (except the 500 ms case) which lasts 300/400 [3] ms.

With reference to the 50 ms transient starting from 80 bar, the delivery and control pressure are plotted in Figure 3. The control saturation is clearly visible, and the

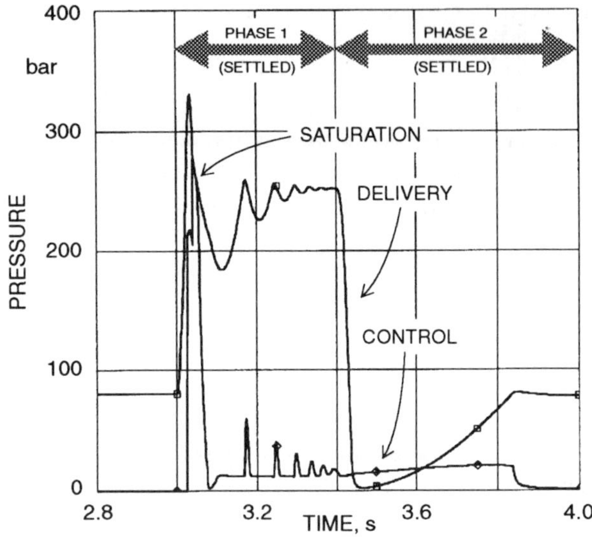


Figure 3 *Time plots of delivery and control pressure in the 50 ms transient (starting delivery pressure $p_1$ = 80 bar)*

small peaks in the control pressure plot are caused by repeated springs of spool 1 from its end stop. The high and low pressure flow consumption during the 500 ms transient starting from 20 bar are plotted in Figure 4 (the opposite sign is for the sake of clarity). In phase 1 the high pressure consumption prevails, whereas in phase 2 the low pressure consumption is higher.

## 4.2 Flow Compensator Testing

The FC control testing implies the analysis of two modes of operation: (a) the regulating mode and its related *margin* pressure (pressure difference across the flow orifice) which depends on the flow orifice setting and its downstream pressure; (b) the *standby* mode, which implies full closure of the flow orifice and null downstream pressure.

Both modes are demonstrated in the sequential test shown in Figure 5, which is divided in three phases (the load orifice is fixed and the load valve is set at 28 bar)

3. Higher values apply to faster transients because of the flow absorption of the bias piston.

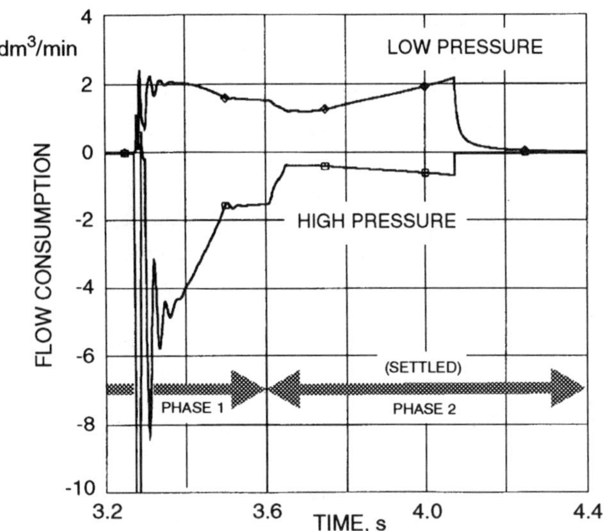


Figure 4 *Time plots of the high and low pressure flow consumption in the 500 ms transient (starting pressure $p_1$ = 20 bar)*

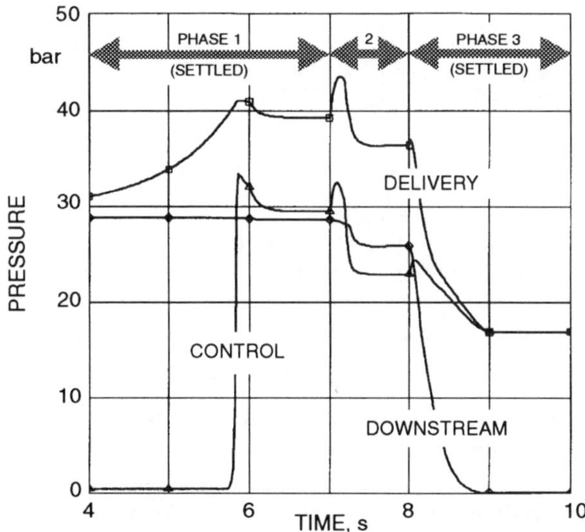


Figure 5 *Time plots of delivery, downstream and control pressure in the sequential flow compensator test*

1) a slow transition (2 s) from the flow orifice wide open (FC inactive) to a high active flow setting (swashplate at about 15 deg). The overshoot of the delivery and con-

trol pressure are partly caused by the timing plate torque (Table 1)

2) a faster transition (200 ms) to an intermediate flow setting (about 7 deg). The margin pressure has here an overshoot due to the overall system dynamics

3) a slow transition (1 s) to the standby mode, where the delivery and control pressure coincide by definition (16.9 bar)[4].

The FC control differs in some way from the PC and SL control because it is expected to be the default pump operation. Consequently, a useful description of the control performance comes from a steady state analysis of its primary indexes, i.e. margin pressure and flow consumption. A contour plot of flow consumption is given in Figure 6 as a function of pump delivery flow (ratioed to 220 dm$^3$/min) and pump delivery pressure (ratioed to 250 bar). The upper right triangle left by the bounding box is not compatible with the engine power. The margin pressure in the same region varies between 10.9 and 15.9 bar, the average being 11.7 bar and the standard deviation 0.77 bar (the nominal setting of valve 2 is 10 bar).

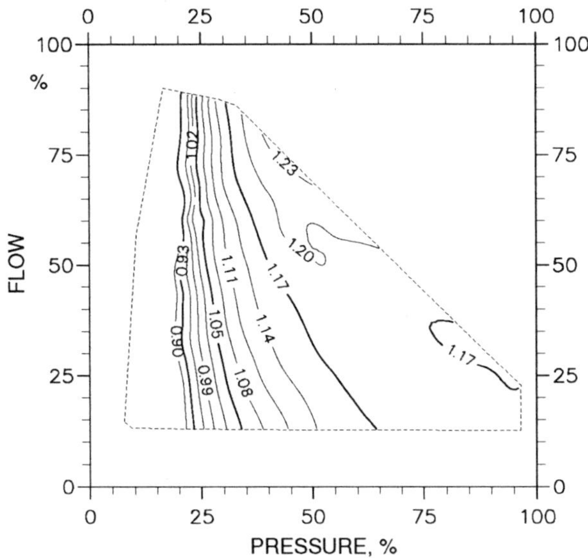


Figure 6 *Contour plot of the flow compensator consumption (levels are shown in dm$^3$/min)*

## 5 Speed Limiter

The SL control testing has to take into account that the unbalance between torque demand (pump) and torque supply (engine) occurs in different ways. For illustration purposes, a limited number of perturbation modes are classified in Figure 7. Their relevant transients are driven by the load orifice, the load valve and the additional

4. This value corresponds to a swashplate angle slightly greater than the lower end stop $\beta = 0$.

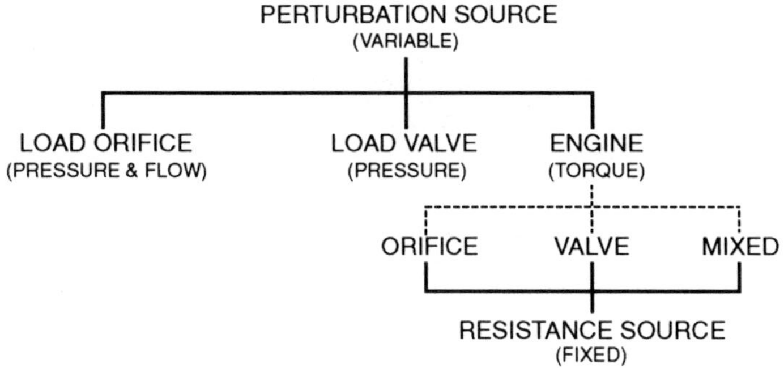


Figure 7 *Perturbation tree to be used in the speed limiter analysis*

engine torque. In the last case, orifice and valve impose independent or combined (mixed) resistance characteristics at the end of the delivery line.

Whereas the PC and SL control are free to interact, the FC control is inactive in all simulations.

## 5.1 Orifice Driven Transient

For a given throttle setting, the simplest steady state interaction between the SL controlled pump and the engine can be demonstrated under two assumptions:

a) the SL orifice (Figure 2) is fixed and in conjunction with the fixed setting of valve 3 produces a fixed *set speed*

b) the load orifice is closed at a sufficiently low rate, and the load valve is inactive.

The engine starts moving along the steep governor curve and then along the flat full injection curve until any further deceleration is opposed by the SL control which reduces the pump displacement. Finally, the PC control overcomes the SL control and speeds up the engine again. The above sequence is plotted in Figure 8 for the

Table 5 *Design parameters of the speed limiting valve*

| | | | |
|---|---|---|---|
| $A_3$ = 71 mm$^2$ | $C_3$ = 0.1 Ns/mm | $K_3$ = 1 bar/mm [a] | $G_3$ = 2 mm$^2$/mm |
| $r$ = 0.1 | $U_3$ = 0.3 mm | $X_{03}$ = 2 mm | $p_{s3}$ = 10 bar [b] |

a) Pressure is referred to the spool active area

b) The by-pass relief is set at 15 bar and the supply pump has 10 cm$^3$/rev capacity

system under study during a transient that lasts 100 s, and starts from 20 bar (delivery pressure). Four phases are in evidence: governor control (1) governor saturation (2) SL control (3) pressure compensator override (4). The throttle is set at 2100 rev/min and the nominal SL set speed (when valve 3 starts moving from its end stop) is 1850 rev/min. The difference between $\omega_s$ and the actual speed limit (hysteresis) is due to the speed sensing subsystem (pump, orifice, and valve 3). The flow consump-

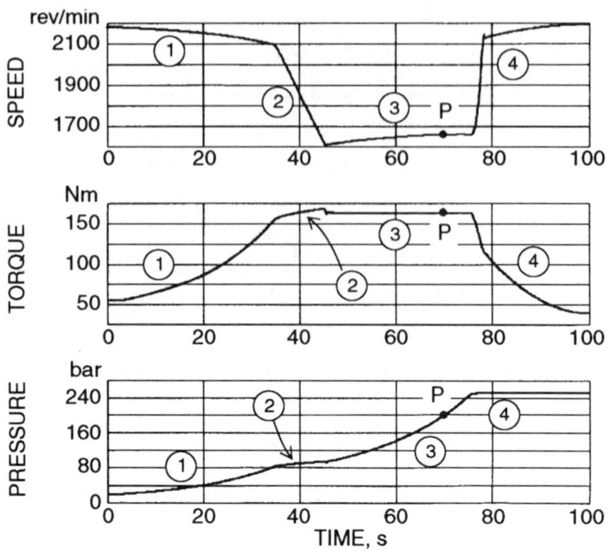


Figure 8 *Time plots of pump speed, torque and delivery pressure during the SL demonstration transient (orifice driven)*

tion of the SL control ranges between 1.9 and 2.1 dm$^3$/min. Some design parameters of valve 3 are collected in Table 5. Their refinement procedure (otherwise similar to PC and FC valves) is *not* general, but specifically applies to the orifice driven transient.

To investigate the SL response during faster transients, the orifice closure can be stopped in any intermediate point of phase 3, e.g. point P in Figure 8 at 200 bar (steady state). The speed droop for various closure times are listed in Table 6. There

Table 6 *Speed droop caused by orifice driven transients*

| closure time, s | 0.5[a] | 0.7[a] | 0.8[a] | 1.0[b] | 1.2[a] | 1.5[a] | 1.8 | 3.0 | 4.5 |
|---|---|---|---|---|---|---|---|---|---|
| speed droop, rev/min[c] | 38 | 113 | 183 | 311 | 287 | 233 | 204 | 113 | 82 |

a) The PC valve opens once while the SL valve is open

b) The PC valve opens twice while the SL valve is open

c) Evaluated from the steady state speed (1663 rev/min)

is a complex interaction with the PC control, but the interesting thing is that the speed droop has a maximum at about 1 s. On the left (less than 500 ms) the speed droop is negligible, and on the right (more than 4.5 s) the asymptotic value in Figure 8 is approached.

## 5.2 Valve Driven Transient

The demonstration transient is the same as in Figure 8, the only difference being in assumption (b) which now refers to the load valve variably set up to 300 bar (the load orifice is closed). The relevant time plots are given in Figure 9. Apart from some obvi-

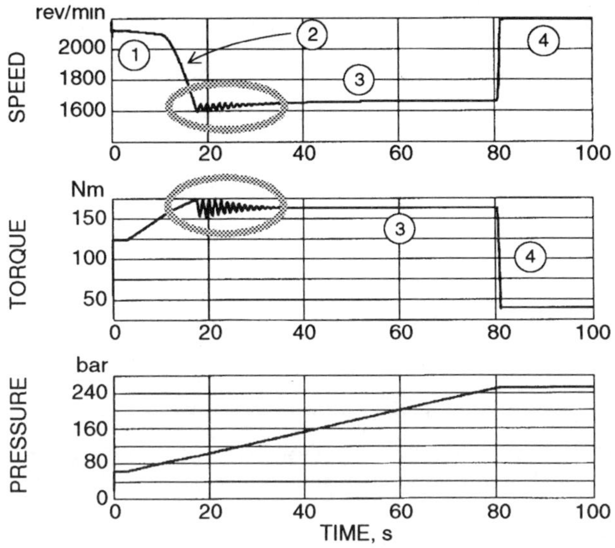


Figure 9 *Time plots of pump speed, torque and delivery pressure during the SL demonstration transient (valve driven)*

ous changes in the general shape of plots, the important difference lies in the speed and torque ripples experienced in the low pressure region of phase 3. A closer analysis, which takes the points of the interesting region as intermediate targets, shows that two things could happen:

a) if the target is close enough to the beginning of phase 3, the swashplate meets at some time its high end stop and the transient gives rise to a limiting cycle. For instance, this occurs at 95 bar

b) if the swashplate never meets its end stop, the system is stable but very lightly damped. At 100 bar, for instance, the settling time is more than 200 s.

The first (and reasonable) attempt to solve the problem goes through the inspection of the root locus and the modification of the valve design parameters accordingly. The effect of spring stiffness, area gain and feedback ratio on the dominant system root are shown in Figure 10, which refers to the 100 bar final set of the load valve. In addition to the fact that the feedback ratio confirms to be necessary in view of stability, the most important result is that the root is effectively shifted by the increase of the spring stiffness and above all the drastic decrease of the valve gain (from twenty to forty times less than in Table 5), which is beneficial to flow consumption but dramatically adverse to speed of response and valve hysteresis[5].

5. Hysteresis is affected by the spring stiffness, too.

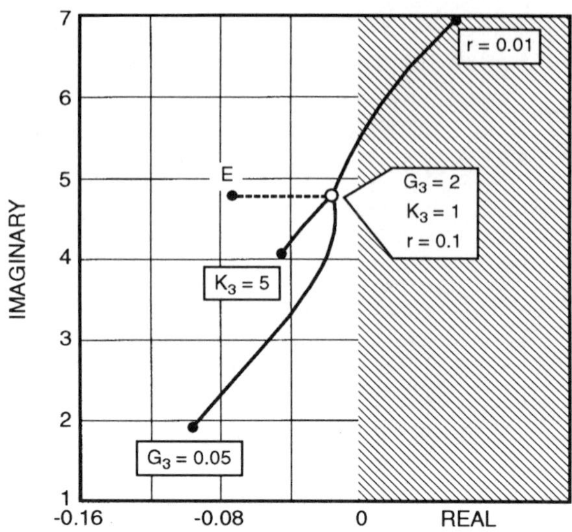


Figure 10 *Position of the dominant system root as affected by three design parameters of valve 3 (valve driven transient)*

An alternative attempt is based on the observation that all problems vanish if the portplate timing torque is withdrawn from the swashplate dynamic model. In fact, the dominant root moves from (-0.013, 4.75) to (-0.9,4.59), much more than acting on the valve parameters. Apart from the opportunity or feasibility of such a pump design, the surprising thing is that the absence of $T_p$ would lead to an *unstable* response to the orifice driven transient. Actually, the dominant root moves to (2.05, 7.68). The consequence is that the role of $T_p$ is contradictory (favourable in one case, unfavourable in another), and the only broad band solution seems to be the reduction of gain and the increase of stiffness. Some tests have been carried out by setting $G_3 = 0.05$ and $K_3 = 3$ and repeating the same experiments as in Table 6. Until 1.2 s the difference is not big, but from 1.5 s onwards the speed droop increases dramatically: 807, 770, 679, 539 rev/min (all measured from 1616 rev/min).

In any case the epilogue of the story is yet to be written, because more facets need consideration. Undoubtedly one of them is on the engine side, as proved in Figure 10 where point E shows the dominant root location if the torque margin of the engine would be changed from 0.11 to 0.2.

## 5.3 Engine Driven Transient

The analysis of the system response when an additional torque is applied to the engine shaft confirms the preceding results. If the SL valve design parameters are as in Table 5 and a 120 Nm torque demand is applied in 5 s, the following happens

a) if the load orifice only is active in the delivery line, the system works well when $T_p$ is included and exhibits a limiting cycle when $T_p$ is disregarded

b) if the load valve only is active in the delivery line, the system works well when $T_p$ is disregarded and exhibits a limiting cycle when $T_p$ is included

c) if both orifice and valve are active in the delivery line, the system exhibits limiting cycles or instabilities, irrespective of the presence or absence of $T_p$.

If the modified design is adopted (low area gain and high spring stiffness) a final steady state equilibrium is reached in *all* case studies, but the preceding time history is not always acceptable.

## 6 Conclusions

Based on the experiments summarized in the paper, and considering that an explicit comparison with the torque limiter is outside the present scope, the following remarks apply to the triple system with PC, FC and SL control:

- the PC and FC control do not have special problems and can be designed in view of the flow consumption. Conversely, the SL control is to be designed in view of stability. In some conditions the PC control helps the SL control
- the behaviour of the particular SL scheme studied in the paper is affected by the swashplate equilibrium internally and the perturbation mode externally. In particular, the portplate timing torque contributes to system stability in some modes, but does the opposite in others
- what seems to be a design compromise of broad validity (low gain and high spring stiffness in the SL valve) has a high price to pay in terms of response to fast perturbations.

## References


[1] Paoluzzi, R., SIDAC Class of Hydrostatic Transmission Controls. In *Preprints of the 5th Bath International Fluid Power Workshop on Circuit, Component & Systems Design*, Fluid Power Centre, Bath, 1992.

[2] Tsai, S.C. and Goyal M.R., Dynamic Turbocharged Diesel Engine Model for Control Analysis and Design, SAE paper 860455, Society of Automotive Engineers, USA, 1986.

[3] Zarotti, G.L. and Nervegna N., Pump Efficiencies Approximation and Modelling. In *Proc. of the 6th International Fluid Power Symposium*, BHRA Fluid Engineering, Cambridge, 1981, pp. 145-164.

[4] Zaichenko, I.Z. and Boltyanskii, A.D., Influence of End Distributor on Regulating Gear of Axial Piston Pumps. In *Russian Engineering Journal*, Vol. L, 1970, No 7, pp. 56-60.

[5] Zarotti, G.L. and Paoluzzi R., Sulle funzioni cicliche delle pompe a piatto oscillante. In *Atti del 47° Congresso Nazionale ATI*, Associazione Termotecnica Italiana, Parma, 1992, Vol.II, pp. 1429-1440.

[6] Harrison, J. et al, *EASY5x Users Guide*, Boeing Computer Services, 1992.


❑❑❑

**WRITTEN DISCUSSION**
**Triple pump controls with speed sensing power management**
**R Paoluzzi, M Puglia & GL Zarotti (CEMOTER, Italy)**

**Question:** **W Backé,**
**IHP, RWTH Aachen, Germany**

In analysing load-sensing systems we have found that they present a connection of a pressure-control (at the pump) and a speed-control (at the consumers) with mutual interactions. That is the reason why load-sensing systems with hydro-mechanical controls have poor dynamic behaviour.

Do you think it possible to decouple these controls by a digital signal circuit with decoupling controllers?

**Answer:**

As a preliminary remark, it should be mentioned that the three control functions within the pump (as addressed in the paper) are mutually exclusive, and they only interact during short transitions. In fact, the flow limiter is the continuous (or default) condition, bounded by the pressure limiter, the speed limiter, or the maximum displacement; in turn, the speed limiter is overcome by the pressure limiter, and so on. If we consider the load-sensing subsystem (flow limiter and flow setting orifice(s)) the following points should be noted:

a) If the simplest case is analyzed, comprising one pump, one control valve and one actuator, it is possible to identify the interaction problem and attempt a solution by electronic means. The question is whether the effort is meaningful or not; in fact such a simple case is useless in practice, because it is easily replaced by a plain displacement setting (at least if the pump speed does not change very much).

b) If complex cases are analyzed and closer to practical applications (one or two pumps, many valves and actuators, one or two main branches with floating or fixed local compensators, and so on), the interaction problems grow more and more critical. But still the question is whether it is wise to attempt the electronic alternative in order to patch a scheme conceived on the basis of hydromechanical controls. Probably, the best solution would a fresh approach driven from the beginning by the electronic "way of thinking".

**Question:** **F Conrad**
**Technical University of Denmark**

Do you envisage any potential for improvements in efficiency?

**Answer:**

In examining the efficiency of triple pump controls, the three functional tasks should be considered separately:

a) The pressure limiter is generally contrasted with a relief valve and its intrinsic energy losses.

b) The flow limiter usually belongs to a class of load-sensing systems and its benefits are well known in terms of efficiency and controllability (the latter property referring to the speed of actuators).

c) With the speed limiter, the point of view is different. Strictly speaking, this control function is not related with efficiency, because it is intended (like the torque limiter) to decouple the size of engine and pump, and counteract the influence of additional loads on the engine shaft. However, the decoupling has some effects on efficiency and/or productivity; in fact, for a given pump a smaller engine is possible (advantages in cost and maintenance), whereas for a given engine a bigger pump is possible (higher costs, but increased machine speed and production).

Finally I would like to remark that in my experience both machine manufacturers and customers do not seem anxious today about efficiency or fuel consumption. Consequently, the best advantage of the triple control scheme is probably to build up an automatic operational envelope for a pump.

**Question:** **W Rampen**
**University of Edinburgh, UK**

Your controller deals with an independently controlled diesel engine. How do you see it changing with electronically-governed diesels such as Lucas EPIC?

**Answer:**

The aim of the paper is to point out some problems arising from the speed limiting function as an alternative of the more popular torque limiting function, all within a conventional application environment. An essential part of such environment is that the diesel engine is taken as an independent unit with its own control system (governor). Though I am not specifically familiar with the EPIC system mentioned, the following points seem appropriate:

a) In principle, the speed limiter has little to do with the governor because it works when the engine moves along its "natural" torque curve. Conversely, the torque limiter (not considered in the paper) would work in conjunction with the governor.

b) Generally speaking, an electronically controlled diesel would become some kind of a "virtual" engine whose characteristics are to be tailored according to the specific requirements of the driven transmission. However, this approach introduces two problems. Firstly, it makes it difficult for pump manufacturers to design general purpose control architectures, unless the use of similar engines is widespread. Secondly, the "virtual" engine would run counter to the basis of speed or torque limiting and the designer would need to consider the pump control right from the beginning. But, are designers ready to use so many degrees of freedom?

# 6. SSP - A Simulation Program for Pneumatics

W Backé *and* R Eschmann

## Abstract

This paper describes the application of simulation to pneumatic circuits which incorporates both open and closed loop control. Details are presented regarding experimental test methods developed to measure seal friction of pneumatic cylinders and valve flow characteristics of continuously actuated electro-pneumatic valves for inclusion in the simulation models. It is shown that simulation results are in close agreement with those obtained from experiment for the cases presented.

## 1. Introduction

Within the scope of the research project "simulation of control pneumatics" the simulation program SSP was developed at the Institute for Fluid Power Transmission and Control at the Aachen University of Technology. This program enables the user to configure a system of pneumatic components and to simulate the combined action of these components.

Concerning the simulation of pneumatic systems, the greatest problem is the description of the non- linear influencing variables. In pneumatics these are the flows through pneumatic resistors and the friction forces at the seals. Therefore, for a simulation it is necessary to know precisely the individual components of the system and to be able to describe them. For the examination of the components included in the simulation, two testing concepts were developed by means of which the non- linear influencing variables can be determined. The tasks here consist of the measuring the components and producing a mathematical description of the measured results for integration into the simulation program.

In addition to the two test benches, a pneumatic drive- system was constructed enabling the comparison of the measured and the simulated movements. This system exclusively consists of components for which the parameters were determined according to the developed methods.

This paper describes the functions of the simulation program, the test benches and the measurement of the non- linear parameters, as well as the drive- systems for the comparison of measurement and simulation. Subsequently some examples will be presented demonstrating the behaviour of a pneumatic cylinder drive both in the open and in the closed loop control as measured and simulated.

## 2. The simulation program "SSP"

The program SSP is a tool for predicting the combined action of pneumatic segments within a system. Figure 1 shows the main menu of the program which is split into five sub- menus. Within the sub- menu "parameters" the user can set the general values needed for the simulation such as length of simulation time or method of integration.

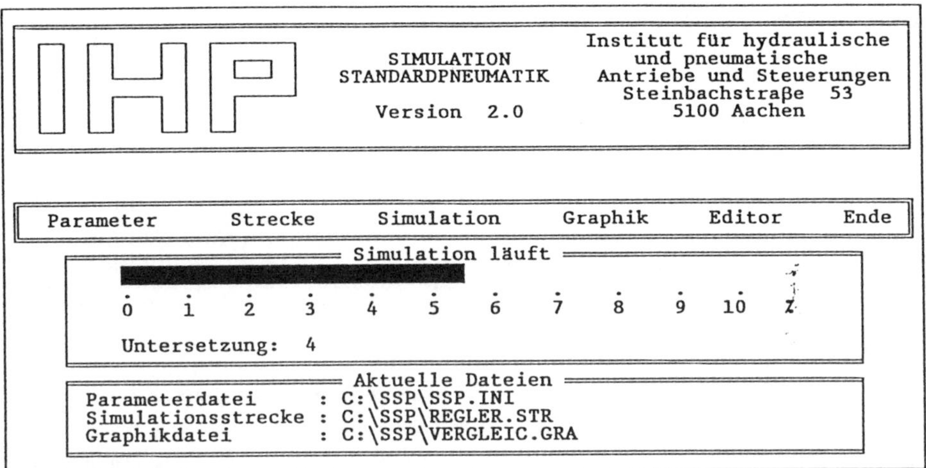


*Figure 1: Main menu of the program*

Within the sub- menu- segment "Strecke" the pneumatic system intended to be simulated is drawn (figure 2). The user can arrange any segment from a collection of programmed pneumatic components. Here different tools can be used to add components to the system or delete them. When adding, the user can select from various kinds and subsequently from different types of components. For each of these components the user can either choose a data record from a library or draw up a new data record. After the selection of the necessary components the user has to join the connections of the components to each other. The letters and figures next to the components (figure 2, on the right) indicate the joints of the connections.

An extension of the simulation program by further components can be carried out by the user with the help of a programmable interface.

The submenu "simulation" of the main menu starts the simulation. During the simulation, a cursor indicates the progress of the calculations. The program installs an output file for each component in which the data of the simulation are written in columns in the ASCII- format. The submenu "graphics" enables the user to present the results of the simulation according to a preferred choice and configuration. In this way it is possible to present up to ten graphs simultaneously as functions of any x- axis. All graphs can be normalized so that a direct comparison of the graphs is possible.

The simulation program has been provided with an "editor" which allows standard data files in ASCII- format to be read in and worked on.

For the simulation of a system, distinctions must be made between pneumatic components that essentially work as a capacity and the components which behave as a resistance.

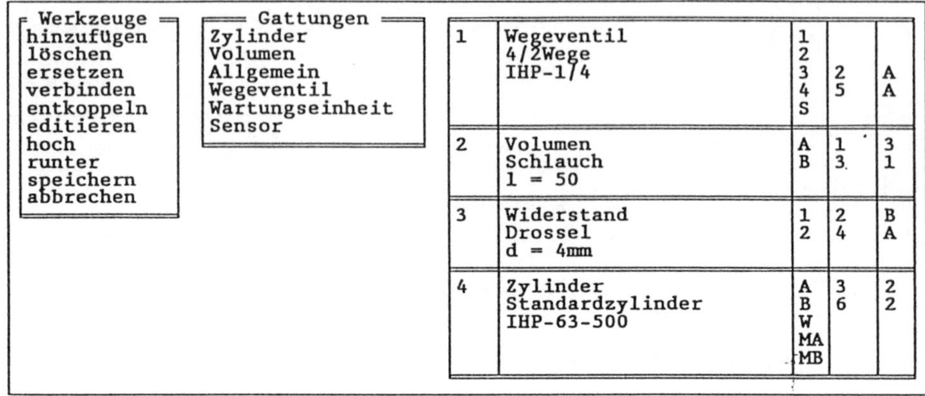


*Figure 2: Drawing up the menu segment*

With a pneumatic resistance, a mass flow rate is calculated from the pressures and temperatures adjacent to the connections and from the parameters describing the resistance, e.g. conductance C and the critical pressure ratio b . In the case of a pneumatic capacity, the input quantities mass flow and temperature are used to determine the overall mass in the capacity which leads to a modification of pressure that is integrated to a current pressure during the simulation step.

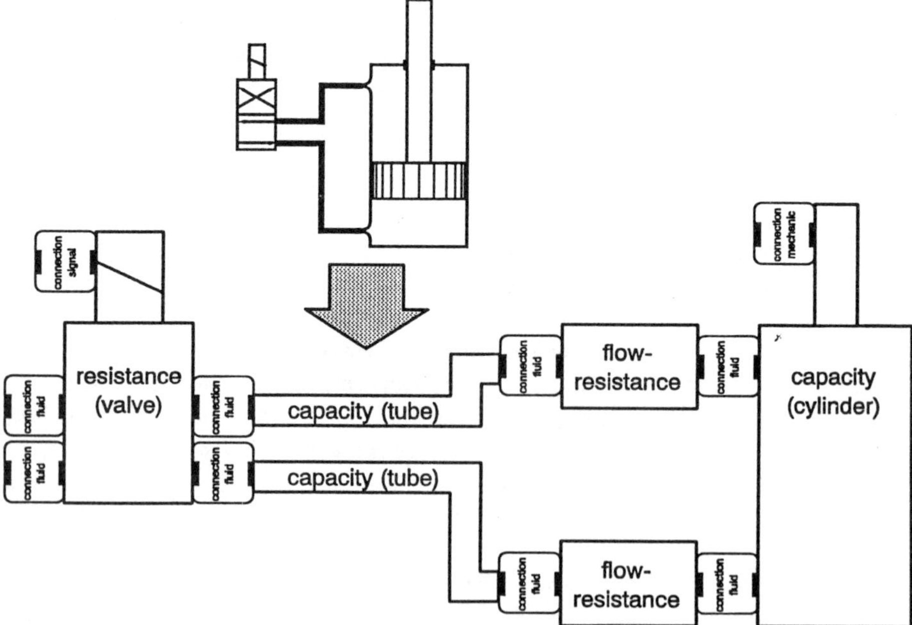


*Figure 3: Transmission of the pneumatic segment into the simulation*

It is obvious that the output quantities of the capacities represent the input quantities of the resistors, likewise the output quantities of the resistors represent the input quantities of the capacities. For this reason all simulation segments consist of an alternating series of connections of capacities and resistors. In figure 3 the component 4/2 directional control valve (resistance) is joined to the cylinder (capacity) by fluid- connections.
With a fluid- connection the parameters pressure, temperature and mass flow are exchanged between the components. A signal connection, however, only consists of a standardized input signal. With a mechanical connection, the quantities position, velocity, acceleration and force are delivered.

## 3. Measuring the friction force at a pneumatic cylinder

In order to measure the friction force at a cylinder drive, it was important to find out the essential parameters which influence the friction force. Distinctions must be made between the characteristic quantities that vary during the movement of a cylinder and those that are to be considered constant for a cylinder drive. The influence of material pairings between tube, lubricant and sealing material is assumed to be constant. Moreover all geometrical quantities are considerd to have a constant effect on the friction force of a cylinder.
Variable quantities influencing the friction force are above all the piston speed and the pressures in the cylinder chambers that press the sealing lips against the tube walls.
In order to be able to determine these factors separately, a test method has been developed forcing a pneumatic cylinder movement with constant pressures in the cylinder chambers and constant piston speed.

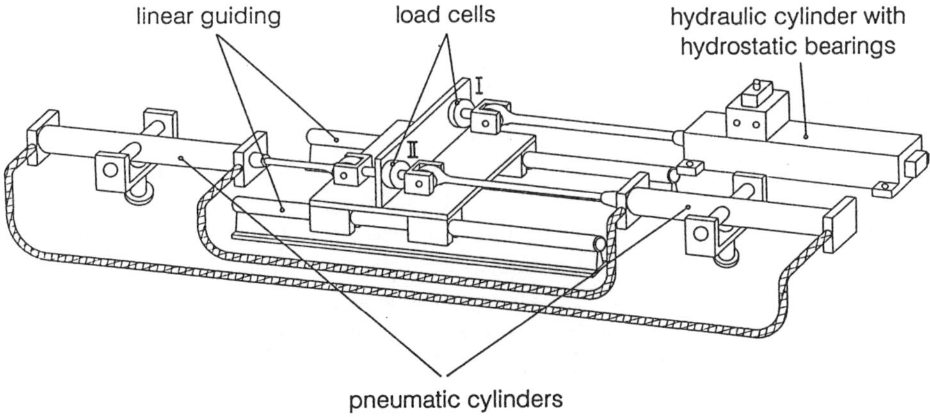


*Figure 4: Test bench for measuring the friction force*

Figure 4 shows the test bench for measuring the friction forces.
With this arrangement, two identical pneumatic cylinders act against each other with the outer and inner chambers being supplied with the same pressure. They are driven by a hydraulic cylinder having hydrostatic bearings at a constant speed. The inner and outer cylinder chambers of the

pneumatic cylinders are connected by lines so that the air volumes and the pressures in the cylinder chambers remain constant. As a result it is possible to measure the friction forces at the load cells for different combinations of speed and cylinder chamber pressures.

Figure 5 shows the measured friction force with speed at different pressures between $\Delta p = 0$ bar up to $\Delta p = 10$ bar. In order to use these results within the simulation, three values of two curves with different $\Delta p$ graphs are required. In the range of $\dot{x} \leq \dot{x}_{grenz}$ the curves are approximated by a parabola and in the range of $\dot{x} \geq \dot{x}_{grenz}$ by a square root function. Between the curves the values for the friction force as a function of the difference of pressure can be determined by interpolation.

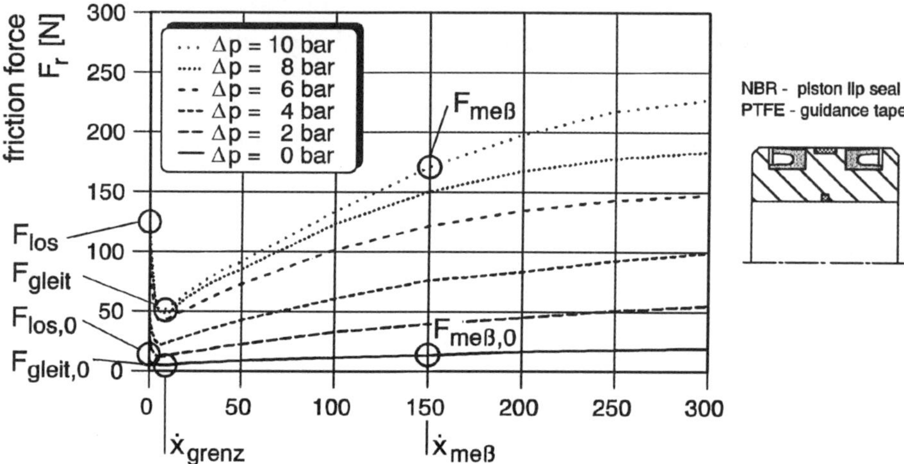


*Figure 5: Friction force in dependence on speed $\dot{x}$ and on the difference of pressure $\Delta p$*

Within the simulating program, the friction force is calculated at every integration step from the current pressure difference at the piston- and the rod seals and the current speed.

# 4. Measuring the flow characteristics of pneumatic resistances

Apart from the friction force, the flow through the valves is the second important influencing quantity for the motional behaviour of a pneumatic cylinder drive. The flow through a pneumatic resistor can best be described by the parameters conductance C and critical pressure ratio b according to ISO 6358. Conductance C is a quantity for the supercritical flow which is given when, at the narrowest cross section in the resistance, sonic velocity has been reached. The critical pressure ratio b describes the value of the pressure ratio at the transition from the subcritical to the supercritical area.

The measurement graph is produced at constant admission pressure with the counterpressure being varied in the outgoing air by a variable resistance (figure 6). Four turbine flowmeters with different flow- ranges can be used to measure the volume flow at constant pressure. The deviations of the inlet- and outlet tubes are laid down by ISO 6358.

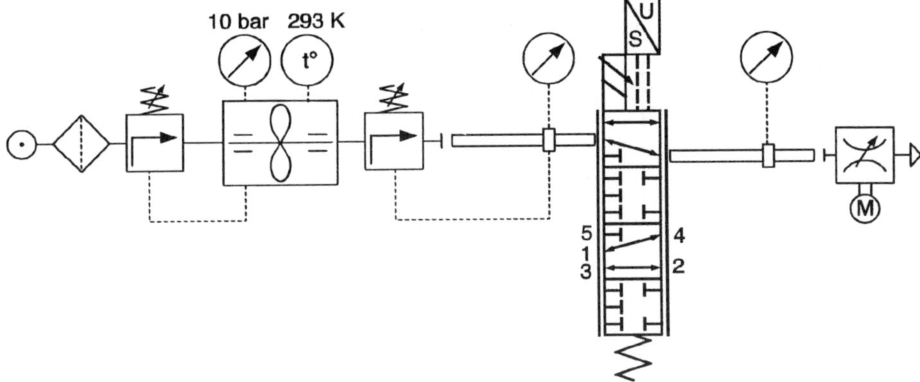


*Figure 6: Flow measurement method*

With a proportional sliding valve this measurement has to be taken at different stroke positions. Figure 7 shows the characteristics of a sliding valve.

In the simulation program flow through a continuously stroked valve is calculated as a function of the actual pressures, of the drive voltage and of the flow parameters b and C. In figure 8, the flow parameters are plotted against the drive voltage U. The flow increase in the range of 0.25 V up to 8 V is nearly proportional. Only in the range >8 V do further resistances in the valve become apparent and the slope of the flow curve decreases. In the range < 0.25 V, flow increases only slowly because of overlap at the metering edges.

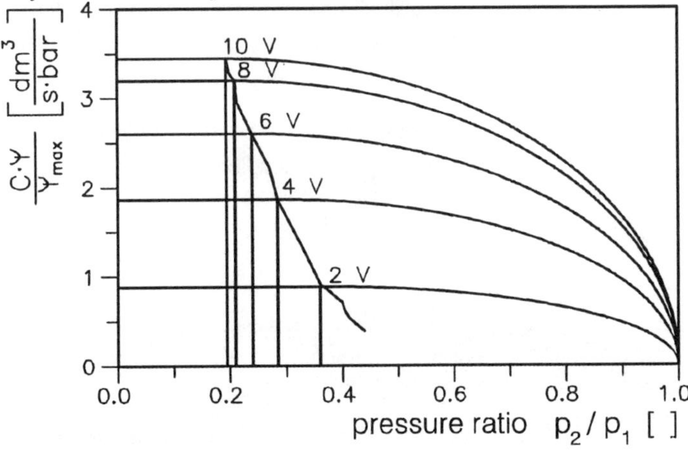


*Figure 7: Flow characteristics of a NG 6 proportional sliding valve*

In this range the critical pressure ratio reaches its highest value, as here the metering edges has an effect on the flow almost like a nozzle. With large valve strokes and - as a result - flows, further resistances become effective and the critical pressure ratio falls off to a value of 0.2 .

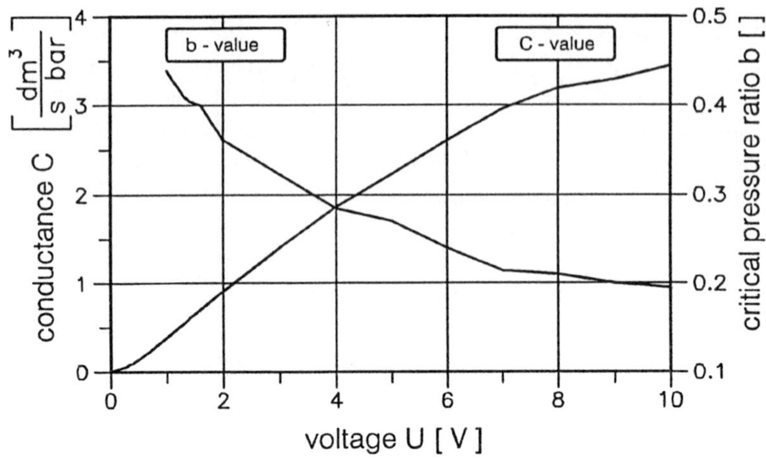


*Figure 8: Characteristics of conductance C and of the critical pressure ratio b dependent on the control voltage U*

## 5. Test bench for the comparison of measurement and simulation

In order to obtain as precise a comparison as possible between the measured and the simulated movements of a pneumatic cylinder drive, a pneumatic drive with continuous valves was constructed. Here it was of particular importance to determine in detail all parameters of this test bench construction. The test bench was equipped with pressure take- ups at the cylinder chambers, at the final position dampers and at the pressure supply, a load cell between the piston rod and the mass that has to be accelerated and with an incremental position sensor.

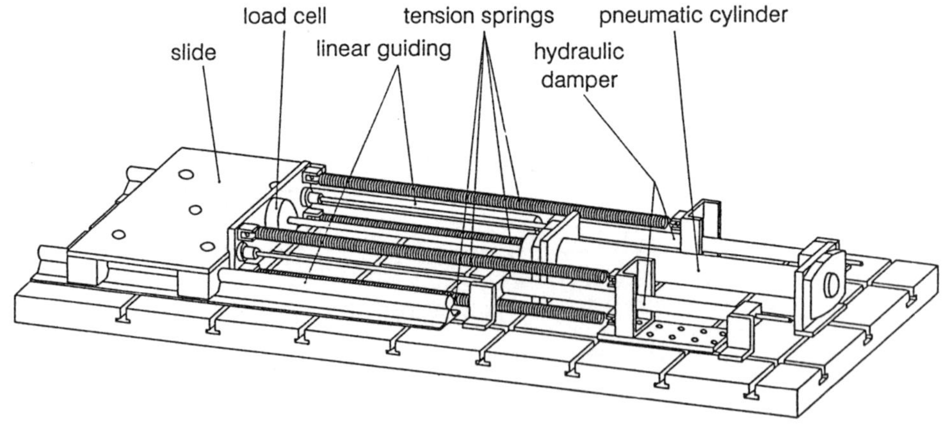


*Figure 9: Test bench for the comparison of measurement and simulation*

Section 3 showed that the friction force of a seal depends to a particularly high degree on the pressure difference acting on it. With a pneumatic drive gear such high differences of pressure at the piston sealing occur, however, only under high loading. The test bench is equiped for various possibilities of load (figure 9). Apart from the possibility of fixing different masses on the slide block of the test bench, four springs were integrated as a stroke- dependent load. To develop a speed- dependent load, two hydraulic dampers - adjustable by means of a throttle - were mounted to the pneumatic cylinder.

The pneumatic cylinder is controlled by a continuous 5/3 directional slide valve NG6 . The supply pressure of the valve is kept almost constant by a 40 litres accumulator.

## 6. Comparison of measurement and simulation

In the following section, some measurements will be demonstrated which were taken using the test bench presented in section 5. The characteristics of the components employed for these investigations were determined as described above. The measured values were integrated into the simulation program and the motional behaviour of the cylinder drive simulated. First the driving system was operated in open, then in closed loop control.

## 6.1 Operation using open loop control

Figure 10 shows the pneumatic system in open loop control. The system consists of a double-acting cylinder, a 4/2- directional control valve, the compressed- air supply and two switches actuating the directional valve when it reaches the end positions. Additionally, a throttle was installed at the operating connection of the valve. In this circuit a dedicated PC undertakes the measurements and provides the starting signal.

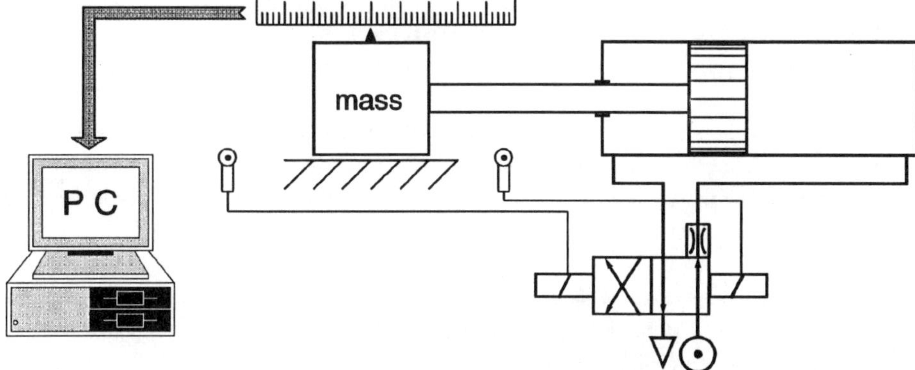


*Figure 10: Test bench as an open loop control*

By throttling the input flow, the released chamber of the cylinder very quickly reaches atmospheric pressure. This means that the level of pressure within the cylinder is low during the motion. As a result the pneumatic spring action is very flexible and the piston is not fixed between two springs. As soon as sufficient pressure has built up at the supply side of the piston to overcome the break- away force of the seal, the piston moves with a jerking motion until the pressure has fallen off and stops again. This results in a stick- slip- movement of the piston (figure 11).

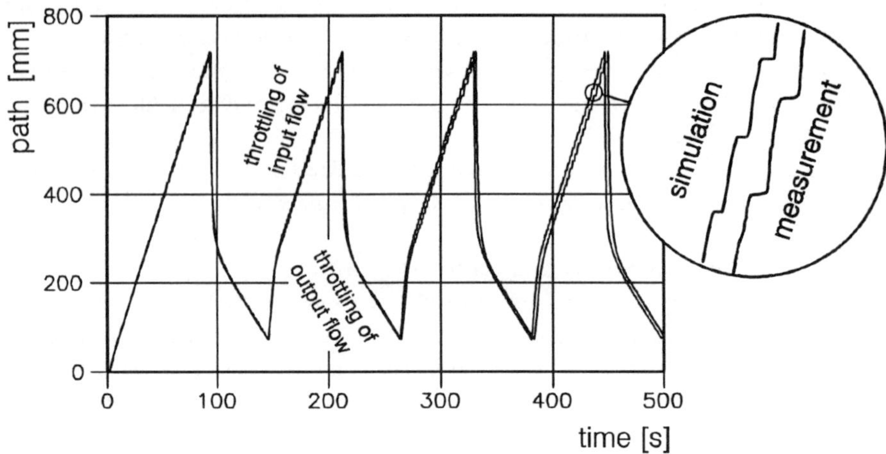


*Figure 11: Measured and simulated movement of the cylinder drive with stick- slip- motions*

When enlarging the stick- slip- motion the small differences between measurement and simulation become obvious. The measurement shows slight irregularities of motion caused by the not completely constant friction conditions. Here above all the influences of the elastic sealing become apparent. During the throttling of the output flow the cylinder is fixed between higher pressures. In this case hardly any stick- slip- motion occurs.

## 6.2 Operation using closed loop control

In contrast to the open loop where two switches actuate the valve either on or off, the position of the piston in the closed loop control is determined by a three- loop control unit (figure 12). Control of the pneumatic positioning system is carried out digitally by the PC. The controller gets its actual position value from the incremental path- measuring devicer. The piston position signal speed $\dot{x}$ and acceleration $\ddot{x}$ are generated by differentiation.The output value y of the controller is then calculated as follows:

$$y = K_x \cdot (w - x) - K_{\dot{x}} \cdot \dot{x} - K_{\ddot{x}} \cdot \ddot{x} \pm K_0$$

$K_0$ is an offset to compensate for the static friction force acting in both directions. The employed valve has an integrated stroke measuring system whereby the adjustment of the valve stroke is carried out by a position control circuit.

Figure 13 compares the step responses measured at a test bench and simulated by the SSP-program. The command variable performs a step from 150 mm to 350 mm. The left part of the diagrams always shows the step responses measured at the test bench and symbolized by a dial gauge; the right part of the diagrams presents the step responses calculated by the simulation program and symbolized by the PC.

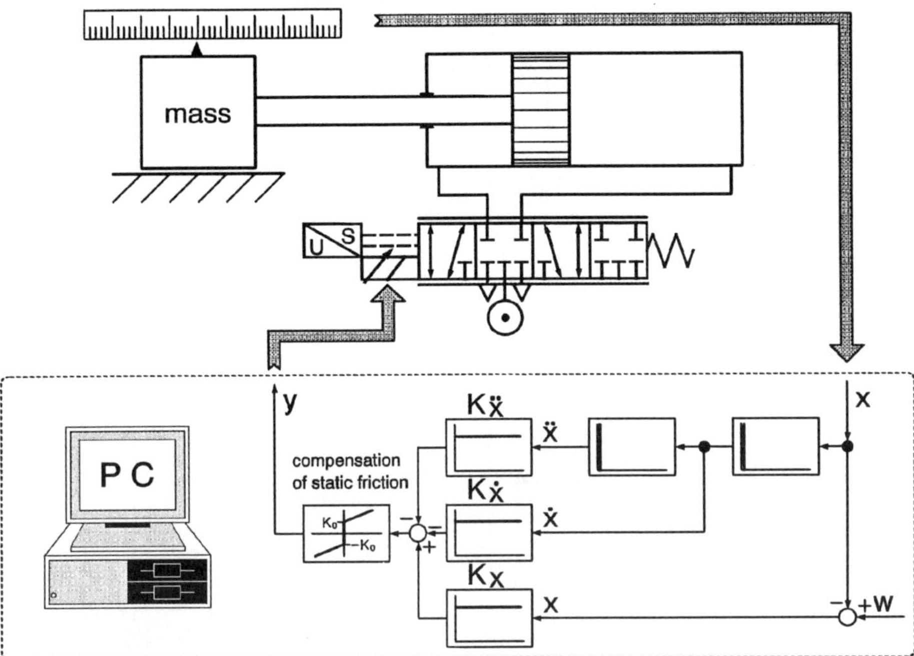


*Figure 12: Test bench within a closed control circuit*

The test bench consists of a cylinder with a piston diameter of 63 mm, a rod diameter of 20 mm and a piston stroke of 500 mm. On the piston a PTFE- cylindrical sealing supported by a NBR- O- ring and two PTFE guidance tapes are installed. The cylinder is operated with oil- free compressed air and initial lubrication. The supply pressure amounts to 10.5 $bar_{abs}$ constantly and the initial pressures in the cylinder chambers are to $p_a$ = 7.3 $bar_{abs}$ and $p_b$ = 8.2 $bar_{abs}$. The cylinder actuates a slide which is loaded with a mass of 30 kg.

For the adjustment of the controller the values $K_x = 0.2225$ V/mm, $K_{\dot{x}} = 0.01$ V s/mm, $K_{\ddot{x}} = 0.032$ V·$s^2$/mm and $K_0 = 0.032$ V were chosen.

From the initial starting position of 150 mm the stroke of the piston x reaches the new nominal position at 350 mm after about 0.5 s . The piston reaches a maximum speed $\dot{x}$ of 1.5 m/s . During the acceleration phase the control unit fully opens the valve, and only when the deviation of position $\Delta x = w - x$ is reduced does the valve close.

Differences between measurement and simulation become most obvious in the deviation of position. The measured deviation $\Delta x$ reached after about 1 s is a very low value near 1 µm. The possibility of reaching this high precision of positioning by means of a pneumatic drives is assisted by the tolerance of the sealing in the groove of the piston and by the elasticity of the sealing. In the simulation an ideally rigid connection between piston and sealing is assumed. Thus - after a standstill - a high break- away force has to be generated at every change of position. In the simulation the increase of the pressure $p_b$ in the lower right diagram can be recognized until the force is sufficient to move the piston (as to be seen in the diagram of the deviation $\Delta x$). In the presentation the deviation of position is normalized to a range of only 0.1 mm in order to demonstrate the precision that can be attained by a pneumatic positioning drive.

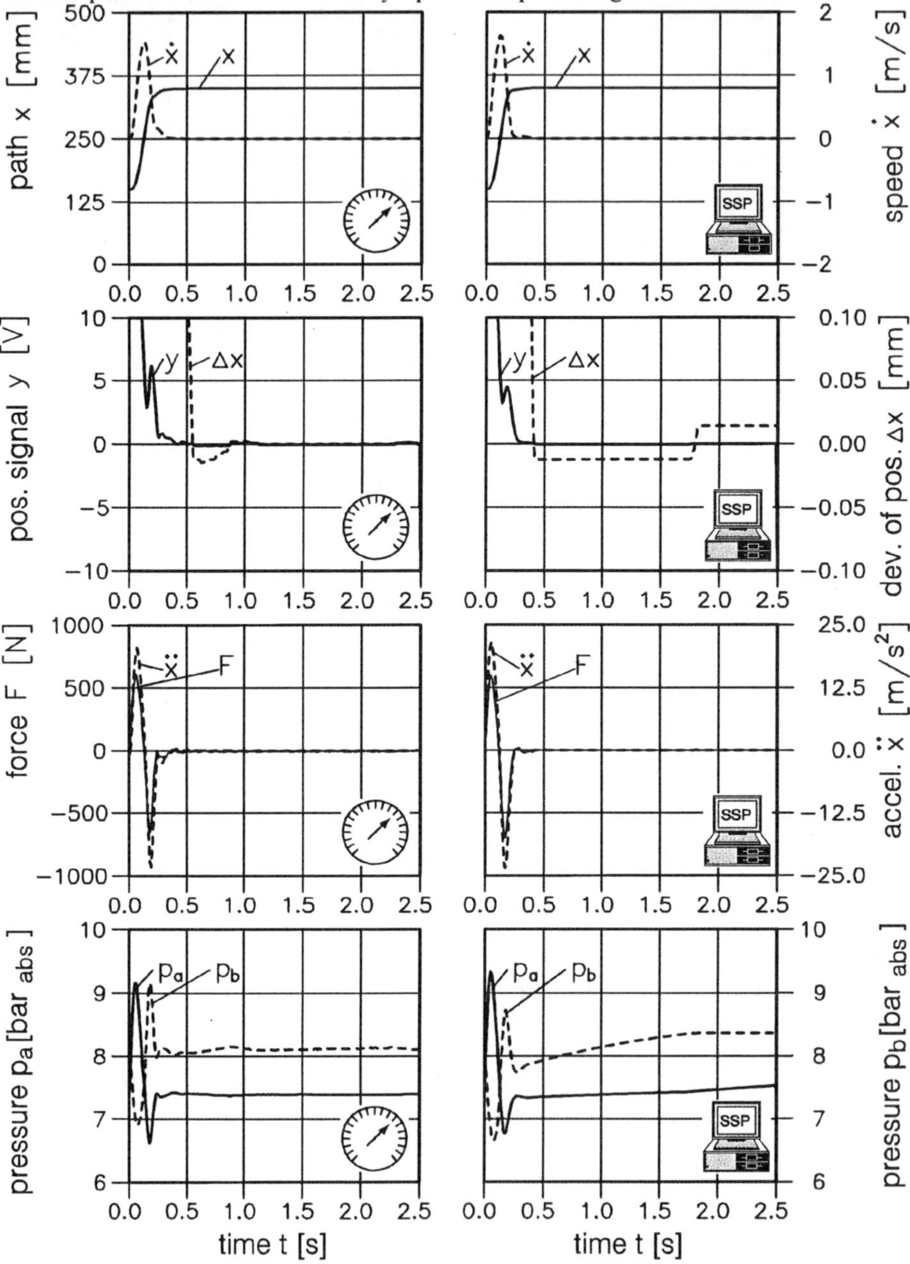


*Figure 13: Comparison of measurement and simulation with optimal positioning*

The function of the force F presented in the illustration represents the force between piston rod and piston, i.e. the force that is required for the acceleration of the mass. Accordingly this force is exactly proportional to the measured acceleration. The friction force of the slide guide is negligible compared with the force of inertia.

In order to prove that with a variation of a control parameter, the measurement and the simulation change in the same way, a variation of the control parameter $K_{\dot{x}}$ was carried out based on the step response represented in figure 13. Figure 14 shows the reaction at $K_{\dot{x}} = \frac{1}{3} \cdot K_{\dot{x},opt}$ and at $K_{\dot{x}} = 3 \cdot K_{\dot{x},opt}$.

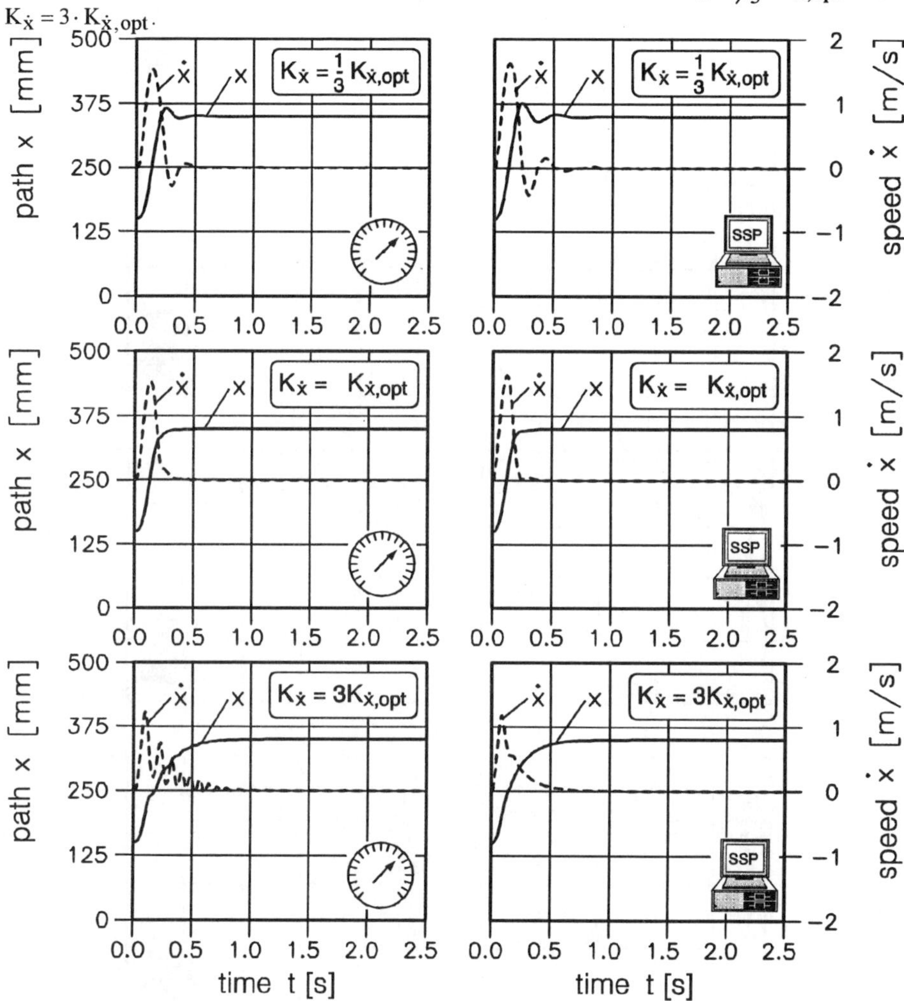


*Figure 14: Comparison of measurement and simulation with variation of the control parameter* $K_{\dot{x}}$

An increase of this factor prolongs the positioning time, as the speed of the drive is reduced by the higher velocity feedback. Additionally, this increases the possibility of the system to oscillate.

If the factor is reduced to a third there will be an overswing of the drive because of the distinctly lower damping.

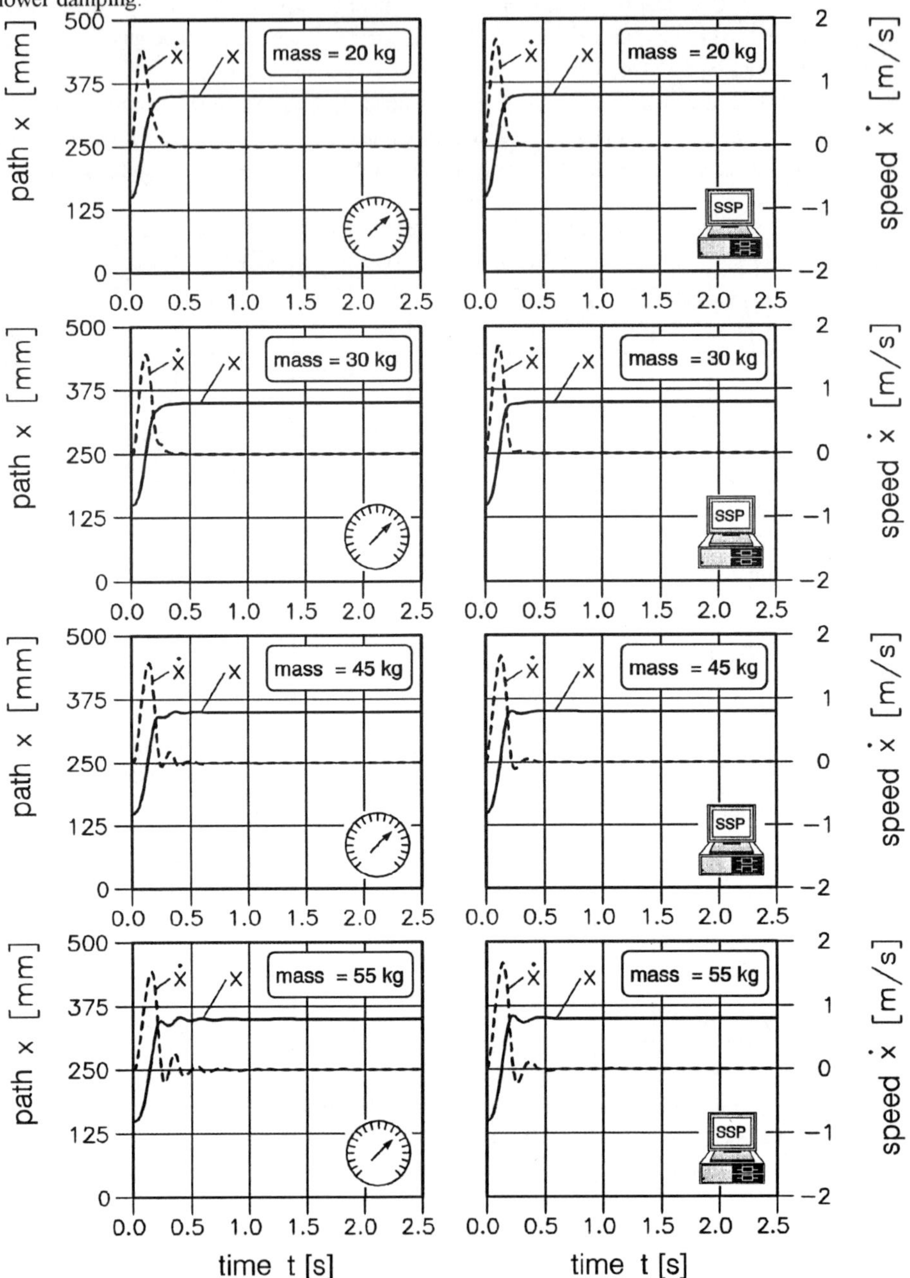


*Figure 15: Variation of the accelerated mass*

As a further variation of the test parameters the accelerated mass was changed on the slide of the drive. Figure 15 shows the different step responses with an unchanged adjustment of the controller. An increase of the accelerated mass reduces the damping of the system, i.e. it leads to overswinging and to a delay in reaching the nominal position. The reduced damping effect of the accelerated mass becomes equally apparent in both measurement and simulation.
The comparisons of measurement and simulation presented here (figures 10 - 15) show it to be possible to simulate the behaviour of a pneumatic system with high precision both in open and in closed loop control. With regard to system design the differences between measurement and simulation are considered to be insignificant. Concerning the closed loop control it is assumed that an optimization of the controller parameters may be possible by means of the simulating program.

## 7 Conclusions

The simulation program "SSP" presents the possibility to combine pneumatic components and to simulate their interactions in a system.
The simulation of pneumatic components in an open- loop or in a closed loop control circuit is possible if the parameters of the individual components required for the simulation are known from exact measurements.
Flow characteristics and the friction force at seals have strong influences on the behaviour of a pneumatic cylinder drive. The paper describes methods for measuring the friction forces and the flow characteristics. The friction force of pneumatic seals is dependent to a high degree on the difference of pressure at the sealing lip and on speed. Concerning the measurement of flow, the dependence of conductance C and of the critical pressure ratio b on the valve stroke is shown for a continuous valve.
By means of a precise reproduction of the friction- forces occurringin reality it is possible to demonstrate both experimentally and by simulation the stick-slip- motion at a pneumatic cylinder drive caused throttling the input flow.
Results obtained from tests and simulation are compared for the positioning of the servo- pneumatic cylinder. There is good correspondence between the tests and simulation for all examined parameters.

## References

[1] **Backé, W.:** Grundlagen der Pneumatik
Vorlesungsmanuskript RWTH Aachen; 7. Auflage 1986

[2] **Eschmann, R.:** Reibkräfte an Pneumatikdichtungen
10. Aachener Fluidtechnisches Kolloquium; März 1992, Band 3 S: 49- 69

[3] **N.N.:** ISO 6358, Oktober 1989
Pneumatik fluid power -Components using compressible fluids -
Determination of flowrate characteristics

[4] **Scholz, D.:** Auslegung servopneumatischer Antriebssysteme
Dissertation, RWTH Aachen, 1990

**WRITTEN DISCUSSION**
**SSP - a simulation program for pneumatics**
**W Backé & R Eschmann (IHP, RWTH Aachen, Germany)**

**Question:** **ND Vaughan**
**Fluid Power Centre, Bath, UK**

What approach is used for the modelling of thermodynamic effects?

**Answer:**

To calculate the power in a volume you have to take the power lost by heat transfer into consideration. The power lost by heat transfer is calculated from the term:

$$P = \frac{k-1}{k} \cdot \alpha \cdot \Delta T \cdot A_{cyl}$$

In this equation $\alpha$ is the heat transfer coefficient, $\Delta T$ the difference between the temperature of the cylinder wall and the air and $A_{cyl}$ the area of the chamber. It is known that $\alpha$ depends on the Reynolds-, Nusselt-, Grashof- and Prandtl-number that means that $\alpha$ depends on the flow conditions of the air in the chambers of a pneumatic cylinder. In experiments, we found that in the simulation $\alpha$ can be taken as constant with a value between 75 and 100 $W/m^2 \cdot K$.

**Question:**

Does the data for valves and cylinder friction vary very much between components?

**Answer:**

In order to get an exact simulation of a pneumatic system, you have to know exactly the characteristics of the valves and seals which are used in the system. A small difference of the conductance C for example may cause big difference between simulated and real behaviour of a pneumatic system. Therefore the characteristics of every type of component have to be known.

Between valves of the same type the characteristics show only small differences of a few per cent. But the characteristics of seal friction can differ much more. These characteristics depend strongly on the lubrication and the wear, which is caused by friction.

**Question:** **J Pu**
**Loughborough University of Technology, UK**

Has any allowance been included for variations in supply pressure when there is a high air consumption?

**Answer:**

Variations of the supply are not taken into consideration. These variations are low because of an additional container with a volume of 40 L. This tank is installed in front of the continuous directional slide valve. Therefore with a supply pressure of 11 bar absolute, the pressure only varies between 0.1 and 0.2 bar. But it would be easily possible to simulate the drop in pressure by adding a flow resistance in front of the continuous directional slide valve.

**Question:** **M Sethson**
**University of Linköping**

What is the performance of the package in terms of simulation time and what type of solver do you use?

**Answer:**

The relationship between the duration of calculation and simulation depends on many different factors and cannot therefore be determined in general. Influencing factors on the duration of calculation are, for example, the speed of the processor and the type and number of components. Since the program works with an automatic step dividing scheme, it may possibly be necessary to employ small time steps when the integration parameters vary too much. This will increase time needed for the simulation. For example the relation of simulation time and the duration of the calculation for the simulations in figures 13 - 15 is 1:100. In these cases the simulations were made with a PC with a 80386/87 40 MHz processor.

For numerical integration the user may choose between Euler's method, improved Euler's method, Heun's method and the method of Runge Kutta. For the simulations shown Runge Kutta's method was used.

# 7. Simulation of Pneumatic Systems using BGSP (Bond Graph Simulation Program)

S Ikeo, H Zhang, K Takahashi *and* Y Sakurai

## Abstract

Recently, with the growing demands for improving the characteristics and reliability of a pneumatic system, it becomes necessary to consider the thermodynamic process for the system in more detail. This paper describes a bond-graph approach for the modelling and simulation of a pneumatic system taking into consideration of the compressibility of air and the heat transfer. On the other hand, BGSP, which has simulated the system using only one-port elements so far, is extended so that it can be used for the simulation of multiport systems. The validity of this method is demonstrated by comparing the simulated results with the experimental ones for a pneumatic meter-out circuit. It was shown that by accounting for heat transfer rather than using a polytropic law, the extended BGSP was able to predict temperature variations accurately.

## 1 Introduction

With growing demands for automation and labour saving, pneumatic systems are being widely used in various fields of industry. In order to design a system with better characteristics and reliability, it is necessary to predict the dynamic behaviour of the system using computer simulation in the design stage. In analysing the dynamic behaviour of pneumatic systems, it is necessary to take into account thermodynamic effects. However, the analysis of a pneumatic system is difficult compared with that of a hydraulic system, because air is a compressible fluid and the thermal, fluid and mechanical system are strongly coupled. Such complexity leads to difficulties in the modelling and simulation of the system. The simulation of such a system is usually made under the assumption that the thermodynamic process is isothermal, adiabatic or polytropic process.

In recent years, the bond-graph method [1], which pays attention to the energy flow through the system, is known as a simple and effective tool for the modelling and simulation of a complex system with many interacting physical components, particularly when several systems are involved. This technique has been applied to different systems such as mechanical, electrical, fluid, thermal and combined systems. On the other hand, many simulation programs such as ENPORT[2] using bond graph have been already developed and obtained fruitful results. In Japan, a simulation program called BGSP[3] was developed ten years ago at Mechanical Engineering Laboratory of Japan, and has been applied to hydraulic systems[4]. From this program, the state equation of the system can be obtained in functional forms and any nonlinear function can be calculated using symbolic manipulation. However, multiport systems have not been used in BGSP so far.

In this paper, a pneumatic system is initially described using bond-graph models which include multiport fields, considering the compressibility of air and the heat transfer. Next, BGSP is extended so that it can simulate multiport systems, and is applied to predict the dynamic behaviour of a pneumatic system. The validity of this method is demonstrated by experiments. The simulated results using this method are compared with those assuming a polytropic change. The results show that the heat transfer must be considered to predict the dynamic behaviour of a pneumatic system precisely.

# 2 Nomenclature

| | | | | | |
|---|---|---|---|---|---|
| $A_e$ | effective area of valve | $n$ | polytropic index | $t$ | time |
| $C$ | capacitance | $p$ | pressure | $U$ | internal energy |
| $c_p$ | specific heat at constant pressure | $Q$ | heat-flow rate | $u$ | specific internal energy |
| | | $R$ | resistance | $V$ | volume |
| $c_v$ | specific heat at constant volume | $R_a$ | perfect-gas constsnt | $v$ | specific volume |
| | | $S_e$ | effort source | $x$ | displacement |
| $E$ | energy | $S_f$ | flow source | 0 | zero-Junction |
| $h$ | specific enthalpy | $S$ | entropy | 1 | one-Junction |
| $I$ | inertance | $s$ | specific entropy | $\kappa$ | ratio of the specific heat |
| $m$ | mass | $T$ | temperature | $\rho$ | air density |

## Subscripts

$_0$: initial value $\quad$ $_1$: inflow $\quad$ $_2$: outflow

$_{cap}$: cylinder cap-end $\quad$ $_{rod}$: cylinder rod-end

# 3 Bond-Graph Approach

Two types of bond graph, the true and the pseudo bond graph[5][6], have been developed for thermal systems. We shall adopt the true bond graph to take the same viewpoint as the hydraulic case, that is, the product of the effort and flow yields power. The temperature and the pressure are taken as efforts, and the entropy flow rate and the volume flow rate as flows, for thermal and fluid systems, respectively. The energy variables are the displacements $V$ and $S$, which are the integrals of the flows.

The constitutive relations and bond-graph representation of some basic elements which are different from the hydraulic case are discussed below.

## 3.1 Connection elements

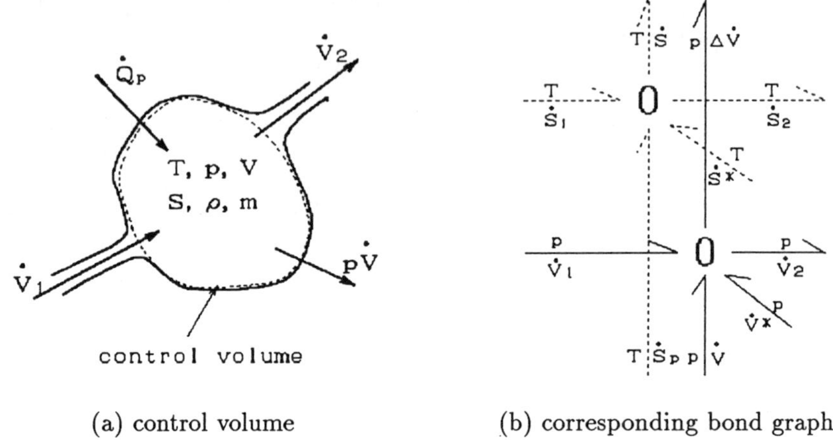


(a) control volume (b) corresponding bond graph

Figure 1: Bond graph of 0-Junctions

The air in the control volume shown in Figure 1(a) exchanges thermal energy $\dot{Q}_p$ per unit time with its surroundings when it is heated or cooled, and it exchanges mechanical energy $p\dot{V}$ per unit time as it changes volume. The state of the air within the control volume is assumed to be homogeneous. Thus, it follows that $U = mu, S = ms, V = mv$. The conservative laws of mass and energy for this control volume can be written as

$$\dot{m} = \dot{m}_1 - \dot{m}_2 \tag{1}$$

$$\dot{E} = \dot{E}_1 - \dot{E}_2 - p\dot{V} + \dot{Q}_p \tag{2}$$

In order to represent the two equations by bond graphs, these equations must be expressed in terms of energy variables $V$ or $S$.

First, concerning equation (1), the state variable $\Delta V$ for the fluid system is defined as

$$\Delta V = \frac{m}{\rho_0} - \frac{m}{\rho} \tag{3}$$

where $\rho_0$ denotes the initial density. $m/\rho_0$ is the volume of the air contained in the control volume at time $t$ reduced to the initial state. Differentiating equation (3) with respect to time $t$ and substituting $\dot{m}$ into equation (1) where $\dot{m}_1 = \rho\dot{V}_1, \dot{m}_2 = \rho\dot{V}_2$, we obtain the following equation:

$$\Delta\dot{V} = \frac{\rho}{\rho_0}\dot{V}_1 - \frac{\rho}{\rho_0}\dot{V}_2 - \dot{V} \tag{4}$$

where $\dot{V}_1, \dot{V}_2$ are the volume flow rates flowing into and out of the control volume under the pressure $p$, respectively. In order to represent easily equation (4) by bond graph, it is written by adding and subtracting $\dot{V}_1 - \dot{V}_2$ in its right hand as

$$\Delta\dot{V} = \dot{V}_1 - \dot{V}_2 - \dot{V} - \dot{V}^* \tag{5}$$

where

$$\dot{V}^* = (1 - \frac{\rho}{\rho_0})(\dot{V}_1 - \dot{V}_2) \tag{6}$$

For a hydraulic system, since the oil compressibility is small, the variation of the density can be neglected. Consequently, $\dot{V}^*$ will disappear. In the case of a pneumatic system, however, because of the air compressibility, $\dot{V}^*$ appears in the continuity equation compared with a hydraulic system.

Secondly, concerning the energy equation (2), the total energy in the control volume is replaced by the internal energy

$$E = U, \tag{7}$$

and assuming that the kinetic and potential energies are negligible, we can express the energy flow rates flowing into and flowing out of the control volume in equation (2) as

$$\dot{E}_1 = \dot{m}_1 h_1, \quad \dot{E}_2 = \dot{m}_2 h_2 \tag{8}$$

The generalized Gibbs's equation is given by

$$dU = TdS - pdV + \varphi dm \tag{9}$$

where the chemical potential $\varphi = u + pv - Ts$. Substituting equations (7)~(9) into equation (2), we obtain the equation of state with respect to $S$[5]

$$\dot{S} = \frac{\dot{m}_1 h_1 - \dot{m}_2 h_2}{T} + \frac{\dot{Q}_p}{T} - \frac{\varphi\dot{m}}{T} \tag{10}$$

Further, we assume that properties such as the temperature and pressure of the air flowing into this control volume are different from those inside the control volume, while those flowing out of the control volume are the same as those inside. Under these assumptions, equation (10) becomes

$$\dot{S} = \dot{S}_1 - \dot{S}_2 + \dot{S}_p + \dot{S}^* \tag{11}$$

in which $\dot{S}_1 = \dot{m}_1 u_1/T$ is the entropy flow rate into the control volume by the convection of air, $\dot{S}_2 = \dot{m}_2 u_2/T$ is that out of the control volume, $\dot{S}_p = \dot{Q}_p/T$ is that due to heat transfer between the air in the control volume and surroundings, and $\dot{S}^* = \dot{S}_1^* + \dot{S}_2^*$ where $\dot{S}_1^* = (Ts - u)\dot{m}/T$ is the entropy flow rate due to the variation of the air mass in the control volume and $\dot{S}_2^* = \dot{m}_1(p_1 v_1 - p_2 v_2)/T$ is that due to the power loss of air flow.

In Figure 1(b), the solid lines represent equation (5) where the flow variable is the volume flow rate, and the dashed lines equation (11) with the entropy flow rate as the flow variable.

## 3.2 Capacitive element

$$\vdash \overset{p}{\underset{\dot{V}}{\rightharpoondown}} C \overset{T}{\underset{\dot{S}}{\leftharpoondown}} \dashv$$

Figure 2: Bond graph of a C-field

In the case of the compressible flow in a pneumatic system, both the thermal and fluid variables interact with each other. This interaction is expressed by the following equations[1] using the equation of state of the ideal gas and the Gibbs's equation.

$$T = T_0 \exp^{\frac{s-s_0}{c_v}} \left(\frac{v}{v_0}\right)^{-(\kappa-1)} \tag{12}$$

$$p = p_0 \exp^{\frac{s-s_0}{c_v}} \left(\frac{v}{v_0}\right)^{-\kappa} \tag{13}$$

The specific volume $v$ is expressed using equation (3) as

$$v = \frac{1}{\rho} = \frac{1}{\rho_0} - \frac{\Delta V}{m} \tag{14}$$

and considering $v_0 = V_0/m_0, s = S/m, s_0 = S_0/m_0$, we obtain equations (15) and (16)

$$T = T_0 \exp^{\left(\frac{S}{c_v m} - \frac{S_0}{c_v m_0}\right)} \left(1 - \frac{m_0 \Delta V}{m V_0}\right)^{-(\kappa-1)} \tag{15}$$

$$p = p_0 \exp^{\left(\frac{S}{c_v m} - \frac{S_0}{c_v m_0}\right)} \left(1 - \frac{m_0 \Delta V}{m V_0}\right)^{-\kappa} \tag{16}$$

These two equations describe the relationships between the efforts$(p, T)$ and the displacements $(\Delta V, S)$. Therefore, a 2-port $C$ element should be used for representing the combined fluid and thermal systems, as shown in Figure 2.

Comparing the constitutive relations (15) and (16) of the $C$ element with the polytropic relation $T/T_0 = (p/p_0)^{\frac{n-1}{n}}$, we obtain the following polytropic index:

$$n = \frac{\kappa}{1 - \frac{s-s_0}{c_v \ln(p/p_0)}} \tag{17}$$

From equations (15) and (16), the following equation is derived:

$$\frac{s - s_0}{c_v \ln(p/p_0)} = (1 - \kappa) + \kappa \frac{\ln(T/T_0)}{\ln(p/p_0)} \tag{18}$$

Combining $\dot{S} = m\dot{s} + s\dot{m}$ with equation (10), we obtain the specific entropy flow rate by

$$\dot{s} = \frac{\dot{m}_1(h_1 - h)}{T} + \frac{\dot{Q}_p}{T} \tag{19}$$

Since during both the inflow and outflow processes of air, the change in temperature has the same sign as the pressure change, and the change in entropy has the reverse sign to the pressure change, the following relation is obtained from equation (18):

$$1 - \kappa \le \frac{s - s_0}{c_v \ln(p/p_0)} \le 0 \tag{20}$$

Substituting inequality (20) in equation (17), we obtain

$$1 < n < \kappa \tag{21}$$

Therefore, the constitutive relations of 2-port $C$ element expressed by equations (15) and (16) mean that the state of air in the cylinder undergoes a polytropic change and the polytropic index $n$ varies as given by equation (17).

## 3.3 Resistive element

**Flow through a valve** Assuming that the air flow is isentropic and the flow through the valve in a pneumatic system is analogous to that through a converging nozzle, we can express the volume flow rate of air upstream of the valve as follows[8]:

$$\dot{V}_1 = A_e\sqrt{2R_aT_1\tfrac{\kappa}{\kappa-1}\{(\tfrac{P_2}{P_1})^{\frac{2}{\kappa}} - (\tfrac{P_2}{P_1})^{\frac{\kappa+1}{\kappa}}\}} \quad \text{for subsonic flow} \tag{22}$$

$$\dot{V}_1 = A_e\sqrt{\kappa R_aT_1\{\tfrac{2}{\kappa+1}\}^{\frac{\kappa+1}{\kappa-1}}} \quad \text{for choked flow} \tag{23}$$

Considering $\rho_1\dot{V}_1 = \rho_2\dot{V}_2$, we obtain the volume flow rate of air downstream of the valve as

$$\dot{V}_2 = A_e\tfrac{\rho_1}{\rho_2}\sqrt{2R_aT_1\tfrac{\kappa}{\kappa-1}\{(\tfrac{P_2}{P_1})^{\frac{2}{\kappa}} - (\tfrac{P_2}{P_1})^{\frac{\kappa+1}{\kappa}}\}} \quad \text{for subsonic flow} \tag{24}$$

$$\dot{V}_2 = A_e\tfrac{\rho_1}{\rho_2}\sqrt{\kappa R_aT_1\{\tfrac{2}{\kappa+1}\}^{\frac{\kappa+1}{\kappa-1}}} \quad \text{for choked flow} \tag{25}$$

On the other hand, since the entropy flow rate upstream of the valve $\dot{S}_1$ is same as the entropy flow rate flowing out of the control volume located upstream of the valve where the temperature is $T_1$,

$$\dot{S}_1 = \frac{\dot{m}_1 u_1}{T_1} \tag{26}$$

$T_1$ $\dot{S}_1$ R $T_2$ $\dot{S}_2$ $p_1$ $\dot{V}_1$ $p_2$ $\dot{V}_2$

(a) flow through a valve

$T_1$ $\dot{S}_1$ R $T_2$ $\dot{S}_2$

(b) heat transfer through a cylinder wall

Figure 3: Bond graph of R-field

Similarly

$$\dot{S}_2 = \frac{\dot{m}_1 u_1}{T_2} \tag{27}$$

The internal energy in equations (26) and (27) is calculated either directly from $u_1 = c_v T_1$ assuming ideal gas, or from $u_1 = \int c_v dT_1$ assuming semi-ideal gas, using the properties of the air upstream of the valve. Simulated results from the above two methods will be shown later.

Equations (22) $\sim$ (27) showed that both the volume flow rate and the entropy flow rate upstream of the valve are different from those downstream because of the compressibility of air and irreversibility due to heat transfer. These equations can be represented simplely by a 4-port $R$ element as shown in Figure 3(a).

**Heat transfer through the cylinder wall** When heat energy does not accumulate in the cylinder wall, the heat-balance equation is written in the form

$$\dot{Q} = HA(T_1 - T_2) \tag{28}$$

where $H$ is the overall heat-transfer coefficient, including the effects of the convection and the conduction, $A$ is the heat transfer area. $T_1$ is the temperature of surroundings and $T_2$ is that of the air in the cylinder. Applying the second law of thermodynamics, we obtain the following equations with respect to the entropy flow rate:

$$\dot{S}_1 = \frac{HA(T_1 - T_2)}{T_1}, \quad \dot{S}_2 = \frac{HA(T_1 - T_2)}{T_2} \tag{29}$$

The corresponding bond graph is shown in Figure 3(b) which uses a 2-port $R$ element.

It should be noted that elements $S_e, S_f$ supply not only fluid power but also thermal power to the system. The corresponding bond-graph models will be shown later in the example. Since there is no $I$ element in thermal systems, the inertial effect of air is represented by a 1-port $I$ element as for hydraulic elements.

# 4 Simulation and Considerations

In order to examine the validity of the bond-graph method developed above, the dynamic behaviour of a simple pneumatic system is simulated using the extended BGSP, and computer predictions are compared with experimental results.

## 4.1 Modelling of a System

Figure 4(a) shows a pneumatic meter-out circuit which is composed of a cylinder, a three-position five-way solenoid valve, and two speed control valves.

Figure 4(b) shows the corresponding bond-graph diagram, where the dashed bonds represent thermal power ports and the solid bonds represent pneumatic and mechanical ones. Two-port source element $S_{e1}$ represents the air source at a certain temperature and pressure, and $S_{e2}$ at atmospheric temperature and pressure. Two-port source element $S_{f1}$ represents the terms $\dot{V}^*, \dot{S}^*$ in equations (5) and (11) for the cap-end chamber of the cylinder, and $S_{f2}$ for the rod-end chambers, because these terms have the same dimensions as the flow variables. The elements $C_1$ and $C_2$ represent the capacitances of the cylinder cap-end chamber and rod-end chamber of the cylinder, $R_1$ and $R_2$ the resistances of the solenoid valve, $R_3$ and $R_4$ the resistances of the speed control valve, $R_5$ and $R_6$ the energy loss due to the heat transfer between the air in the cylinder and surroundings. One-port elements $I_1$, $R_7$ and $S_{e3}$ connected to 1-Junction represent the inertia, viscous friction and Coulomb friction force of the piston-rod assembly, respectivitly, as in hydraulic systems.

Comparing this bond graph with that for an equivalent hydraulic system, the following differences can be seen:

1) Elements $C, R, S_e, S_f$ produce a multiport field instead of a 1-port field.

2) The thermal bond graphs, which are plotted by dashed bonds in parallel with the pneumatic ones, are added to model the effect of heat flow induced by the air flow.

3) In order to model the effect of heat transfer, 2-port elements $R_5$ and $R_6$ are inserted between two 0-Junctions, one of which represents the cylinder chamber and the other the atmosphere.

4) In order to model $\dot{V}^*$ in equation (5) and $\dot{S}^*$ in equation (11), 2-port elements $S_{f1}$ and $S_{f2}$ are connected to 0-Junctions which represent the cylinder chambers.

5) Since the absolute pressure is used in a pneumatic system, it is necessary to add a 0-Junction representing the atmosphere.

## 4.2 Extension of BGSP

The BGSP consists of two programs: one a pre-processing program PRE.F which generates the state equations for physical systems, and the other a nu-

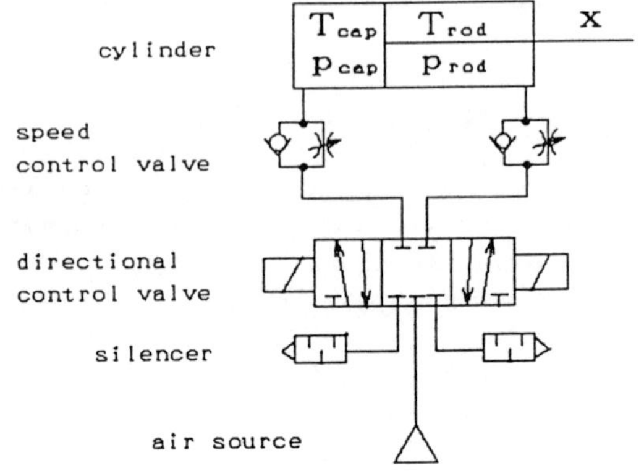


(a) meter-out circuit

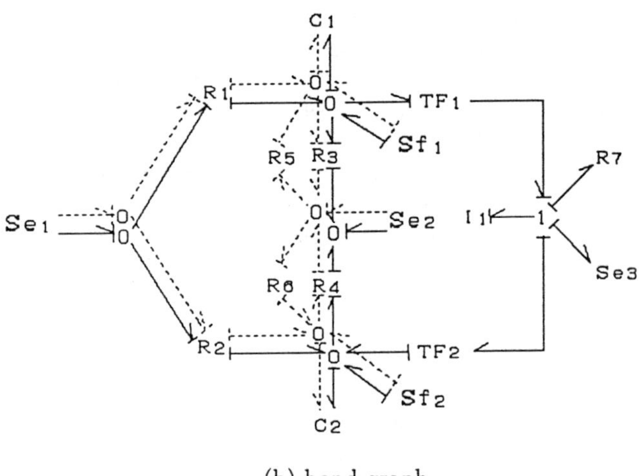


(b) bond graph

Figure 4: Pneumatic system and corresponding bond graph

merical program BGS.F which solves a set of state equations by a Runge-Kutta fourth-order method. The BGSP can solve nonlinear equations using symbolic manipulation.

The BGSP has been applied to analyse the properties of various hydraulic systems, where elements such as $C, R, I, S_e, S_f$ are 1-port ones and the relations of variables are one-to-one. For pneumatic systems, it is necessary to use multi-port elements, and the relations between variables become n-to-n, because heat transfer must be taken into consideration.

## 4.3 Comparison of the Simulated Results with Experiment

The temperature and pressure variations in the cap-end and rod-end chambers of the cylinder and the displacement of the piston were obtained from the bond-graph diagram shown in Figure 4 using the extended BGSP. The simulated condition included a pressurised cap-end chamber and a movement of the piston-rod assembly from left to right after the solenoid has been activated.

In numerical simulation, the effective area of the valve and the overall heat transfer coefficient between the air and the cylinder wall[7] were determined from the measured pressure waveforms in the inflow and the outflow processes. They are considered to be important parameters influencing the dynamics of the pneumatic system.

Under the assumption that the air is a semi-ideal gas, the specific heats are calculated from the following empirical equations[8]:

$$c_v = \frac{D}{M}(1 + \xi T), \quad c_p = \frac{D}{M}(k + \xi T) \tag{30}$$

where $D = 18.994, M = 28.96, \xi = 0.0002885, k = 1.438$. Thus, the internal energy is expressed as

$$u = \frac{D}{M}(T + \frac{\xi}{2}T^2) \tag{31}$$

Figure 5 shows the simulated results and experimental ones. The air temperature was measured using a $25.8\mu m$ copper-constantan thermocouple that was inserted to the cylinder through the cylinder tube in the radial direction. The air pressures in the cap-end and the rod-end chambers were measured with absolute pressure transducers, and the piston displacement with a linear potentiometer(0.3m).

It can be seen that the simulated results agree well with the experimental ones. The waveform of the measured air temperature fluctuates slightly. This results from the circulation caused by a nonuniform density distribution of the air around the thermocouple junction due to forced and free convections.

The simulated results, assuming the specific heat to be constant, are also shown in Figure 5. It can be seen that the temperatures calculated assuming varying specific heat are lower during inflow process and higher during outflow

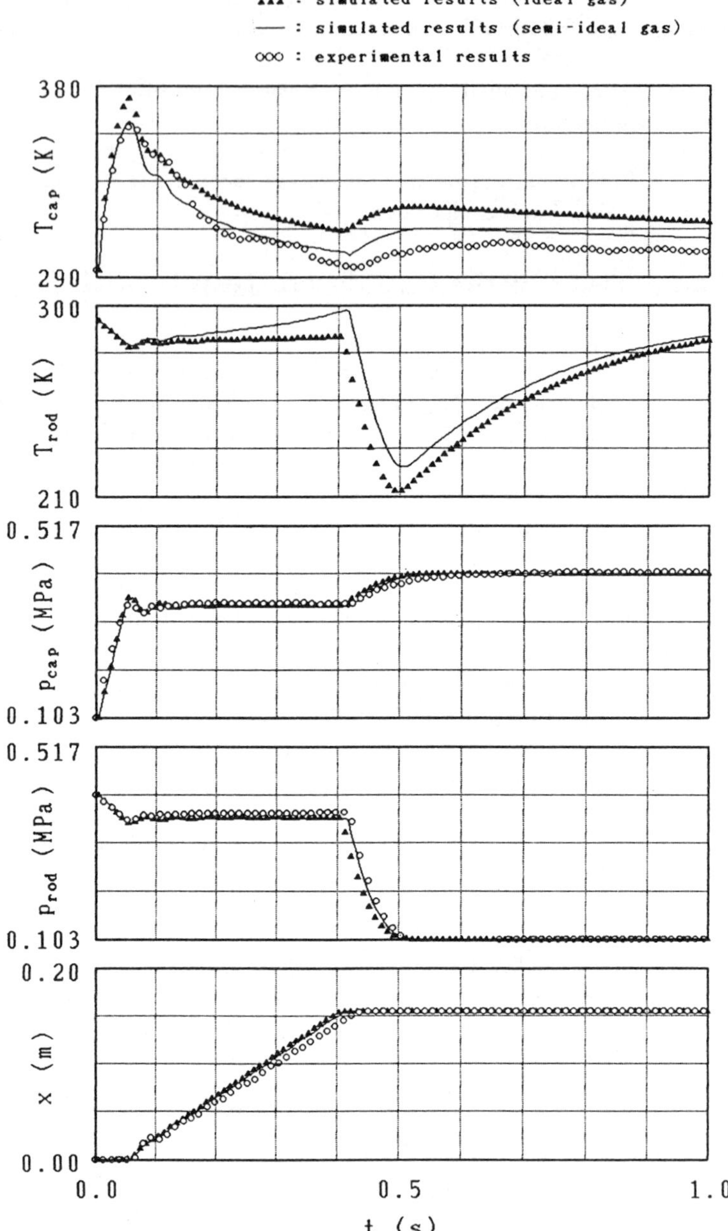


Figure 5: Simulated results and experimental results

process than those assuming constant specific heat. It can be seen from Figure 5 that the former shows better agreement with the experimental results.

## 4.4 Comparison with Polytropic Process

In most applications, the change of state of the air in the cylinder is assumed to be polytropic during the cylinder-actuating process because of some heat exchange. By doing so, the modelling and simulation of a pneumatic system is simpler, and the elements are represented as 1-port when the bond-graph model is used[9]. However, it is accompanied by a great error in the prediction of thermodynamic behaviour as shown later.

Figure 6 shows the simulated results when polytropic index was varied between 1.0 and $\kappa$. From these results, the following inferences can be drawn:

1) The pressures and displacements calculated by this method agree with the experimental results as shown in Figure 5, although the response becomes faster with an increase in $n$,

2) When an appropriate polytropic factor $n$ is selected as $n = 1.1$, the waveforms of pressure and displacement agree well with the experimental results,

3) The waveforms of temperature change in a similar manner to the pressure. However, the temperature does not return to the initial state after the pressure reaches an equilibrium. In experiments, it returns to the initial state due to the heat transfer through the cylinder wall.

Therefore, when the change of state of the air is assumed to be polytropic, and a suitable polytropic factor between 1 and $\kappa$ is selected, it is possible to predict the pressure and the displacement accurately, but the temperature cannot be predicted accurately using this method.

# 5 Conclusions

1. In order to analyse the dynamic behaviour of a pneumatic system, it is necessary to take into consideration the compressibility of the fluid and the heat transfer. The use of multiport elements in producing a bond-graph diagram greatly facilitates the simulation of the system. The power state variables used are temperature and entropy flow rate for thermal variables and pressure and volume flow rate for fluid variables. The bond-graph diagram for a pneumatic system can be constructed similarly to that for a hydraulic system with certain additions.

2. The BGSP was extended to be able to be applied to the simulation of multiport systems, and applied to a simple pneumatic system to predict temperature, pressure and displacement. The results agree fairly well with the experimental ones.

3. In the simulation of a pneumatic system, parameters such as the effective area of valve, the heat transfer coefficient need to be determined by experiment

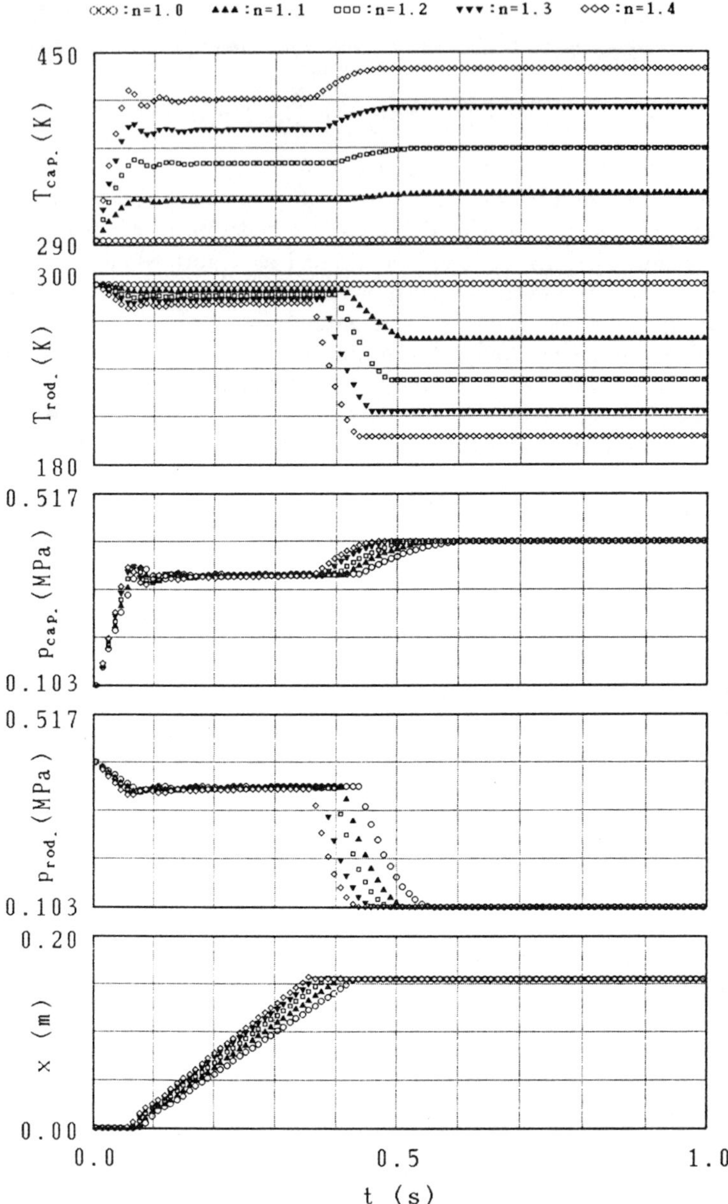


Figure 6: Simulated results based on polytropic process

and the influence of temperature on the thermodynamic properties such as specific heat must be considered.

4. It was shown that by accounting for heat transfer rather than using a polytropic law, the extended BGSP was able to predict temperature variations accurately.

# References


[1] **Karnopp, D. C., and Rosenberg, R. C.,** "System Dynamics: A Unified Approach", Wiley, New York, 1975

[2] **Rosenberg, R. C.,** "A User's Guide to ENPORT-4", John Wiley and Sons, Inc., USA, 1974

[3] **Kohda, T., Nakada, T., Kimura, Y. and Mitsuoka, T.,** "Simulation of Bond Graphs with Non-Linear Elements by Symbolic Manipulation", Bull. Mech. Eng. Lab. ISSN 0374-2725, No. 49, 1988

[4] **Sakurai, Y., et al,** "Simulation of the Dynamic Characteristics of an Injection Modeling Machine by the Bnod Graph Method", Proc. JHIPS International Symp. Fluid Power Tokyo, 49-54,1989

[5] **Shoureshi, R., and McLaughlin, K.,** "Application of Bond Graphs to Thermofluid Processes and Systems" ASME Journal of Dynamic System, Measurement, and Control, Vol. 107, No. 12, 1985, pp. 241-245

[6] **Karnopp, D. C.,** "State Variables and Pseudo Bond Graphs for Compressible Thermofluid Systems", ASME Journal of Dynamic System, Measurement, and Control, Vol. 101, No. 9, 1979, pp. 201-204

[7] **Kagawa, T.,** "Dynamics of Pneumatic RC Circuits" (in Japanese), J. Jpn. Hydraulics and Pneumatics Soc, Vol. 17, No. 3, 1986, pp. 205-212

[8] **Nishikawa, K., et al,** "Application Thermodynamics" (in Japanese), Koronasha, 1983

[9] **Araki, K., Liu, X. P., Ishino, Y.,** "The Pneumatic System Analysis with Bond Graph Method"(in Japanese), 3rd Joint Symposium on Fluid Control and Measurement, Tokyo, 1991

**WRITTEN DISCUSSION**
**Simulation of pneumatic systems using BGSP (Bond graph simulation program)**
**S Ikeo, H Zhang, K Takahashi & Y Sakurai (Sophia University, Japan)**

**Question:** **ND Vaughan**
**Fluid Power Centre, Bath, UK**

Does the technique give advantages with pneumatic system simulation which are not evident with hydraulic systems?

**Answer:**

With pneumatic systems, the temperature variation in the cylinder chamber due to the rapid expansion of compression should be considered in the analysis. This temperature variation affects the dynamic behaviour of the system and also sometimes causes the water condensation in the system.

The effects of the compressibility of air and the heat transfer through the cylinder wall on the dynamic behaviour of pneumatic systems have been extensively investigated by Kagawa (7), (10) - (12) . Using the proposed BGSP system, these effects can be easily simulated. The problem of condensation can also be analyzed by the program package. Figure 7 shows an example of simulated results with represents the effect of initial humidity on the condensation (13).

[10] Kagawa, T. "Heat transfer effects on the frequency response of a pneumatic nozzle flapper". Trans ASME Jnl of Dynamic Systems Measurement & Control, Vol 107, No 4, 1985, pp 332-336

[11] Kagawa, T. and Ohligschlager, O. "Simulations modell fur pneumatische zylinderantriebe". Olhydraulik und Pneumatik, Vol 34, No 2, 12990, pp 115-120

[12] Kagawa, T. Shimizu, M and Ishii, Y. "Air temperature change of pneumatic cylinder with meter-out control and its effect on the velocity". Proc of 3rd Triennial International Symposium on Fluid Control, Measurement and Visualization, FLUCOME '91, pp 543-548

[13] Zhang, H., Ikeo, S., Takahashi, K. and Sakurai, Y. "On condensation of water in pneumatic cylinder" (In Japanese). Proc of JHPS spring meeting, 1993

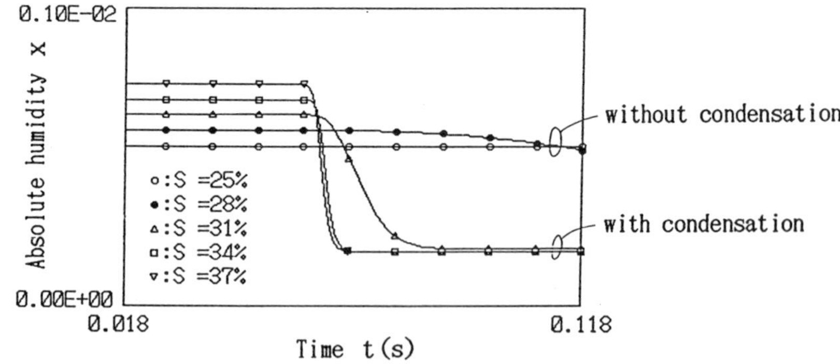


FIGURE 7 Effect of initial humidity on condensation

# 8. Computer Simulation of Human Interaction with Underwater Breathing Equipment

S P Tomlinson, J K W Lo *and* D G Tilley

Abstract

A mathematical model of the human respiratory system is being developed to simulate manned diving operations using various types of breathing equipment. In addition to the diving industry [1],[2], substantial benefit is expected in a number of other fields involving the human using breathing equipment such as rescue, aerospace [3],[4] and respiratory medicine [5],[6]. The model includes a representation of the nasal passageways, tracheobronchial tree, alveolar and pleural compartments, blood supply and neurogenic control action by using appropriate differential and algebraic equations. At surface conditions, the model gives good agreement with observed behaviour and is able to predict time transient variations in lung volume, flow rate and alveolar and pleural pressures during a maximum inspiratory-expiratory manouevre and during quiet continuous breathing. It also predicts time transient variations in constituent gas concentrations and partial pressures throughout the airways, alveolar compartment and blood supply to a tissue compartment.

A simulation study is presented of a human diving to a depth of 50 metres, using a semi-closed re-breathing system [7], in which the majority of the breathing gas is re-circulated continuously between the diver and apparatus. Two different oxygen flow supply settings are examined; a 32.5% oxygen concentration which results a safe dive and a 60% setting which indicates a dangerous diving condition. Typical respiratory model input data is illustrated for the lung, together with dimensional details for the re-breathing counterlung. The simulation results show variations in alveolar and pleural pressure and gas concentrations and partial pressures in the equipment and human respiratory system.

## 1 Introduction

Software for the elements of the respiratory system is being developed in a modular manner, using a fluid power simulation program BATH*fp* [8], produced at the Fluid Power Centre, University of Bath, U.K., in which the user can assess the behaviour of a human using breathing equipment. To perform computational procedures, the software employs a cursor, which is positioned on a schematic of manned breathing equipment, appearing on a screen. These include:

(i) supplying respiratory parameters and equipment dimensions prior to a simulation.

(ii) inspecting respiratory and breathing equipment parameters on completion of a simulation.

(iii) changing the breathing equipment when necessary to establish an alternative simulation.

The equations for the items of breathing equipment and each respiratory element exist in modular subroutines which are connected together. In this manner, differing levels of complexity or competing theories may be tested for the behaviour of the equipment and the various parts of the human respiratory system.

## 2 Nomenclature

A area. $m^2$
B bulk modulus. bar
C flow or discharge coefficient
CV closing volume. L
f viscous friction coefficient. N/(m/s)
FRC functional residual capacity. L
k stiffness. N/m
k diffusion coefficient. L/(min/bar)
M mass. kg
M gas production/consumption. kg/min
P pressure. mm Hg or bar
Q volumetric flow rate. L/min
n polytropic index
R gas constant. kJ/(kg.deg.K)
RV residual volume. L
T temperature. deg. K
V volume. L
V mass flow rate. kg/min
x displacement. m
U gas production. kg/min
d/dt time derivative
$\alpha$ volumetric concentration. L/L
$\Sigma$ summation

### SUBSCRIPTS

a airway or alveolar compartment
al alveolar
$CO_2$ carbon dioxide
b pulmonary blood pool
bA alveolar blood
d discharge coefficient
dv dead space
e extra-pulmonary airway
i intra-pulmonary or constituent gas
l lung wall
m mass flow
o mouth or lung wall
$O_2$ oxygen
pa pulmonary artery
pl pleural compartment
sh physiological right-to-left shunt
t tissue
th thoracic
$\alpha$ atmospheric conditions
2 extra-pulmonary
3 intra-pulmonary

## 3 Simulation of the human respiratory system and underwater breathing equipment using the BATH*fp* package.

Fig.1 is a BATH*fp* simulation schematic of a human using a semi-closed [7],[9] re-breathing system.

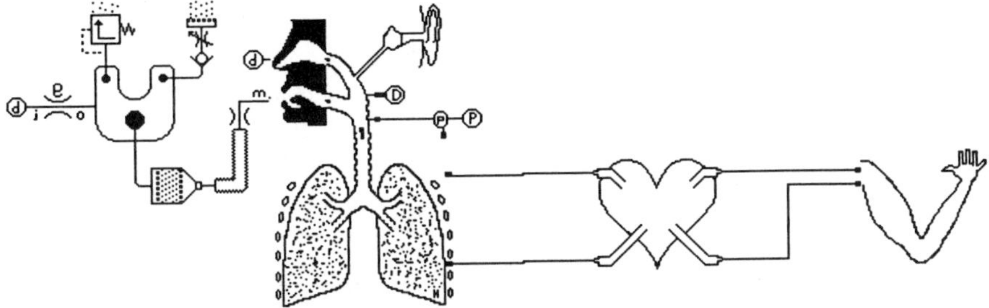

Figure 1. *BATHfp simulation schematic of a human using a semi-closed re-breathing system.* The system employs a counterlung (re-breathing bag) and mass-flow unit to supply a constant mass flow of mixed gas containing a fixed oxygen fraction to the counterlung, irrespective of depth. The model of the human respiratory system comprises elements including:

(i) mouth
(ii) tracheobronchial airways
(iii) ear and Eustachian tube
(iv) alveolar compartment
(v) pleural compartment
(vi) cardiovascular system
(vii) metabolic processes

The breathing equipment includes elements for the counterlung, carbon dioxide removal scrubber, buoyancy control valve, pressure relief valve, breathing tube and mouthpiece.

### 3.1 MOUTH

When using breathing equipment, a significant pressure loss occurs across the mouth and measured data are shown (fig.2) of dimensionless pressure loss coefficients [10] for the mouth to alveoli and the central airways as a function of Reynolds number.

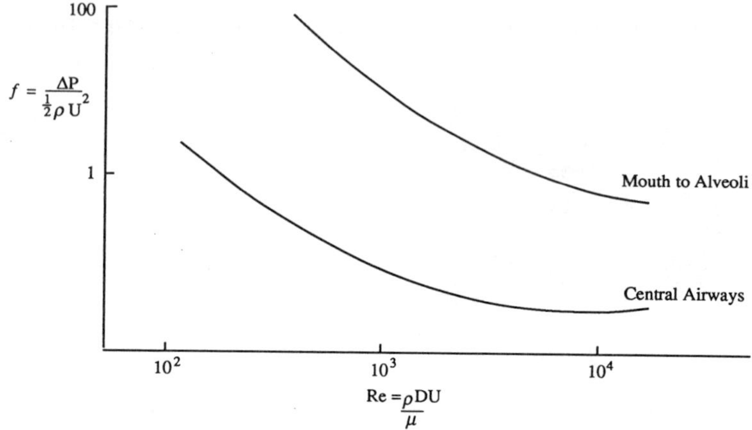


Figure 2. *Dimensionless pressure loss coefficients for the mouth to alveoli and the central airways as a function of Reynolds number.*

The mass flow rate of gas $\dot{V}_1$ through the mouth of area $A_o$, in terms of atmospheric inlet pressure $P_a$ and temperature $T_a$, is determined using the standard orifice compressible flow equation.

$$\dot{V}_1 = \frac{C_m A_o P_\alpha}{\sqrt{T_\alpha}} \tag{1}$$

where $C_m$ the net mass flow coefficient, is derived from fig.2.

## 3.2 TRACHEOBRONCHIAL TREE

The model of the tracheobronchial tree, extending from the mouth to the alveolar compartment has been simplified by divided it into three distinct regions (fig.3); the extra-thoracic zone, the intra-thoracic extra-pulmonary zone and the intra-thoracic intra-pulmonary zone.

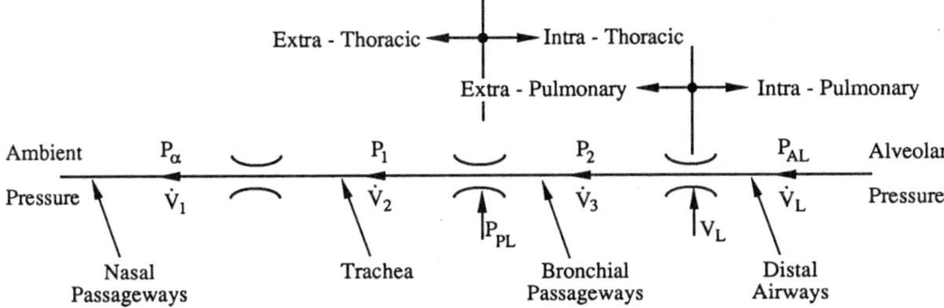


Figure 3. *The extra-thoracic zone, the intra-thoracic extra-pulmonary zone and the intra-thoracic intra-pulmonary zone.*

A uniform (but time varying) pressure is assumed to exist in each zone, allowing lumped parameter theory [11] to be used in the model.

## 3.3 EXTRA-THORACIC ZONE AND INTRA-THORACIC EXTRA-PULMONARY ZONES

The extra-thoracic zone includes the downstream side of the nasal passageways, leading to the trachea. This leads to the intra-thoracic extra-pulmonary zone, which extends to generation 16 [12] of the airways. The zone pressure $P$ is dependent on the net flow rate $\sum \dot{Q}$ . It is given by:

$$\frac{dP}{dt} = \frac{nRT}{V} \sum \dot{Q} \tag{2}$$

where $n$ is the polytropic index, $R$ the gas constant, $T$ the gas temperature and $V$ the volume of the zone.

The intra-thoracic extra-pulmonary airways comprise elastic walls (area $A_a$) which distend under the action of airway pressure $P_a$, acting to open and pleural pressure $P_{pl}$, acting to close the airway. The smooth muscle surrounding a variable area orifice (fig.4) is idealized as a mass-spring-damper system acted on by the transmural force described.

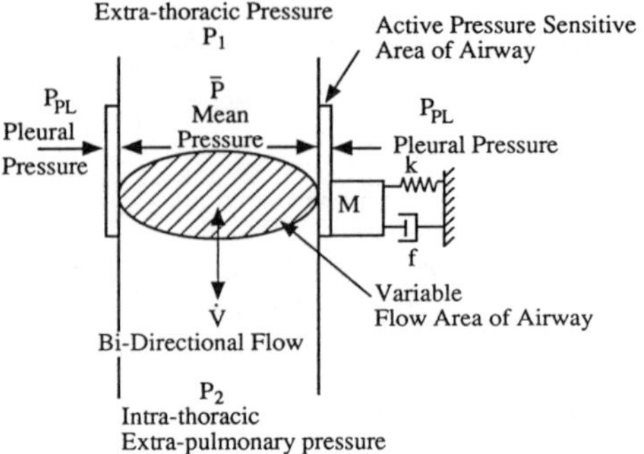


Figure 4. *A mass-spring-damper system acted on by a transmural force.*

It is assumed that the nominal displacement of the airway $x_e$ is the radius at the equal pressure point [13],[14],[15] and [16] when the transmural pressure force equals the spring recoil of the airway. The force acting on the wall of a single airway is given by

$$M\frac{d^2x}{dt^2} = (P_a - P_{pl})A_a - f\frac{dx}{dt} - k(x - x_e) \tag{3}$$

where $M$ is a representative mass term for the moving part of the airway and $d^2x/dt^2$, $dx/dt$ and $x$ are respectively the acceleration, velocity and displacement of the airway wall. The dimensions of tracheobronchial airways used in the model are based on the work of numerous authors including Fung [12] and Fredberg [17]. Successive integration of acceleration $d^2x/dt^2$ gives the velocity $dx/dt$ and displacement $x$ of the airway wall respectively. The average effective flow area of the airways is thus determined from the equation:

$$A_e = n_e \pi x^2 \tag{4}$$

where $n_e$ is the number of extra-pulmonary airways.

### 3.4 INTRA-THORACIC INTRA-PULMONARY ZONE

The region close to the alveolar compartment is termed the intra-pulmonary zone as the airway openings are directly influenced by the degree of inflation of the lung [18],[19]. The walls of these airways begin to close at the lung closing volume *CV* [20],[21],[22] and [23] and completely close off at residual volume *RV*. The flow is assumed to be laminar and is given by:

$$\dot{V}_3 = C_d\, n_e A_i\, (P_2 - P_{al}) \tag{5}$$

where $n_e$ is the number of intra-pulmonary airways of average area $A_i$.

## 3.5 BLOCK DIAGRAM OF TRACHEO-BRONCHIAL TREE

The interactions of the mouth and the tracheobronchial tree leading to the alveolar compartment are summarised in the block diagram shown in fig.5.

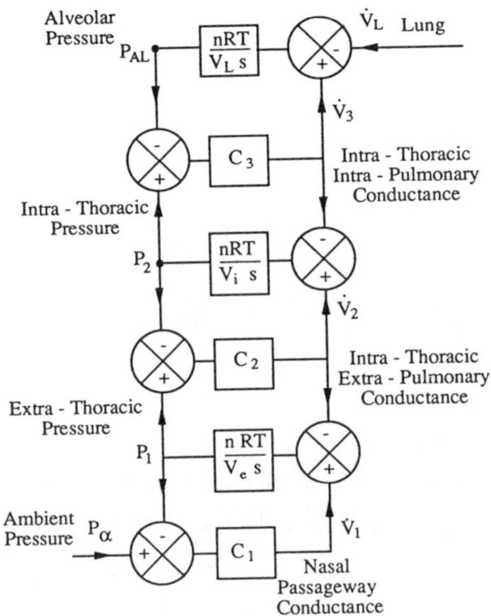


Figure 5. *Block diagram of tracheo-bronchial tree.*

In this diagram, the crossed circles indicate that quantities are being added (+ sign) or subtracted (- sign) and quantities in boxes are gains, where the term *1/s* denotes an integration process. For instance, the net flow into the extra-thoracic zone is indicated by subtracting the flow into the extra-thoracic zone $V_2$ (from the lung) from the flow supply through the mouth $V_1$. This difference is then multiplied by the gain $nRT/V_e$ and integrated to obtain extra-thoracic pressure. Block diagrams are particularly helpful in understanding complicated interactions between components in systems and are often used in Control System analysis. They also indicate the variables that need to be quantified and relationships which need to be established in order to form a complete model.

## 3.6 LUNG AND PLEURAL COMPARTMENT

The model of the lung and pleural compartments is based on that of pistons acting together (fig.6) with spring and friction resistance effects [2],[13], moving at approximately 90 degrees to the spinal column.

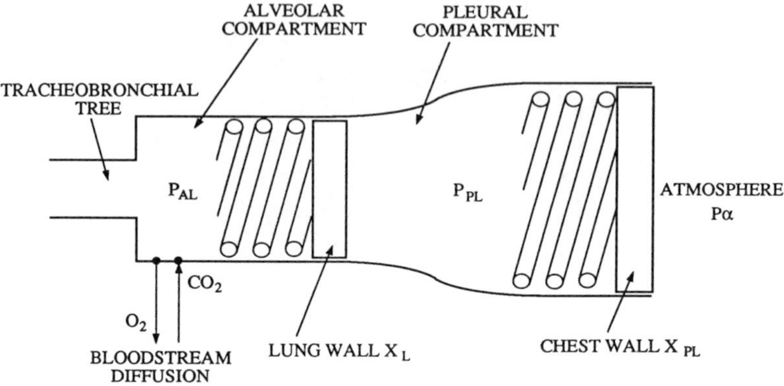


Figure 6. *Model of the lung and pleural compartments.*

Motion of the lung wall of area $A_l$ is due to alveolar and intrapleural pressures. The stiffness *(k)* of the moving lung and chestwall system comprises two distinct effects; one due to the lung and surrounding tissue and the other due to the ribcage, diaphragm and pleural compartment. These vary considerably [24],[25] over the working range of the lung from residual volume to total lung capacity. The net force on the lung wall is zero at functional residual capacity *FRC*, when the lung wall displacement is $x_o$. Viscous *(fdx/dt)* and Coulomb resistance *($F_c$)* [11],[26] forces are also present and the equation of motion of the lung wall is given by:

$$M\frac{d^2x_l}{dt^2} = (P_a - P_{pl})A_l - f\frac{dx_l}{dt} - k(x_l - x_o) - F_c \tag{6}$$

where $M$ is the mass of the lung and $d^2x_l/dt^2$, $dx_l/dt$ and $x_l$ are respectively the acceleration, velocity and displacement of the lung wall face. Typical values for the mass of a lung are given by Cotes [23], 0.5 kg being typical for a normal male.

The alveolar compartment is regarded as a control volume $V_l$ into which air flows during inhalation and from which air flows during exhalation. The rate of change of alveolar lung pressure is given by:

$$\frac{dP_a}{dt} = \frac{nRT}{V_l}(\dot{V}_3 - \dot{V}_l - \dot{V}_{O_2} + \dot{V}_{CO_2}) \tag{7}$$

where for air, $n$ is the polytropic index, $R$ is the gas constant and $\dot{V}_{O_2}$ and $\dot{V}_{CO_2}$ are respectively the oxygen flow into and carbon dioxide flow from the bloodstream.

The model of the pleural compartment is similar to that of the lung except that the compartment is filled with liquid and not gas. The face of the pleural compartment is idealized as a plane piston of area $A_{pl}$ which moves under the action of forces generated by the ribcage and diaphragm, intrapleural pressure and atmospheric pressure $P_a$. These pressures which, together with the intercostal and diaphragm thoracic forces $F_{th}$ (due to neurogenic action), act to change the compartment volume. Viscous $fdx_{pl}/dt$ and Coulomb resistance $F_c$ forces are also present and the equation of motion of the chest wall is:

$$M_{pl}\frac{d^2x_{pl}}{dt^2} = F_{th} + (P_{pl} - P_a)A_{pl} - f_{pl}\frac{dx_{pl}}{dt} - k_{pl}x_{pl} - F_c \tag{8}$$

where $k_{pl}$ is the stiffness and $M_{pl}$ the mass of the moving part of the chest wall and $d^2x_{pl}/dt^2$, $dx_{pl}/dt$ and $x_{pl}$ are respectively the acceleration, velocity and displacement of the chest wall.

The pleural compartment is regarded as a closed control of trapped pleural fluid of volume $V_{pl}$ which compresses during inhalation giving an increase in pleural pressure and expands during exhalation giving a decrease in pleural pressure. The rate of change of pleural pressure is dependent on the motion of the lung wall and pleural compartment wall and is given by:

$$\frac{dP_{pl}}{dt} = \frac{B}{V_{pl}}\left[\frac{-A_{pl}dx_{pl}}{dt} + \frac{A_l dx_l}{dt}\right] \tag{9}$$

where $B$ is the effective bulk modulus.

## 3.7 BLOCK DIAGRAM OF LUNG AND PLEURAL COMPARTMENT

A linearised representation of the interactions of the lung and pleural compartment in generating air flow is shown in fig.7.

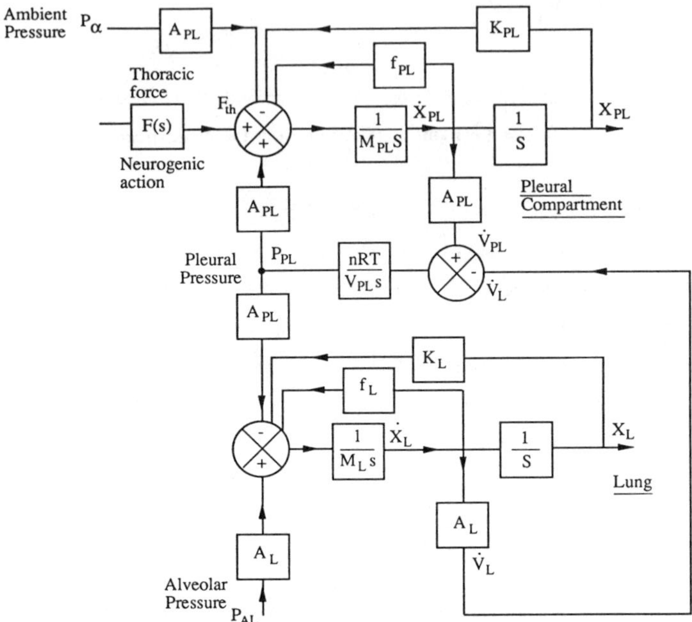


Figure 7. *Block diagram of lung and pleural compartment.*

This block diagram depicts these elements as mass-spring-damper systems, connected together and driven by the thoracic force, generated by the diaphragm and muscles in the ribcage. This force is neurogenic and depends on a number of factors including lung volume (sensed by stretch receptors), and the partial pressures of oxygen and carbon dioxide [2],[5] in the blood.

## 3.8 HEART AND BLOOD SUPPLY

Fig.8 is a simplified hydraulic circuit equivalent of the blood flow system. The heart is a dual pump, comprising four chambers: two receiving chambers called atria and two pumping chambers called ventricles. The right atrium and ventricle pump blood in the pulmonary circuit to the lungs and the corresponding left pair pump blood to the rest of the body. The direction of flow is maintained by a series of non-return valves in the heart and veins.

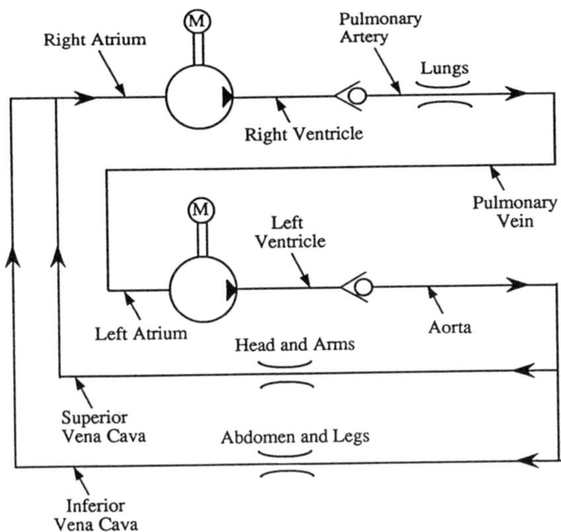


Figure 8. *A simplified hydraulic circuit equivalent of the blood flow system.*

### 3.9 GAS EXCHANGE IN BLOOD

Oxygen carriage in blood is largely a molecular process involving haemoglobin, although a certain amount of oxygen is dissolved physically and this obeys Henry's law. The variation of percent saturation of oxygen in haemoglobin with oxygen partial pressure is complex, being above 90% in the region of the lung and falling to 60-70% at tissue sites to facilitate the necessary oxygen supply. Carbon dioxide carriage in blood is also largely molecular although like oxygen, some is physically dissolved. The main chemicals involved are bicarbonates and protein carbaminos. Both the variation of oxygen and carbon dioxide concentration in blood with corresponding partial pressure are included in the model as empirical relationships.

Constituent gas flow takes place through three distinct phases:

(i) Mouth to alveolar compartment

(ii) Diffusion across alveolar membranes

(iii) Carriage of molecular gas in blood to and from tissues

These effects are accounted for in models of the equipment, mouth and tracheobronchial airways, alveolar membrane, pulmonary vein and artery, heart, systemic veins and arteries and tissue (including capillaries).

### 3.10 CONSTITUENT GAS FLOW THROUGH MOUTH AND TRACHEOBRONCHIAL AIRWAYS

The model accounts for four gas constituents at present; oxygen, nitrogen, helium and carbon dioxide. This is easily extendable to other gases. The alveolar compartment is divided up into two zones: one which allows gas equilibrium with the pulmonary circulation and the other, termed the alveolar dead space, which does not. The rate of change of gas constituents in the alveolar compartment is given by:

$$\frac{dm_i}{dt} = \frac{f_{dv}\, m_i}{\sum m_i}\dot{Q} - \dot{U}_{o_2} + \dot{U}_{co_2} \qquad (10)$$

and in the alveolar dead space by:

$$\frac{dm_i}{dt} = \frac{(1 - f_{dv})\, m_i}{\sum m_i}\dot{Q} \qquad (11)$$

### 3.11 DIFFUSION OF GAS FROM ALVEOLAR COMPARTMENT TO BLOOD

The gas exchange across the alveolar compartment is given by Graham's law of diffusion. The volumetric flow rate is:

$$\dot{V}_i = k_i\,(P_{ai} - P_{bi}) \qquad (12)$$

where $P_{ai}$ and $P_{bi}$ are respectively the partial pressures of gas constituents 'i' in the alveolar compartment and alveolar blood and $k_i$ is the diffusion coefficient. For oxygen, $k_{o2}$ = 23.4 L/min/bar [25] and for carbon dioxide, $k_{co2}$ = 467.4 L/min/bar [25].

## 3.12 CARRIAGE OF CONSTITUENT GASES BY BLOOD

The rate of change of oxygen concentration is given by:

$$\frac{dC_{o_2}}{dt} = \frac{\alpha_{o_2}(\dot{Q}_1 C_{1o_2} - \dot{Q}_2 C_{o_2})}{V} \tag{13}$$

where $V$ is the total volume of the arteries or veins [5],[6]. For carbon dioxide, the equation is:

$$\frac{dC_{co_2}}{dt} = \frac{\alpha_{co_2}(\dot{Q}_1 C_{1co_2} - \dot{Q}_2 C_{co_2})}{V} \tag{14}$$

The steady state concentrations of these gases are functions of their respective partial pressures [25].

## 3.13 METABOLIC PROCESS IN TISSUE COMPARTMENT

The metabolic tissue model generates oxygen consumption rate and carbon dioxide production rate, the rate fractional oxygen concentration change in the tissue blood pool 't', $dC_{2t}/dt$ in relation to inlet mass flowrate $\dot{Q}_1$ from compartment 1 of fractional concentration $C_{1i}$ being given by:

$$\frac{dC_{to_2}}{dt} = \frac{[\dot{M}_{o_2} + \alpha_{o_2}(\dot{Q}_1 C_{1o_2} - \dot{Q}_2 C_{t_{o_2}})]}{V_t} \tag{15}$$

where $\dot{M}_t$ is the consumption rate of oxygen, $V_t$ is the volume of the tissue compartment, $\dot{Q}_2$ is the outlet mass flowrate and $\alpha_t$ is the volume of oxygen in L/L of blood, nominally 0.2 [5],[6].

The corresponding rate of change of carbon dioxide concentration in the tissue blood pool, $\frac{dC_{2t}}{dt}$ in relation to compartment 1 of concentration $C_{1co2}$ is given by:

$$\frac{dC_{tco_2}}{dt} = \frac{[\dot{M}_{co_2} + \alpha_{co_2}(\dot{Q}_1 C_{1co_2} - \dot{Q}_2 C_{tco_2})]}{V_t} \tag{16}$$

where $\dot{M}_t$ is the production rate of carbon dioxide $\dot{Q}_2$ and $\alpha_t$ is a factor to give the volumetric flow rate of carbon dioxide.

### 3.14 ARTERIES AND VEINS

The arterial and vein systems are modelled as flexible pipes, the pressures obtained by solving the continuity differential equation:

$$\frac{dP}{dt} = \frac{B}{V}(\dot{Q}_1 - \dot{Q}_2) \tag{17}$$

where $\dot{Q}_1$ and $\dot{Q}_2$ are respectively the inlet and outlet blood flow rates to the artery or vein.

### 3.15 BLOCK DIAGRAM OF GAS EXCHANGE IN BLOOD

The time transient gas exchange between an artery, a tissue compartment and a vein can be visualized using the block diagram shown in fig.9.

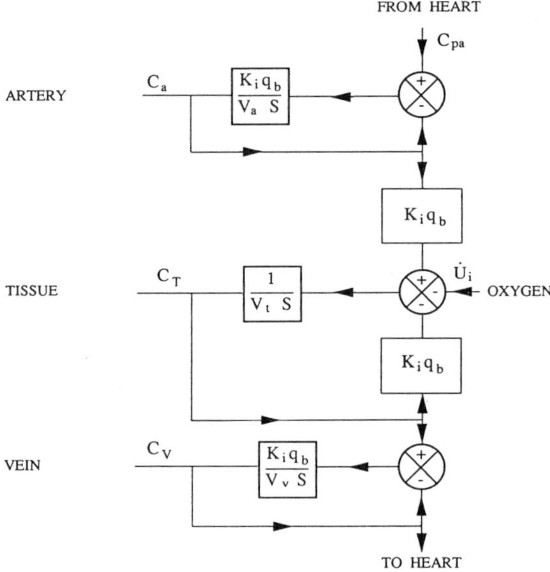


Figure 9. *Block diagram of gas exchange in blood.*

The fractional concentrations of gas constituents are denoted by the variables *C*. From the diagram, it is clear that in steady state, the downstream concentrations of gas are dependent on the upstream concentrations and any consumption or production of gas at a tissue.

### 3.16 GAS EXCHANGE IN BREATHING EQUIPMENT

The simulation of a manned diving operation involves the overall gas exchange process from atmosphere to tissue and it is necessary to include the effect of the breathing equipment. This has been described in detail by Tilley et al [7],[9] and Tomlinson et al [26] is included in the overall model of the diving operation.

## 4. Simulation

The model (fig.1) was used to simulate a diver descending from surface conditions to a depth of 50 metres. A breathing frequency of 25 breaths per minute, 2 litre tidal volume and a metabolic oxygen consumption rate of 2 L/min and carbon dioxide production rate of 1.6 L/min were assumed. The simulation was used to show the effect on alveolar oxygen partial pressure of two different oxygen flow settings (a) 32.5% and (b) 60% oxygen from a constant supply mass flow unit. Table 1 is a sample of the parametric data used in the lung model, based on tests in a clinical laboratory. Parametric data relating to the counterlung has been described previously by Tomlinson et al [26].

| | |
|---|---|
| Alveolar pressure | 1.013 bar |
| Lung face displacement | 0.06 m |
| Lung face velocity | 0.0 m/s |
| Mass of lung | 0.5 kg |
| Tidal volume | 0.5 L |
| Residual volume | 2.71 L |
| Functional residual capacity | 3.61 L |
| Total lung capacity | 8.46 L |
| Fractional dead space | 0.25 |

**Table 1. Dimensional details of the lung**

## 4.1 SIMULATION RESULTS

The variations in mouthpiece oxygen partial pressure during descent to a depth of 50 metres are shown in fig.10.

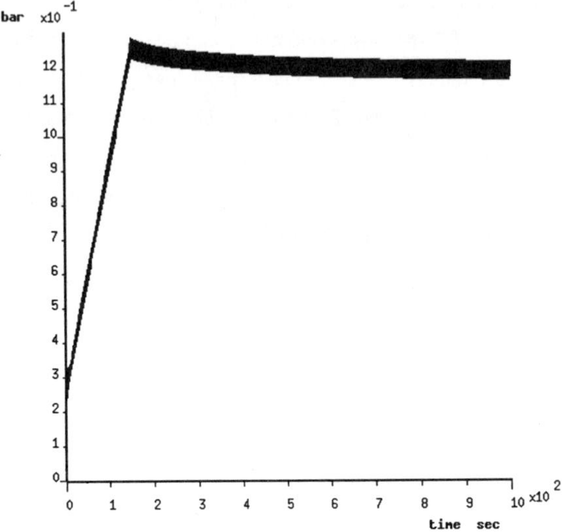


Figure 10. *Variations in mouthpiece oxygen partial pressure during descent to a depth of 50 metres.*

With a 32.5% oxygen setting, the mean level rises from approximately 0.21 bar at surface conditions to 1.2 bar at 50 metres depth. The corresponding variations in alveolar compartment gas and blood pool oxygen partial pressures are shown in fig.11(a).

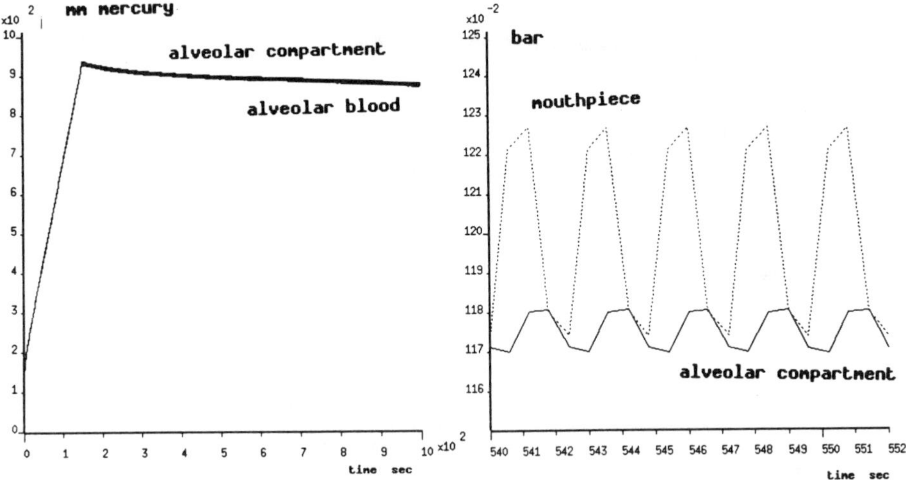


Figure 11(a) and (b). *Variations in alveolar compartment gas and blood pool oxygen partial pressures.*

The mean level rises from approximately 180 mm Hg at surface conditions to 890 mm Hg at 50 metres depth. Fig.11(b) shows the mouthpiece and alveolar compartment oxygen partial pressure at depth. Gas concentrations vary with the respiratory cycle.

When the oxygen mass flow setting is increased to 60%, the oxygen partial pressure at the mouthpiece and alveolar compartment rises above 2 bar as shown in fig.12, which implies an unsafe diving condition [2].

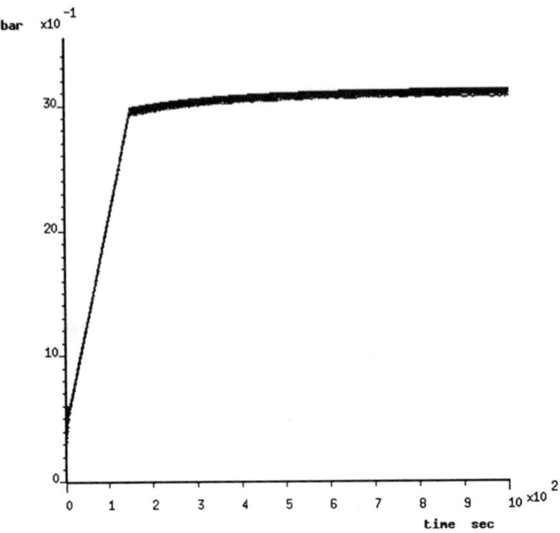


Figure 12. *Oxygen partial pressure at the mouthpiece and alveolar compartment*.

The simulation indicates the importance of using an appropriate oxygen setting with this equipment to maintain the partial pressure within acceptable limits.

## 5 Conclusions

The research to date indicates the potential for using computer simulation to investigate the interaction between the human respiratory system and underwater diving equipment. Computer software enables circuits to be configured and changed, thus enabling human and equipment aspects to be assessed. The simulations are non-linear and discontinuous, involving the numerical solution of sets of differential and algebraic equations which describe time transient and steady-state behaviour. Using such techniques, it is possible to assess different viewpoints on the behaviour of respiratory elements using two representational criteria: an empirical criterion to determine how well a model compares with observed behaviour and a theoretical criterion which determines how consistent a model is with accepted physiological, physical and chemical theories.

The software enables the effects of variations in parameters such as equipment settings or respiratory characteristics to be assessed. In the model, the effects of using two different breathing equipment oxygen flow settings were shown. A 32.5% oxygen flow setting was shown to give adequate alveolar partial pressure whereas a 60 % setting indicated an excessive partial pressure and consequent danger to the diver.

The model has the potential to predict many important parameters in underwater breathing equipment which are at present only avaible due to experimental manned equipment testing. In future, it is intended to use the model in the diving industry to:

(a) give estimations of oxygen consumption and carbon dioxide production when a diver is working at depth in order to assess diver effort and the requirements of equipments.

(b) give estimations of humidity factors in order to assess the requirements of underwater diving eqipment.

(c) assess extremes of temperature on equipment and the human respiratory system.

(d) develop a data base of equipment models to assess human interaction with a wide range of underwater breathing equipment.

## 6. Acknowledgements

We are grateful to Professor D.M.Denison for his advice on the model concept and allowing the use of facilities at Brompton Hospital to obtain pulmonary data to compare with the respiratory model.

We are grateful to Dr.G.J.Gibson of Freeman Hospital, Newcastle-on-Tyne for providing lung compliance data incorporated in the model.

The research described in this paper has been jointly funded by the Science and Engineering Research Council and the Ministry of Defence.

## References

[1] **Sterk W.** Respiratory Mechanics of Diver and Diving Apparatus. PhD thesis. University of Utrecht. Druukerij, Elinkwijk, Utrecht, Holland. 1973;5:105-116.

[2] **Bennet P B, Elliot D H.** "The Physiology of Medicine and Diving". 3rd edition. Bailliere Tindall. London. 1982. 55-99.

[3] **Kleinman D L, Baron S, Levison W H.** An Optimal Control Model of Human Response. Part 1: Theory and Validation. *Automatica.* 1970. 6;357-369.

[4] **Sutton R.** "Modelling Human Operators in Control System Design". John Wiley & Sons Inc. 1990. 169-180.

[5] **Sarhan N A S, Leaning M S, Saunders K B, Carson E R.** Development of a complex model: breathing and its control in man. *Biomed Meas Infor Contr.* Vol.2 1987-88;2:81-100.

[6] **Dickinson C J.** "A Computer Model of Human Respiration (MacPuf)". MTP Press Ltd. 1977. 3-251.

[7] **Tilley D G, Tomlinson S P, Livesey J.** Computer Simulation of a semi-closed-circuit breathing system. *Proc. Instn. Mech. Engrs. Journal of Systems and Control Engineering.* 1991;Vol 205. 163-173.

[8] **Richards C W, Tilley D G, Tomlinson S P, Burrows C R.** BATH*fp* - A Second Generation Simulation Package for Fluid Power Systems. *9th International Conference on Fluid Power.* Cambridge UK. 25-27 April 1990.

[9] **Tilley D G, Livesey J, Tomlinson S P, Himmens IA.** An approach to the modelling of underwater breathing apparatus. *NOAA/UHMS workshop on 'Physiological and Human Engineering Aspects of Underwater Breathing Apparatus'*. University at Buffalo, USA, April 1989. 125-137.

[10] **Sullivan K J, Chang H K.** "Flow Dynamics of the Nasal Passage". Department of Biomedical Engineering, University of Southern Carolina. Los Angeles, California. Springer Verlag. *Respiratory Biomechanics Symposium of the First World Congress of Biomechanics.* La Jolla. California. USA. August-September 1990. 98-105.

[11] **Tomlinson S P.** "The Hydraulic Automatic Simulation Package. Modelling and Simulation Aspects". *Ph.D thesis,* University of Bath. United Kingdom. 1987. 60-64.

[12] **Fung Y C.** "Biomechanics. Motion, Flow, Stress, and Growth". Springer Verlag. 1990. 229-237.

[13] **Denison D M, Waller J F, Turton C W G, and Sopwith T.** Does the Lung Work? Breathing in and Breathing Out. *British Journal of Diseases of the Chest.* No.5. Vol.76. October 1982. 237-252.

[14] **Dawson S V, Elliott E A.** Wave-speed limitation on expiratory flow - a unifying concept. *J.Appl.Physiol.* Vol.43. No.3. 1977. 498-515.

[15] **Lambert R K, Wilson T A, Hyatt R E, Rodarte J R.** A computational model of expiratory flow. *J.Appl.Physiol.* Vol.52. No.1. 1982. 44-56.

[16] **Pedersen O F, Ingram jr R H.** Configuration of maximum expiratory flow-volume curve: model experiment with physiological implications. *J.Appl.Physiol.* Vol.58. No.4. 1985. 1305-1313.

[17] **Fredberg J J.** Spatial considerations in oscillation mechanics of the lung. *Biomedical Engineering and Respiratory Physiology Fed Proc.* Vol.39. No.10. August 1980. 2747-2754.

[18] **Hyatt R E, Schilder D P, Fry D L.** Relationships Between Maximum Expiratory Flow and Degree of Lung Inflation. *J.Appl.Physiol.* Vol.13. No.3. 1958. 331-336.

[19] **Stubbs S E, Hyatt R E.** Effect of increased lung recoil on maximal expiratory flow in normal subjects. *J.Appl.Physiol.* Vol.32. No.3. 1972. 325-331.

[20] **Travis D M, Green M, Don H F.** Expiratory flow rate and closing volumes. *J.Appl.Physiol.* Vol.35. No.5. 1973. 626-630.

[21] **Leblanc P, Ruff F, Milic-Emili J.** Effects of age and body position on "airway closure" in man. *J.Appl.Physiol.* Vol.28. No.4. 1970. 448-451.

[22] **Denison D M, Du Bois R, Sawicka E.** Does the Lung Work? 5. Pictures in the Mind. *British Journal of Diseases of the Chest.* No.1 Vol.77. January 1983. 35-50.

[23] **Cotes J E.** "Lung Function. Assessment and Applications in Medicine". Fourth edition. Backwell Scientific Publications. 1979. 166-169.

[24] **DuBois A B, Brody A W, Lewis D H, Franklin Burgess jr B.** Oscillation Mechanics of Lungs and Chest in Man. *J.Applied Physiology.* Vol 8. 1956. 587-594.

[25] **Mines A H.** "Respiratory Physiology". Second Edition. Raven Press. 1986. 24-28.

[26] **Tomlinson S P, Livesey J, Tilley D G, Himmens I.** Computer Simulation of Counterlungs. *Undersea and Hyperbaric Medicine.* Vol.20. No.1. 1993. 63-73.

**WRITTEN DISCUSSION**

**Computer simulation of human interaction with underwater breathing equipment**

**SP Tomlinson, JKW Lo & DG Tilley (University of Bath, UK)**

**Question:** **JB Gamble**
**Vickers Systems Division, Trinova Ltd, UK**

Bearing in mind that someone's life may depend on equipment developed using this model, please comment on the sensitivity of the model to its parameters, and the variation between different humans. Will these variations prevent the model from being used?

**Answer:**

It is accepted that respiratory behaviour between different humans is widely different and even more important that significant variations occur in the respiratory behaviour of similar human subjects. However, the human being is an example of a complex set of simultaneous engineering systems involving air flow, lung motion, blood-gas exchange and neurogenic control which are amenable to simulation using ordinary differential equations. Thus, allowing for the dissimilarities that occur, we intend to develop the model and investigate the best uses to which the model can be put. These are likely to be respiratory medicine, diving and aerospace.

# 9. Computing the Dynamic Response of Hydraulic Systems using Knowledge-Based Programming

G Grossschmidt *and* J Pahapill

Abstract

For computing transient or frequency responses as well as static and stationary characteristics of hydraulic systems, especially hydraulic and electrohydraulic drives, a program package is developed using the knowledge-based programming system CPRIZ. The program package contains a knowledge base in the form of computational models (modules) of functional elements, physical properties of fluids, disturbances and graphical output. Multi-port models of hydraulic system functional elements are used. Mathematical models of the elements are composed with the help of signal-flow graphs. Used forms of multiport models of some functional elements of hydraulic drives are given. The modules of hydraulic chains: tube, hydraulic resistance , capacitance and tee-fitting are also reviewed. The instrumental programming system CPRIZ grants an automatic synthesis of computing programs on the basis of the description of conditions of the task. It is not necessary to give an explicit algorithm of the solution, the system provides it automatically.

The program package is realized on a computer that is provided with UNIX or MS-DOS operating system. Computational models are written in the C language. Program modules are formed in the input language of system CPRIZ.

For calculating characteristics of hydraulic systems a two-level method is used. In case of the first level, characteristics of separate functional elements are calculated. Any method of calculation may be used. The connection of variables between the functional elements is carried out with the help of an iteration procedure. Examples of calculating the dynamic response of hydraulic chains and comparing results to experimental data are given.

## 1 Introduction

Hydraulic drives hold a firm place in modern technology. Creating hydraulic drives of high quality would be unwise without the analysis of their dynamic properties. Such an analysis, due to the complicated systems, is possible only with the help of computers. In order to compute the dynamic responses of hydraulic drives it would be expedient to possess a unitary computing system, based on one compilation method, which is open to everyone for introducing alterations and additions. In this way it would be possible to connect the efforts of elaborators of computing systems with those of consumers into a wider collaboration and more effective exploitation of dynamics analysis when observing and designing hydraulic systems. This work presents approaches, which could be useful for creating such a unitary calculation system.

## 2 Mathematical models

Mathematical models of hydraulic systems must observe the following:

- necessary causal dependences;
- evidently, take into account direct disturbances as well as reactions;
- adequacy of functional dependencies to the described physical processes;
- figurative compiling of the model which considers all dependencies and forms the right equations;
- possibility to modify the model.

The mathematical models used in the program package are made up according to the following principles. The hydraulic system dealt with is divided into functional elements (FE) according to the functional or assembling scheme. Such FEs could be simple, or complicated links, tubes, hydraulic pumps and hydraulic motors, various hydraulic components and control system equipment, etc.

Functional elements are described as multi-port elements [1,2]. The system of notation of multi-port models of FEs is given in Figures 1, 2 and 3. Letter I denotes any single input, letter O any single output (Fig.1).

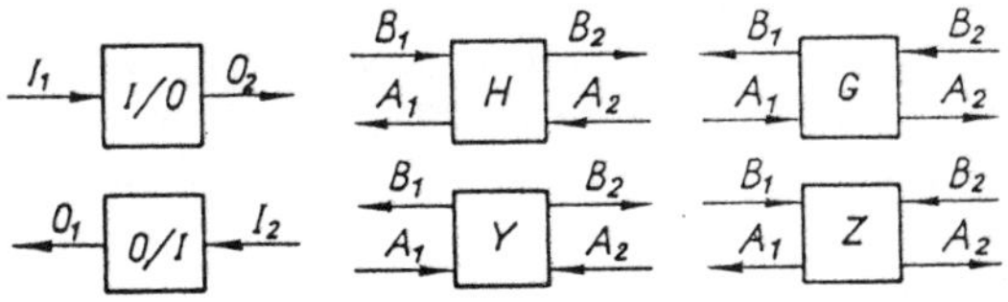

Fig.1. Notation of two and four-port models of functional elements depending on input and output variables.

Left ports are denoted by odd numbers, right ports by even numbers. Letter A denotes potential variables (pressure, force, torque, voltage) and letter B flowrate variables (volumetric flowrate, displacement, speed, angle, current) which come up in pairs. Letters C and D denote the pairs of variables (Fig.2). Depending on the existence of input and output variables four-port models are denoted by the letters H, G, Y and Z (Fig.1), as in electrotechnics. If a multi-port model has a four-port part, it will be denoted by the corresponding sign (Fig.2 and 3). The rest of the ports will be marked separately with corresponding letters, while the left and right ports are separated with a sloping line.

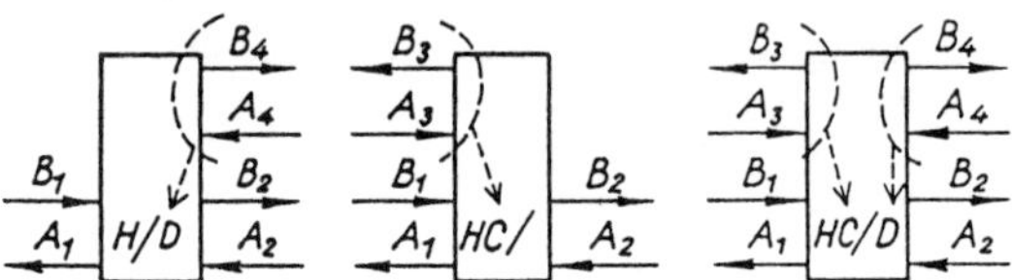


Fig.2. Notation of six- and eight-port models of functional elements according to the scheme of variable pairs.

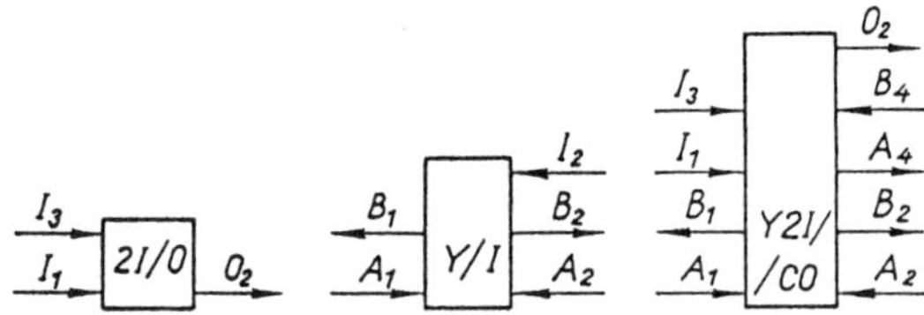


Fig.3. Notation of multi-port functional elements, which have singular inputs and outputs.

Notations and forms of models of some functional elements of hydraulic system can be found in Table 1.

Every FE may have several mathematical models differing in compilation and estimated factors. The multi-port models of dynamics of FE are compiled provisionally as linear models of dynamics in Laplace form, where variables are deviations from the statics (or from the stationary movement). These provisional linear models are transformed into non-linear ones, introducing respective non-linear dependencies. All linkages between variables are established by compiling a corresponding signal flow graph.

The main stage in solving a given task is to put together a block scheme using the multi-port FE [1]. The block scheme is general for calculating the statics (or stationary motion) and frequency responses. Frequency responses are calculated using the method of harmonic linearization of nonlinearity. The shift of the medium value of oscillation is considered. To calculate transient responses there is, as a rule, a necessity for another block scheme which considers the process of integration in the models of dynamics of the elements, allowing a higher precision of calculation than the procedure of differentiation. Building up a block scheme out of multi-port elements should proceed as follows:

- the given input and output variables;
- the table of multi-port models of functional elements being used;
- the recommendations on the selection of the forms of multi-port models;
- the possible variants of connecting the multi-port models;
- the block schemes of typical hydraulic systems.

Table 1. Notation and used forms of multi-port models of some functional elements of hydraulic systems.

| Nomination | Notation | Used forms of multi-port models |
|---|---|---|
| Piston | PI | H/D2O G/D2O Y/C2O |
| Cylinder without piston | CY | Y2I/O |
| Volume elastance a chamber of cylinder | VECY | HI/ GI/ ZI/ |
| Hydraulic cylinder with piston and volume elasticities | HCY | H/C G/C Y/C Z/C |
| Hydraulic motor | HM | H/D G/D Y/C |
| Hydraulic pump | HP | HC/ GC/ YD/ |
| Pressure control valve | VP | H Y |
| Pressure-reducing valve | VR | G |
| Flow control valve without spill-off | FC2 | Y |
| Flow control valve with spill-off | FC3 | H/D |
| Electromechanical transducer | ET | Y/I |
| Nozzle-flapper hydraulic amplifier | NF | YC/2C |
| Spool:displacement hydraulical,feedback to flapper | SP | Y2C/ |
| Sliding spool resistor | RS | H/C G/C Y/C HD/ GD/ YD/ |
| Pairs of sliding spool resistors | RS13,RS24 | G/DC Y/DC, HCD/ YCD/ |
| Feedback:potentiometer, comparison,amplifier | FB | 2I/O |

## 3 The program package

The system part of the program package is based on the instrumental programming system CPRIZ [3,4]. CPRIZ is an environment for developing application programs based on knowledge processing. The system CPRIZ supports a method of solving problems which is called 'conceptual programming' [5]. Programs for solving concrete tasks can automatically be synthesized with the help of the system using common and particular knowledge in the problem domain. The description of the

problem domain is a complex of concepts which are used for specifying tasks. A concept is a computational model containing a complex of objects (variables) and the relations between these objects.
Such a model base of knowledge enables any inquired calculation. In the system CPRIZ the description of a task changes into a description of conditions of the task. The advantage of such a system is that it is not necessary to give an explicit algorithm of the solution, the system provides it automatically.

## 4 Feedback

When describing the system with the help of multi-port models of the elements closed cycles between separate elements are formed (Fig.4).

Fig.4. Formation of a loop when connecting four-port models forms Y and H.

The graphic interpretation of the feedback system is a signal flow graph with loops. For synthesizing a program, using the iteration method, it is indispensable to break all loops. In order to break the loops the block of feedback FB is introduced into the system (Fig.5) [3].

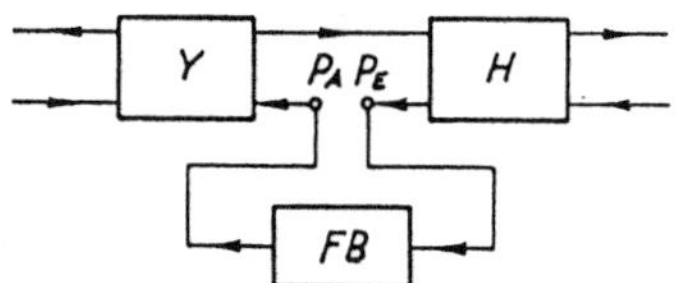


Fig.5. The scheme of iterational calculation of a chain with four-port model forms Y and H with the help of introducing the FEEDBACK-block.

The feedback block must solve the equation

$$|p_E - p_A|/p_E \leq \varepsilon ,$$

i.e. it must guarantee the equality of $p_A$ and $p_E$. It is sufficient to connect one variable of a pair, for example a potential variable, even when calculating parallel and feedback chains.

## 5 Calculation

The program package is compatible with a computer that is provided with the UNIX or MS-DOS operating system. Computational models are written in the C programming language. Program modules are formed in the input language of the CPRIZ system.
For calculating characteristics of hydraulic systems the two-level method is used. In case of the first, FE level, characteristics of separate FEs are calculated. Any method of calculation may be used. The calculation of dynamic responses on the FE level in case of numerical non-stiff continuous functions is carried out by the standard Runge-Kutta method. Numerical stiffness problems may be avoided using mathematical models with procedure of differentiation. Then for computing FE transient responses the Newton's iteration method is used. Discontinuities should be transferred for continous functions in short time intervals. The connection of variables between the FEs is carried out with the help of Wegstein iteration procedure. If no less than 100 points are calculated for one period of oscilla-tion, no more than two iterations for connecting the modules are necessary, as a rule .

## 6 Some fundamental concepts of hydraulic systems (modules)

### 6.1 Module LIQUID

It is understood that the following values of physical characteristics of liquids are used: density $\varrho$, compressibility figure of the mixture of the liquid and air $\beta_M$ and kinematic viscosity $\vartheta$ . The values of physical characteristics of liquid will be determined at every step of calculation, depending on the pressure at the given temperature. At present, there are 20 fluids used in a module for which density $\varrho_T$ at t=15°C and kinematic viscosity at different temperatures with the interval of 10°C from -60°C to +60°C are given in tabular form.
Example of a liquid description:
AMG10 liquid type='AMG-10' theta=50 air=0.02
where AMG10 - identifier; liquid - nomination of the module; type - mark of the liquid; theta - temperature of the liquid °C; air - relative content of undissolved air.

### 6.2 Module TUBE

In the equations of dynamics of mathematical models of tubes with lumped parameters members of hydraulical resistance R, inertia L and liquid volume elastance C are used. Simultanuously, hydraulic resistance taking into account the linear as well as square dependence of the pressure drop from the volumetric flowrate is used. It enables the use of universal equations for laminar, as well as turbulent flow of liquid. In addition, local hydraulical resistances that directly depend on the tube (bends, connections, elbows, inputs to a vessel, outputs from a vessel, etc.) are taken into account. The changes of hydraulic linear and square-law resistance depending on the flow speed increment are introduced. In order to correct these members for dynamical resistance, corresponding coefficients $k_{RL}$ and $k_{RT}$ have been introduced in accordance with the experimental data (see the equations below).

Depending on the input and output variables the mathematical models of tubes may occur as fourport forms H, G, Y and Z. In this paper mathematical models of tubes with lumped parameters are presented, and different types must be chosen, depending on the required natural frequency f, tube length l, as well as the necessary form of the four-port model (Table 2).

Table 2. Fourport model forms of TUBE and the areas of usage

| Four-port model form | Areas of usage | | | | | |
|---|---|---|---|---|---|---|
| | Stationary process | f≤10Hz | for rigid tubes | | | |
| | | | l≤250/f | l≤500/f | l≤1000/f | l≤1500/f |
| | | | for bending hoses | | | |
| | | | l≤200/f | l≤400/f | l≤800/f | l≤1200/f |
| | Notation of model forms | | | | | |
| H | HA | | H | 2ZY | 4ZY | 6ZY |
| G | GA | | G | 2YZ | 4YZ | 6ZY |
| Y | YA | | Y | | 3YZY | |
| Z | - | ZA | Z | | 3ZYZ | |

Four-port models forms HA, GA and YA take into account only resistance R, but form ZA takes into consideration only volume elasticity C of the liquid. The model of form H has the structure C-R-L, the model of form G L-R-C, the model of form Y L/2-R/2-C-R/2-L/2 and the model of form Z C/2-R-L-C/2.

With the help of coefficient k the values L and C are corrected in order to provide correspondence of natural frequency of the tube with lumped parameters to natural frequency of the tube with distributed parameters. In case of form H and G of the four-port model, coefficient k equals to $k=(2/\pi)^{(1/n)}$, where n is the number of joint similar tube parts. For the forms Y and Z of four-port model it is indispensable for correcting natural frequences to use coefficients $k_1$ and $k_2$: $k_1=(2/\pi)^{(1/2n)}$ and $k_2=(2/\pi)^{(1/n)}$. For four-port model form Y it is necessary to introduce corrections in the mode $k_1L$ and $k_2C$, for four-port model form Z corrections in the mode $k_1L$ and $k_2C$.

For a rigid tube with length l≤500/f the model of the four-port form

H is achieved with the help of two elements, model forms Z and Y, being in sequence. The model of fourport form G using elements with models Y and Z sequentially is achieved. For a rigid tube with length l≤1000/f and l≤1500/f models are used where the separate four-port forms are compiled by connecting a greater number of model forms Y and Z (models 3YZY, 3ZYZ, 4ZY, 4YZ, 6ZY, 6YZ). In case of longer tubes, it is indispensable to use still more elements of form Y and Z sequentially, or tube models with distributed parameters.

The block-scheme of the Y-type fourport model of a tube, which consists of elementary L, R and C links and corresponding signal flow graph is represented on Fig.6.

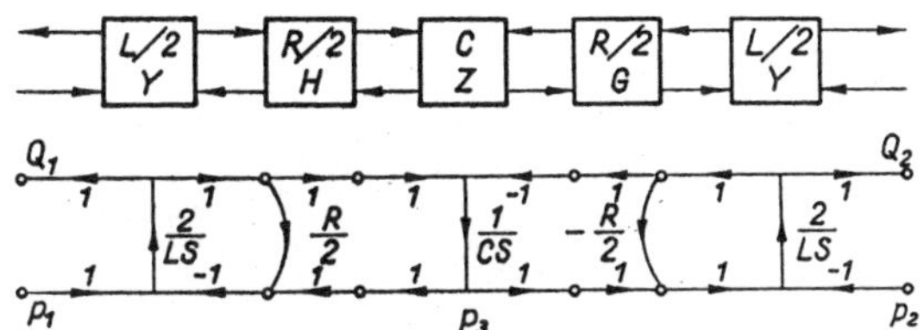

Fig.6. Block scheme of four-port model form Y for tube and corresponding signal-flow graph.

The system of non-linear differential equations for Y-type fourport model of a tube is used in form:

$$\frac{dQ_1}{dt}=\frac{2}{k_1L}(p_1-p_3-\frac{1}{2}(R_L+R_T|Q_1|Q_1)-\frac{1}{2}(k_{RL}R_L+k_{RT}R_T|Q_1|)*(Q_{1j}-Q_{1(j-1)})\frac{\tau}{A}),$$

$$\frac{dp_3}{dt}=\frac{1}{k_2C}(Q_1-Q_2),$$

$$\frac{dQ_2}{dt}=\frac{2}{k_1L}(p_3-p_2-\frac{1}{2}(R_L+R_T|Q_2|Q_2)-\frac{1}{2}(k_{RL}R_L+k_{RT}R_T|Q_2|)*(Q_{2j}-Q_{2(j-1)})\frac{\tau}{A}),$$

where $R_L$ - linear hydraulic resistance; $R_T$ - square hydraulic resistance; $Q_{1j}$, $Q_{1(j-1)}$ - volumetric flowrates in present and in former time moment; A - passage area; $\tau$ - inverse quantity of the time step.

Example of a tube description:

```
t1 tubeH fluid=AMG10 s=0.2e-2 l=1 d=0.01 AL=75 E=2.1e11
        lambda=0.04 dzeta=2 n=1 kRL=1e2 kRT=1.5e2 q1.s=500e-6
```

where s - thickness of the tube walls; l - length; d - inner diameter; AL - coefficient of hydraulic friction of laminar flow; E - elastic modulus of the tube material; lambda - coefficient of hydraulic friction of turbulent flow; dzeta - coefficient of local hydraulic resistance; n - number of similar connected tubes; kRL, kRT - coefficients of correcting dynamic linear and square resistances; ql.s - fixed volume flow rate at the left end of the tube.

## 6.3 Module RR (resistor)

The models of local hydraulic resistance can be given in four-port models H, G and Y. Local hydraulic resistances can be described with linear $R_L$, square $R_T$ and inertial resistance L. Thus, the equations for all local hydraulic resistances can be similar, with different quantities $R_L$, $R_T$ and L for each particular local resistance. At present, there are 15 resistors used in a module.

## 6.4 Module CA (capacitance)

Tube models ZA, H, G and Z are used as mathematical models of liquid volume elasticity. Tube model ZA takes only volume elasticity C into consideration. With the models H, G, and Z hydraulic resistance and liquid inertia are also considered.

## 6.5 Module IE (interface element)

In the package, the interface element takes the form of a tee-fitting. Basic mathematical model of tee-fitting expresses continuity equation and pressure change conditions in local resistance of the tee-fitting. There are six different forms of the mathematical model of the tee-fitting (Fig.7).System CPRIZ performs automatic transformation of the basic model for every concrete model of needed form.
The coefficients of local hydraulic resistance in a tee-fitting are determined, depending on the sign of volumetric flowrate in the branches, their quantities are given tabularly.

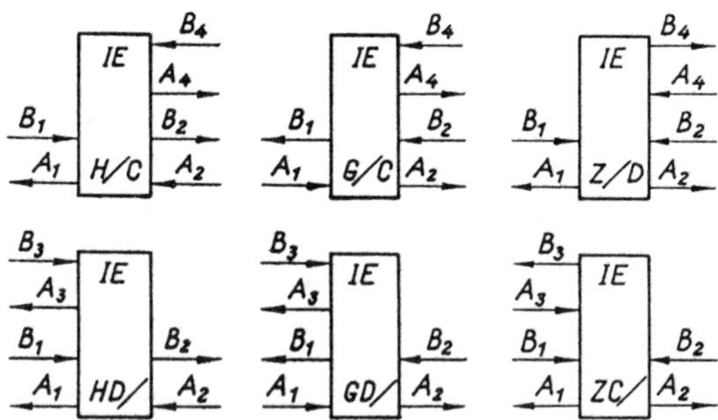


Fig.7. Variants of hydraulic tee-coupling models (A-potential, B-flowrate variable).

## 6.6 Module DIST (disturbance)

The disturbances in practice are 'jump' (step disturbance), 'sinus' (sinusoidal disturbance), 'rectangle' (rectangular disturbance) and 'imp' (impulse disturbance).

Example of a disturbance description:

dp dist type='jump' media=50e5 delta=5e5 deltat=0.001 result=t1.p2.s

where media - the value of the variable before disturbation; delta - step of the variable; deltat - duration of the step; result - nomination of the variable being disturbed.

A disturbance may simultaneously be applied to several inputs. Every input may also be given several disturbances.

## 6.7 Module DISPLAY

This module organizes the calculation process and graphic output of dynamic response on the display screen.

Example of an output description:

d display eps=1e-5 time=0.02 step=0.00005 res=t1.p1.s

where eps - permitted relational calculation accuracy on the iteration level of intermodule variables; time - duration of the calculated dynamic responses; step - time step of the calculations; res - the variable being displayed.

Stationary values or dynamic responses to all of the variables are simultanuously calculated. It is advisable to point out the variables, the values and processes that must appear on the screen.

# 7 Examples

Example 1. Dynamic response in a tube using model 2ZY.

Model of tube t1 form Z and model of tube t2 form Y are placed sequentially in this dynamic response model. Stationary fluid flow variables that are required as initial data for dynamics models forms Y and Z, are calculated using model 2H, where two tube model forms H are placed sequentially.

Fluid being used: type='AMG-10' ($\vartheta_{50}$=10 mm$^2$/s), theta=50°C, air=0.02.

Tube parameters: d=0.01 m, general length l=7.2 m, s=0.003 m,
E=2.1($10^{11}$) N/m$^2$.

Flow parameters: AL=75, lambda=0.04, dzeta=0.

Coefficients that take into consideration fluid flow resistance depending on the flow speed increment: kRL=0, kRT=0.

Disturbance parameters: type='jump', media=500($10^{-6}$) m$^3$/s,
delta=5($10^{-6}$) m$^3$/s, deltat=0.001 s.

Disturbed variable: result=t1.q1.s (volumetric flowrate at the left end of tube t1).

Constant input: t2.p2.s=50($10^5$) Pa (pressure at the right end of tube t2).

Output: t2.q2.s (volumetric flowrate at the right end of tube t2).

Step and relative error of the calculation: step=0.0001 s, eps=1($10^{-5}$). Result dynamic response (Fig.8) differs little from a damped sinusoid.

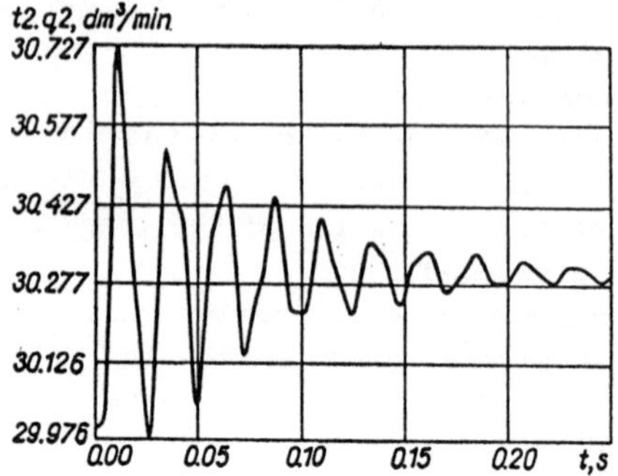


Fig.8 Dynamic response in a tube that has been calculated according to the conditions of Example 1.

Example 2. Dynamic response in tube using model 4ZY.
In case of this dynamic response model, we have tube models t1...t4 forms Z, Y, Z and Y in sequence. Initial data is the same as in Example 1.
Calculated dynamic response is shown in Fig.9. In this dynamic response the fourth harmonical oscillation comes out.

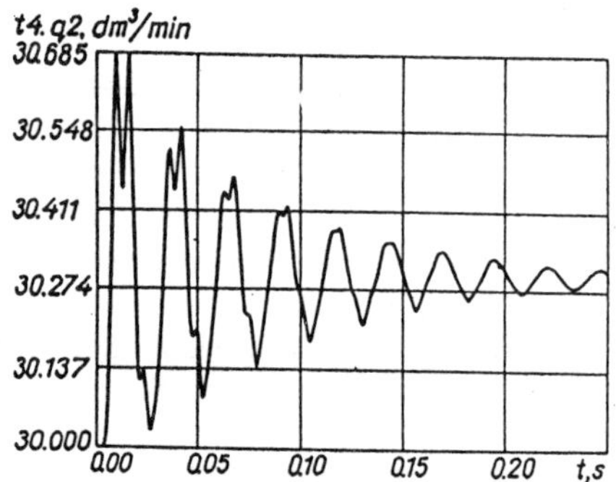


Fig.9. Dynamic response in a tube that has been calculated according to the conditions of Example 2.

Example 3. Comparing calculation results of dynamic response of hydraulical chain to experimental data.
The scheme of experimental device for examining dynamic response in a tube is shown in Fig.10. Dynamic response in a tube is compiled using quick valve opening at the end of the tube.

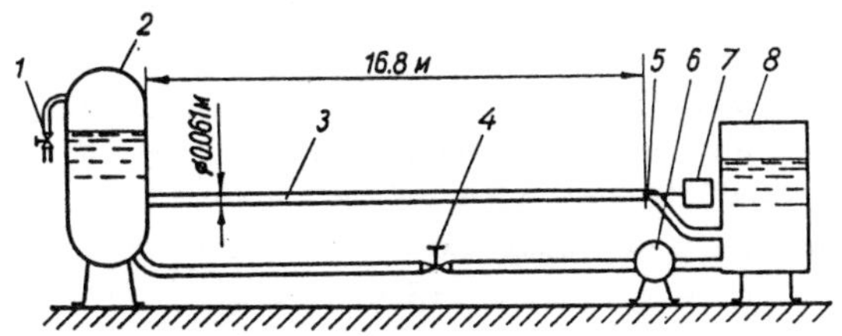


Fig. 10. The scheme of an experimental device for observing dynamic response of fluid flow in a tube: 1 - tap for letting air in and out; 2 - reservoir; 3 - tube; 4 - tap; 5 - quickly opening valve; 6 - fluid reserve pump; 7 - valve opening drive; 8 - receipt vessel.

Mathematical model which corresponds to the experimental conditions includes the following functional elements: local hydraulical resistance rr1 at input to the tube; tube presented as consisting of 6 equal parts t1...t6; valve rr2 that evokes a disturbance; channel rr3 behind the valve with resistance and inertia of output into the vessel.

For calculating stationary flow model RRcG is used for element rr1, model tubeG for elements t1...t6, model RRcY for element rr2 and model RRaH for element rr3. For calculating dynamic response different from the latter for elements t1, t3 and t5 models tubeY and for elements t2, t4 and t6 models tubeZ are used.

Initial data is the following.

Liquid: type='H2O' (water), theta=20°C, air=0.00015 (found with testing method).

Parameters of model RRcG: d=0.061 m, l=0.1 m, dzeta=0.5, eps=1($10^{-6}$).

Parameters of tubes t1...t6: d=0.061 m, l=2.8 m, s=0.006 m, E=2.1($10^{11}$) N/$m^2$.

Parameters of fluid flow in tubes: AL=64, lambda=0.025, dzeta=0, kRL=-3($10^{-3}$), kRT=-3($10^{-3}$), (these coefficients also founded with testing method).

Parameters of model RRcY: d0=0.005 m, l=0.04 m, dzeta=1, eps=1($10^{-6}$).

Parameters of model RRaH: d=0.061 m, l=0.6 m, AL=64, lambda=0.025, dzeta=0.5

Disturbance parameters: type='jump', media=0.005 m, delta=0.056 m, deltat=0.001 s.

Disturbance variable: result=rr2.d (diameter of the valve passage).

Constant inputs: rr1.p1.s=1.5($10^5$) Pa, rr3.p2.s=0.1($10^5$) Pa.

Output: t5.p2.s

Step and relative calculation error: step=3($10^{-5}$) s, eps=1($10^{-5}$).

According to the model of stationary flow for using in dynamics model, the following initial values are calculated.

For rr1: rho0 (density); for t1, t3 and t5: q0 (volumetric flowrate ); for t2, t4 and t6: p10, rho0, nu0 (kinematic viscosity); for rr2: rho0, q0; for rr3: rho0, nu0.

Calculated dynamic response of a pressure at the distance of 14 m from the input to the tube and experimental dependence at the distance of 12.6 m from the input to the tube are expressed in Fig.9. Experiments were carried out in the Department of Hydro- and Aeromechanics of Tallinn Technical University by professor U.Liiv. The equipment used did not allow register of pressures below atmospheric pressure.

Comparing the results, the mathematical models and calculation methods used reflect the processes that actually occur sufficiently well.

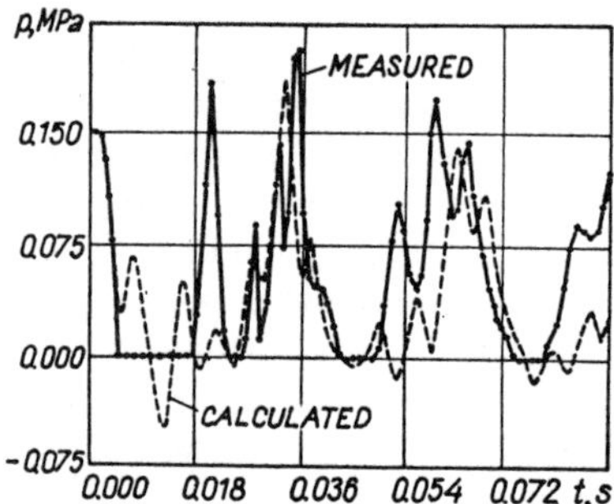


Fig. 11. Calculated and measured dynamic response in a hydraulic chain, according to Example 3.

## 8 Conclusions

For computing transient or frequency responses, as well as static and stationary characteristics of hydraulic systems, the model building block concept coupled with the instrumental programming system CPRIZ provides the possibilities of the following:

- obvious development of mathematical models of functional elements, as well as the complete system;
- application of complicated nonlinear mathematical models of the hydraulic system functional elements, models with distributed parameters and variable structure models;
- utilization of various mathematical procedures in different models of functional elements;
- performing high-speed automatic synthesis of computing program for the whole system based on the description of functional element models, connections between them and formulation of the problem.

The approach described above excels in the methodical and universal compiling of mathematical models. Automatic synthesis of a computing program compiled from prepared program modules makes this computing system very flexible and convenient for the user.

## References


[1] **Grossschmidt G, Vanaveski J, Pahapill J.** *Compiling block schemes of hydromechanical systems out of multiport models of functional elements (in German)*, 7th Conference of hydraulic and pneumatic, Papers - Vol 1, Vol 2, Dresden, pp 138-146, 1987

[2] **Grossschmidt G.** *The principles of compiling mathematical models of hydromechanical chain systems (in Russian)*, Proc. of Tallinn Techn. Univ. no 632, pp 3-13, 1987

[3] **Pahapill J.** *Compiling calculating models of functional elements of hydromechanical systems (in Russian)*, Proc. of Tallinn Techn. Univ. no 632, pp 37-45, 1987

[4] **Pahapill J, Grossschmidt G.** *Program package for modelling hydromechanical systems (in Russian)*, Proc. of Tallinn Techn. Univ. no 632, pp 47-56, 1987

[5] **Tyugu E.** *Knowledge-based programming*, Addison-Wesley Publishing Company, Amsterdam, 243 p, 1988

**WRITTEN DISCUSSION**
**Computing the dynamic response of hydraulic systems using knowledge-based programming**
**G Grossschmidt (Tallinn Technical University, Estonia) & J Pahapill (Estonian Academy of Sciences)**

**Question:** **F Conrad**
**Technical University of Denmark**

Does the program include the capability to simulate different types of digital controller?

**Answer:**

In order to simulate hydraulic drives operating in cooperation with a digital control system, we need to include modules which describe the analogue-to-digital and digital-to analogue conversion processes. The mathematical relations needed can be found in the following, for instance:

1. G. H. Hostetter. Digital Control System Design. Holt, Rinehart and Winston, Inc. 1988

2. C. Kuo. Digital Control Systems. Second Edition, Saunders College Publishing, 1992

**Question:** **ND Vaughan**
**Fluid Power Centre, Bath, UK**

Your conclusions mention "high speed automatic synthesis". The "automatic" part was not clear to me; would you identify the automatic aspects of the simulation package?

**Answer:**

The program approach used is termed conceptual programming. Programming is done using the terms of the subject and an automatic synthesis of programs takes place to calculate the given output variables, proceeding from the description of the task using the parameters in the input language of the system together with the input variables. The whole scheme that is calculated (as well as every partial scheme) has input and output variables. In synthesizing the programs automatically, the calculation system CPRIZ first checks the determination of all the inputs and parameters of the partial schemes, starting from the beginning of the scheme. If the system finds a method of solution, in which the set of output variables can be found from among the given set of input variables and parameters, the established task can be solved. Succeedingly all necessary subprograms (modules) are collected and a program is compiled and calculations performed accordingly.

# 10. An 'Expert' Simulation Software Package for Electro-Hydraulic Control Systems

R B Walters *and* D Harrison

**Synopsis**

**A design tool for the modelling of hydraulic and electro-hydraulic control systems is described. The programs are structured to be user-friendly and are written in conversational mode. The package incorporates integrity checks, operates in both frequency and time domains and includes comprehensive component databases covering major hydraulic equipment manufacturers.**

1. INTRODUCTION

The creation of a viable analytical tool for the design and analysis of complex multi-disciplinary control systems is a time consuming task, which has to be measured in years rather than months. The essential steps leading to the successful conclusion of such an undertaking are as follows:

1. Formulation of **Concept**
2. Derivation of the **Mathematical Model**
3. Setting up a **Computer Program Structure**
4. Formulating **Algorithms**
5. **Writing** of programs
6. **Debugging** of programs
7. Preparation of the **Operating Manual**
8. Conducting **Feasibility Studies**
9. Conduct and evaluate **Correlation Studies**
10. Receiving **Feedback** from users
11. **Revision** of concept by looping 9 to 1
12. **Updating** programs by looping 10 to 2

The omission of any one of these stages can seriously jeopardise the desired outcome of the set objectives.

By way of illustrating the use of this evolutionary process, a brief outline and demonstration of the use of the package HYDRASOFT will be given in this paper by means of a worked example. It is not intended to include a description of the mathematical model or that of the various algorithms deployed since these have been covered in some depth in [1] and [2].

2. PROGRAM FEATURES

The 'knowledge based' package comprises a series of modules to perform a design study. These are briefly described:

**The 'S' Module** performs two quite distinct functions which can be summed up as:

(i) Investigating the integrity of the hydraulic transmission

(ii) Formulating a transfer function which describes its dynamic behaviour.

Input parameters, e.g. cylinder configuration and sizing, valve type, flow rating, supply pressure, mass, loading etc. are entered in practical units via a simple questionnaire.

**The 'C' Module** performs five functions:

(i) Provides Hydraulic Equipment Manufacturers Component Databases
(ii) Provides a means of entering new Component Data
(iii) Automatic Component selection
(iv) Lists steady-state system performance parameters
(v) The ability to create a customised database

A typical manufacturers database will display as shown in Table1.

```
VICKERS SYSTEMS COMPONENT DATA

Servo Valves                          Proportional Valves
V1:SH4-3090                           V8:KDG4V-3
V2:SH4-3220                           V9:KDG4V-5
V3:SM4-10                             V10:KFDG4V-3
V4:SM4-15                             V11:KFDG4V-5
V5:SM4-20                             V12:KFDG5V-7
V6:SM4-30                             V13:KFDG5V-8
V7:SM4-40                             V14:KHDG5V-5
                                      V15:KHDG5V-7
                                      V16:KHDG5V-8
                                      V17:
Feedback Transducers
F1:LVDT 100
F2:FT 07

Components selected:KFDG5V-7
For component update enter YES(Y)?
```

Table 1

**The 'F' and 'P' module** perform four basic functions:

(i) Derivation of the overall system transfer function
(ii) Optimisation of system loop gain
(iii) Analysis of system stability
(iv) System dynamic performance prediction

**The 'T' module** investigates system performance in the time domain. It is a single purpose module whose objective can be summed up as 'The prediction of system transient response to a given input stimulus'. The input stimulus can range from a simple step demand to a compound duty cycle.

Additional features include:
Automatic looping to check performance trends as a function of various independent variables.
The ability to apply system enhancement, i.e. passive networks, flow feedback etc.
The provision of in-built models of circuit configurations to simplify modelling procedure.

## 3. CASE STUDY

Three alternative solutions to a typical problem are now illustrated

| | |
|---|---|
| Ref. WE1 | Servo valve without system enhancement |
| Ref WE1A | Servo valve with PID control |
| Ref WE2 | Proportional valve without system enhancement |

Let us assume a problem such as a table moving in a horizontal plane being required to oscillate over a stroke of 100 cm at an operating frequency of 25 cpm by means of a double ended cylinder. The total mass of the table is 2000 kg and the actuator is subject to an external resisting force of 10 kN, a coefficient of sliding friction of 0.1, and a coefficient of viscous friction of 2kp per cm/sec.

It is required to determine dynamic and steady-state system characteristics for the given operating condition, and establish the limiting operating frequencies and stroke compatible with a maximum positional error at either end of the stroke of 1 mm, and establish comparative performances of systems controlled by servo and proportional valves. See Fig.1

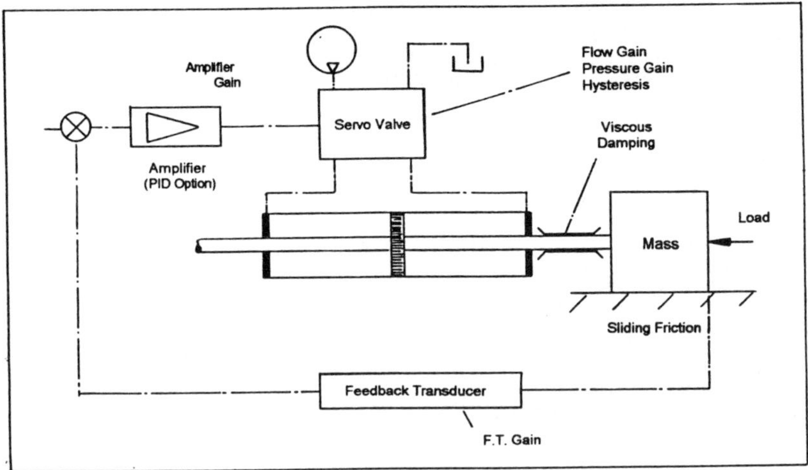


Fig. 1 Circuit Diagram

Two alternative system configurations were investigated:

- System at optimised loop gain
- System with PID at optimised loop gain

Two alternative wave forms were also investigated:

- Sinusoidal
- Triangular

The triangular wave form was found to be the most suitable, giving the highest operating frequencies for the specified positional accuracy. (Not illustrated in this paper).

With the servo valve, the maximum permissible operating frequency for system WE1 is 180 cpm at a maximum stroke of 14 cm; whereas for system WE1A, operating frequency can be increased to 250 cpm at a max stroke of 10 cm.

In the case of the proportional valve, WE2, a comparison of the dynamic performance characteristics of the two systems i.e. band width, natural frequency and step response time using tables 4 and 10 shows that the reduced dynamic performance rating of the proportional valve coupled with the lower supply pressure required, degrades dynamic system performance by a factor of 3. This has the effect of reducing the maximum permissible operating frequency compatible with the specified positional accuracy of 1 mm to 60 cpm at a max stroke of 42 cm. Ref. Fig 11

Due to the lower loop gain and higher valve flow gain, steady-state errors are increased. However due to the lower valve pressure drop required by the proportional valve in relation to the servo valve pressure losses, power efficiency of comparative power supply units is doubled, and power dissipation reduced by a factor of 41.

3.1 The 'S' Module

This is the first module to be completed, and is common to systems WE1 and WE1A. See table 2

```
system reference:WE1          Closed loop system
valve operated system without flow feedback
4 way valve
Double ended Cylinder
 8 bore x  4 rod dia. x  100 stroke (cm)
inlet cylinder area= 37.69911 cm}, outlet cylinder area= 37.69911 cm}
mass = 2000 kg
coefficient of viscous damping =  2 kp/cm per sec
load = 15 kN
inlet/outlet orifice area ratio= 1
equivalent valve flow rating= 188.4876 L/min
actuator velocity= 83.32999 cm/sec
inlet flow = 188.4876 L/min
supply pressure = 160 bar,  valve pressure drop= 113.6747 bar
port diameter= 1.5 cm
trapped volume= 6.769911 L ( 3.769911 :Actuator +  3 :Piping)
actuator shunt coefficient = 0 L/min per bar
bulk modulus= 13793 bar
WHO= 12.23hz
DFO=  0.74
KO/K= 1.019073
S,F,T,P,L,H,C,E,D,R,M,Q,O,G,V,X
```

Table 2

Select the cylinder configuration and dimensions. Make an initial estimate of rod diameter to satisfy the conventional Euler crippling criteria. In the example a standard 8x4x100 cm cylinder is selected. It is interesting to note however that the cylinder bore could be varied, although dynamic performance would be adversely affected by the wrong selection. The selection can be tested in the 'F' module and the dimensions optimised.

The program provides prompts for all input parameters necessary to complete table 2 which includes both directly entered and derived values.

The Load is a summation of the external resisting force plus the effects of sliding and viscous friction and has been rounded to 15 kN.

The Actuator velocity is calculated from the stroke and frequency.

The computed valve pressure drop is the pressure drop available over and above the load pressure drop, and takes into account pressure losses in the valve. It will be necessary to set the supply pressure, i.e. the sum of the load and valve pressure drops, at a level such that the valve pressure drop and equivalent flow rating can be satisfied by a valve in the selected component database.

3.2 The 'C' module

Select a servo valve from the chosen manufacturers database that meets the required specification computed in the previous module i.e. 190 L/min at a valve pressure drop not exceeding 114 bar.

The dynamic characteristics will be extracted from the database and automatically loaded into the 'F' module. In this particular case the valve selected has a natural frequency of 41.25 Hz. and a damping factor of 1.31. This can be seen in table 3

### 3.3 The 'F' and 'P' module

Having assembled the transfer functions of the liquid spring (S module) and the control valve (C module), table 3, the feed-back transducer not being considered in this example, Loop Gain Optimisation is now selected to obtain the optimised system transient performance characteristics Table 4.

```
system reference:WE1            Closed loop system
Components selected:SM4-40
OPEN LOOP SYSTEM PARAMETERS
Frequencies in hz:
WH0= 12.23
WH1= 41.25
WH2=%100000.00
WH3=%100000.00
Damping factors:
DF0=  0.74
DF1=  1.31
DF2=  0.00
DF3=  0.00
Time constants in seconds:
t0= 0          t1= 0          t2= 0          t3= 0          t4= 0
free integrator:SINGLE
loop gain= 26.95  1/sec
'C' to continue 'any key' for frequency domain access
```

Table 3 Summary of overall system transfer function (F module)

All variables are generated within the program and are for reference only

*0 and *3 are reserved for the hydraulic transmission (derived in 'S' module)

*1 is reserved for the control valve (derived in 'C' module)

*2 is reserved for the feed-back transducer (derived in 'C' module)

```
Natural frequency WR=  5.41hz                CL Damping factor DFR=  0.37
Band width at 4 dB attenuation WC=  8.65hz
Max.Overshoot at  0.099sec = 28.91%
Step response time=  0.057sec
system reference:WE1
S,F,T,P,L,H,C,E,D,R,M,Q,O,G,V,X
```

Table 4 System transient performance characteristics (P module)

The effect of introducing PID control can be investigated by adding the appropriate compensation, in which case it is necessary to re-run the 'F' module, selecting the three term controller at the screen prompt. In the example, the integral and differential terms have each been set at 0.5. The effect of changing these values can be easily observed by re-calculation.

We again obtain the system transient performance characteristics table 5, and as expected it will be seen by a comparison with table 4 that natural frequency and band width have been substantially improved together with a significant reduction in step response time.

```
Natural frequency WR=  8.89hz                CL Damping factor DFR=  0.35
Band width at 4 dB attenuation WC= 13.65hz
Max.Overshoot at  0.060sec = 30.52%
Step response time=  0.034sec
system reference:WE1A
S,F,T,P,L,H,C,E,D,R,M,Q,O,G,V,X
```

Table 5 Enhanced system transient performance characteristics (P module)

A Closed Loop Bode Diagram and Nichols Charts have been plotted to check the dynamics. For convenient reading the Bode Diagrams of both systems have been combined.

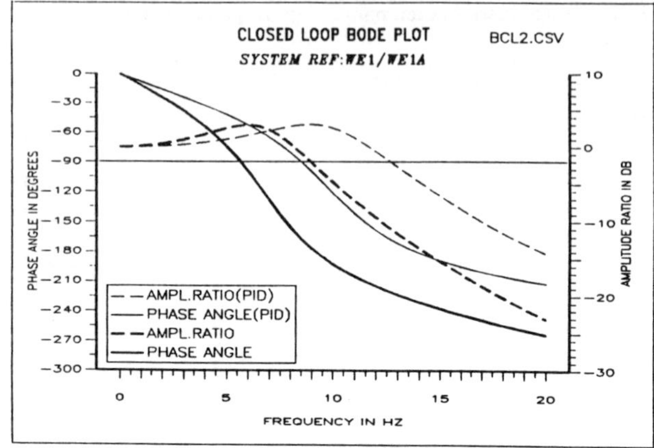


Fig. 2 Closed loop Bode Diagram showing system band width and natural frequency. See also Tables 4 and 5.

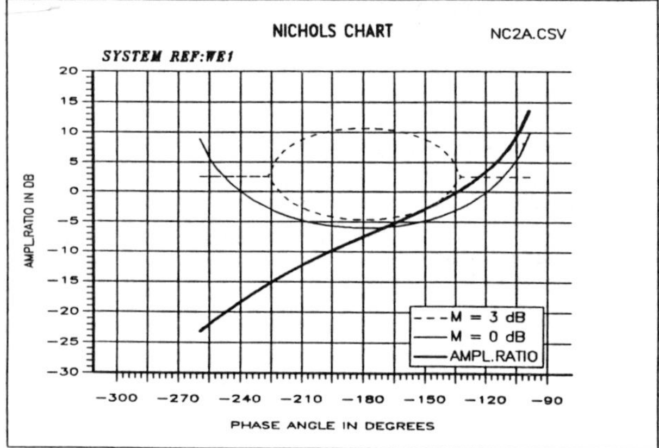


Fig. 3 Nichols Chart of non-enhanced system showing system stability margins (gain and phase)

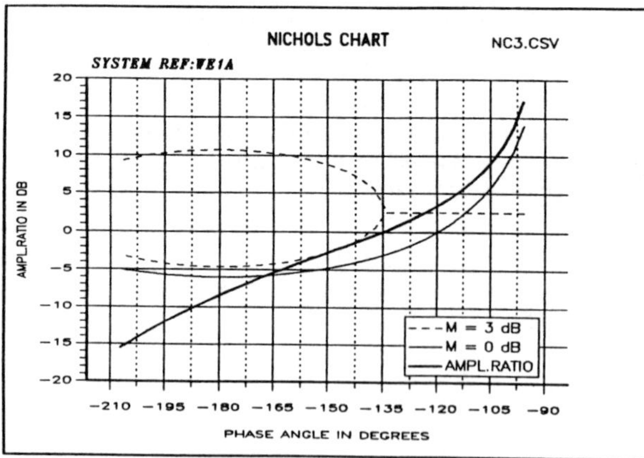


Fig. 4 Nichols Chart of enhanced system (PID control) showing system stability margins (gain and phase)

3.4 The 'T' Module

Having determined the stability limits of the system, we are now able to examine the selected system in the time domain, in order to check for the desired positional accuracy. By plotting the Triangular Response Parameter Fig 5, we can determine the minimum ramp time that can be achieved without exceeding the specified error. The graph can be 'zoomed' for greater accuracy Fig 6.

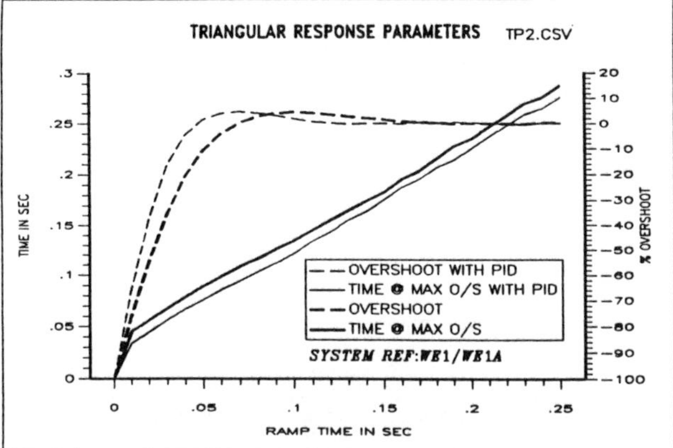


Fig. 5 Triangular Response Parameter

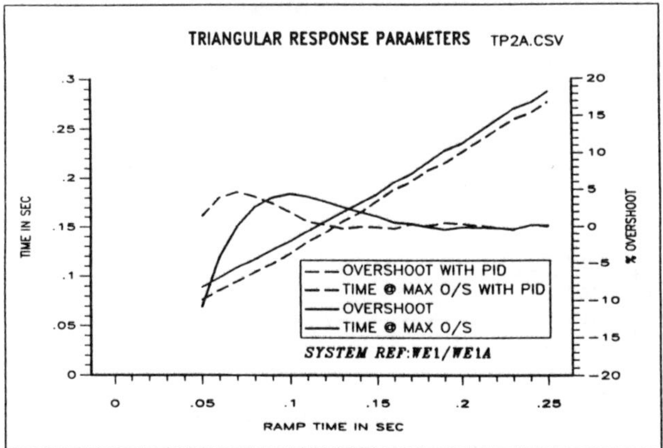


Fig. 6 (with overshoot 'zoomed')

It will be seen that the minimum ramp time that can be achieved by the non-enhanced system (WE1) is **0.17 secs***. As this is the half cycle, this determines the max. reciprocating frequency, say 180 cpm.

The amplitude (stroke) is the time x velocity = 83.33 x 0.17 i.e. **14 cm*** (rounded).

*These two figures are carried forward to the Transient Response and determine the values of the Triangular Input (Forcing Function).

A Transient Response plot ref. Fig. 7 confirms that the amplitude error is within acceptable limits i.e. 1 mm in 14 cm for the non PID system. Again it is beneficial to zoom the graph (amplitude) to confirm the accuracy ref. Fig. 8.

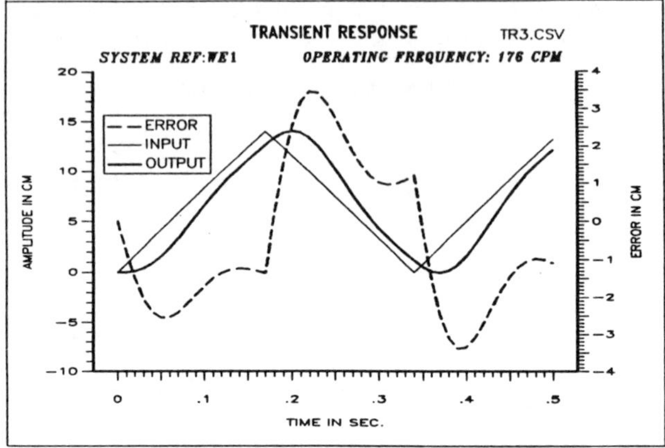


Fig. 7 Transient Response of non-PID system

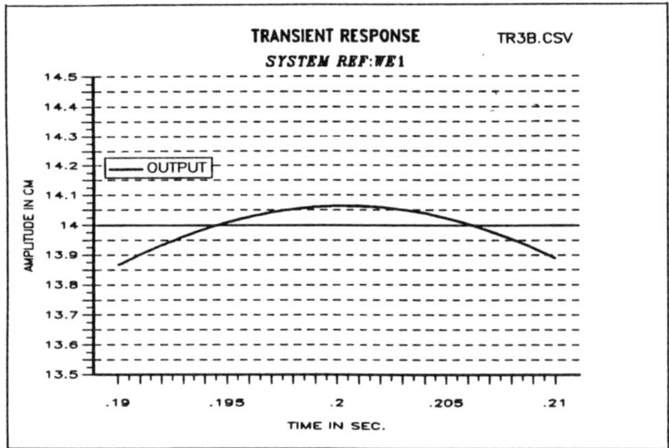


Fig. 8 Transient Response showing 'zoomed' amplitude

The effect of adding PID control (Example WE1A) can now be illustrated by repeating the procedure. The corresponding Figs. 9 and 10 show the improved performance and confirm that it is within the set positional tolerance.

Minimum ramp time has now been reduced to 0.12 seconds (from Fig. 5) resulting in a higher maximum frequency of 250 cpm at 10 cm amplitude.

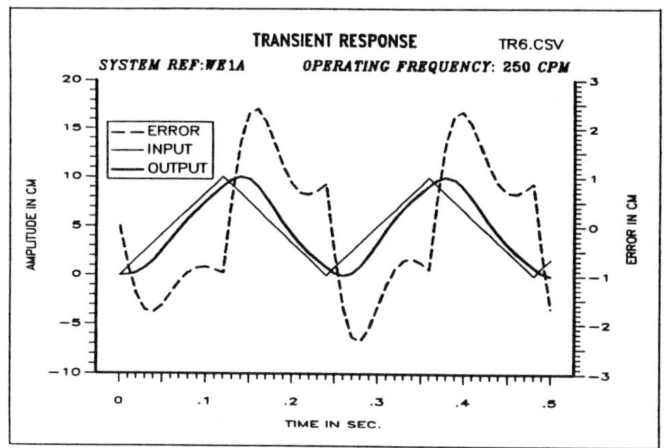


Fig. 9 Transient Response of enhanced (PID) system

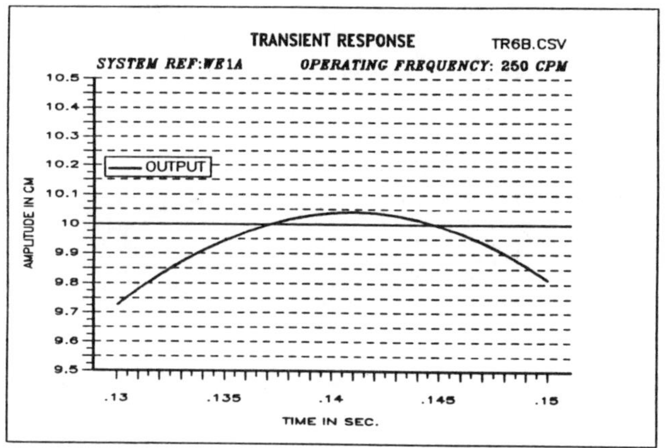


Fig. 10 Transient Response of PID system showing 'zoomed' amplitude

### 3.5 Steady-state Errors

These are obtained via the 'C' module by selecting the appropriate command. The valve parameters i.e. flow gain, pressure gain and hysteresis for the selected valve are retrieved automatically from the component database.

Displays for systems WE1 and WE1A are shown in tables 6 & 7. As may be expected, the errors associated with the PID system are shown to be significantly lower than those of the non-enhanced system.

```
STEADY-STATE ERRORS
SPECIAL FEATURES
VALVE PARAMETERS MODEL:SM4-40
Flow Gain= 1.924243 L/min per % of command signal
Pressure Gain= 48 bar per % of command signal
Hysteresis= 2 % of command signal units
Enter YES(Y) for mode update?
POSITION CONTROL
Load Error @  15 kN= .2616805 mm
Hysteresis Error= .6313679 mm
Velocity Error @  83.32999 cm/sec= 30.92255 mm
VELOCITY CONTROL
Load Error @  15 kN= .2616805 mm/sec
Hysteresis Error= .6313679 mm/sec
Acceleration Error @  83.32999 cm/sec)= 30.92255 mm/sec
To obtain Amplifier Gain, enter Feedback Transducer Gain in Volt/cm for posi
 control and Volt per cm/sec for velocity control?
```

Table 6 Steady-state errors WE1

```
STEADY-STATE ERRORS
SPECIAL FEATURES
PID Controller
VALVE PARAMETERS MODEL:SM4-40
Flow Gain= 1.924243 L/min per % of command signal
Pressure Gain= 48 bar per % of command signal
Hysteresis= 2 % of command signal units
Enter YES(Y) for mode update?
POSITION CONTROL
Load Error @  15 kN= .1550296 mm
Hysteresis Error= .3740465 mm
Velocity Error @  83.32999 cm/sec= 9.159852 mm
VELOCITY CONTROL
Load Error @  15 kN= .1550296 mm/sec
Hysteresis Error= .3740465 mm/sec
Acceleration Error @  83.32999 cm/sec)= 9.159852 mm/sec
To obtain Amplifier Gain, enter Feedback Transducer Gain in Volt/cm for posi
 control and Volt per cm/sec for velocity control?
```

Table 7 Steady-state errors WE1A

**SYSTEM WE2 (PROPORTIONAL VALVE COMPARISON)**

The procedure adopted is identical to that for the servo valve and is quantified in the following illustrations. For brevity, further explanation is restricted to the highlighting of salient points of difference.

To meet the requirements of 188 L/min, a two-stage valve has been selected. The supply pressure can now be reduced to 70 bar to match the lower pressure rating of the proportional valve, table 8.

By comparing the dynamic performance characteristics of the two non-enhanced systems i.e. servo v. proportional, tables 4 & 10, in terms of system band width, natural frequency and step response times, the reader will notice that the figures differ by a factor of 3. It can be concluded that the reduced dynamic performance rating of the proportional valve coupled with the lower supply pressure required, degrades dynamic system performance by a factor of 3:1.
If a comparison is made with the PID system, tables 5 & 10, it will be seen that the corresponding degradation is 4:1.

This has the effect of reducing the maximum permissible operating frequency compatible with the specified positional accuracy of 1 mm, to 60 cpm at a maximum stroke of 42 cm. Comparison of the three systems is shown graphically in Fig. 11. Using an associated program - Power Efficiencies - not covered in this paper, it can also be shown that due to lower valve pressure losses when compared to a servo valve, power efficiency of comparable power units is doubled, and power dissipation reduced by a factor of 4:1.

```
system reference:WE2          Closed loop system
valve operated system without flow feedback
4 way valve
Double ended Cylinder
 8 bore x  4 rod dia. x  100 stroke (cm)
inlet cylinder area= 37.69911 cm}, outlet cylinder area= 37.69911 cm}
mass = 2000 kg
coefficient of viscous damping =  2 kp/cm per sec
load = 15 kN
inlet/outlet orifice area ratio= 1
equivalent valve flow rating= 188.4876 L/min
actuator velocity= 83.32999 cm/sec
inlet flow = 188.4876 L/min
supply pressure = 70 bar,    valve pressure drop= 26.68301 bar
port diameter= 1.75 cm
trapped volume= 6.769911 L ( 3.769911 :Actuator +  3 :Piping)
actuator shunt coefficient = 0 L/min per bar
bulk modulus= 13793 bar
WHO= 12.59hz
DFO=  3.04
KO/K= 1.081256
S,F,T,P,L,H,C,E,D,R,M,Q,O,G,V,X
```

Table 8 'S' module screen for proportional valve

```
system reference:WE2          Closed loop system
Components selected:KFDG5V-7
OPEN LOOP SYSTEM PARAMETERS
Frequencies in hz:
WHO= 12.59
WH1= 21.00
WH2=%100000.00
WH3=%100000.00
Damping factors:
DFO=  3.04
DF1=  0.71
DF2=  0.00
DF3=  0.00
Time constants in seconds:
tO= 0          t1= 0          t2= 0          t3= 0          t4= 0
free integrator:SINGLE
loop gain= 13.20  1/sec
'C' to continue 'any key' for frequency domain access
```

Table 9 Summary of overall system transfer function for propl. valve (F module)

```
Natural frequency WR=  2.02hz              CL Damping factor DFR=  0.39
Band width at 4 dB attenuation WC=  3.00hz
Max.Overshoot at  0.269sec = 26.35%
Step response time=  0.153sec
system reference:WE2
S,F,T,P,L,H,C,E,D,R,M,Q,O,G,V,X
```

Table 10 System transient performance for proportional valve (P module)

Steady-state errors are also increased due to the lower loop gain and higher valve flow gain as illustrated in Table 11.

```
STEADY-STATE ERRORS
SPECIAL FEATURES
VALVE PARAMETERS MODEL:KFDG5V-7
Flow Gain= 4.083734 L/min per % of command signal
Pressure Gain= 21 bar per % of command signal
Hysteresis= 2 % of command signal units
Enter YES(Y) for mode update?
POSITION CONTROL
Load Error @  15 kN= 2.591074 mm
Hysteresis Error= 2.735073 mm
Velocity Error @  83.32999 cm/sec= 63.11961 mm
VELOCITY CONTROL
Load Error @  15 kN= 2.591074 mm/sec
Hysteresis Error= 2.735073 mm/sec
Acceleration Error @  83.32999 cm/sec}= 63.11961 mm/sec
To obtain Amplifier Gain, enter Feedback Transducer Gain in Volt/cm for pos
 control and Volt per cm/sec for velocity control?
```

**Table 11 Steady-state errors of proportional valve system**

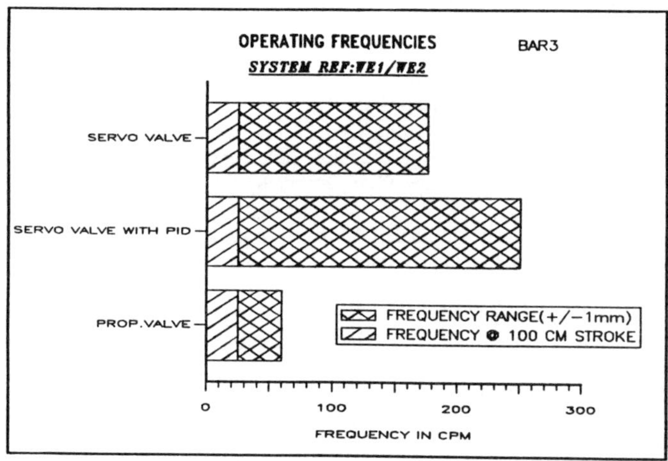


Fig. 11 Graphical comparison of the three alternative solutions

## 4. CONCLUSION

The use of a computerised performance prediction package such as HYDRASOFT can considerably reduce the risks associated with the design and development of a new control system.
However the programs have to be sufficiently user-friendly to provide a suitable tool for the Application Engineer who lacks any specialised knowledge of Control Theory, whilst at the same time meeting the requirements of the experienced system designer for a versatile and comprehensive design aid.

## 5. REFERENCES

1. WALTERS, R : Modular Optimised System Simulation, 8th International Symposium on Fluid Power, Birmingham 1988 (BHRA)

2. WALTERS, R., Hydraulic and Electro-Hydraulic Control Systems 1991 Elsevier Applied Science Publishers

## 6. BIBLIOGRAPHY

THALER G.J., and BROWNE, R.G., Analysis and Design of Feedback Control Systems. McGraw-Hill, New York (1960)

HADEKEL, R., 'Hydraulic And Pneumatic Servos', Automation (1954-1955)

AREFIN, K.M.M., 'Seal Characteristics of a Hydraulic Ram and Effects on Servo System Performance, School of Mech.Eng. Thames Polytechnic, (1975)

WALTERS, R., Hydraulic and Electro-hydraulic Servo Systems 1967 London. Iliffe Books

WALTERS, R : Electrically Modulated Actuator Controls 4th International Fluid Power Symposium, Sheffield 1975 (BHRA)

WALTERS, R., Mathematical Model of a new Actuator Control System, Pneumatics-Hydraulics Conference, Gyor, Hungary, (1975)

WALTERS, R., Electrically Modulated Actuator Controls, International Symposium on Oil Hydraulics, Hannover, (1977)

WALTERS, R., Desk Top Computer Performance Prediction Package, I.Mech.E. Seminar on Computer Aided Design, London, (1983)

WALTERS, R., Proportional Control Systems and Applications. FLUMEX 84 Conference, Birmingham 1984

WALTERS, R., State of the Art Lecture: Electro-Hydraulic Proportional Controls, I.Mech.E. London 1986

WALTERS, R., Electro-Hydraulic Proportional Controls, Oxford University (I.Mech.E.) 1987

**WRITTEN DISCUSSION**
**An 'expert' simulation software package for electro-hydraulic control systems**
**RB Walters & D Harrison (Flotron Limited, UK)**

**Question:** **W Backé**
**IHP, RWTH Aachen, Germany**

You told us that you can reach higher efficiency with a proportional valve in comparison with a servo valve. How is this possible, because both are "resistance" - controls?

**Answer:**

Although both servo and proportional valves are "resistance" controls, higher efficiency with proportional valves is achieved by reducing the valve pressure drop which is the cause of power dissipation. In that sense proportional valves can best be regarded as oversized servo valves.

By applying load-sensing to a fixed displacement pump (pressure-matching) or to a variable displacement pump (power-matching), appreciable improvement in power efficiency can be achieved, as shown in the car chart.

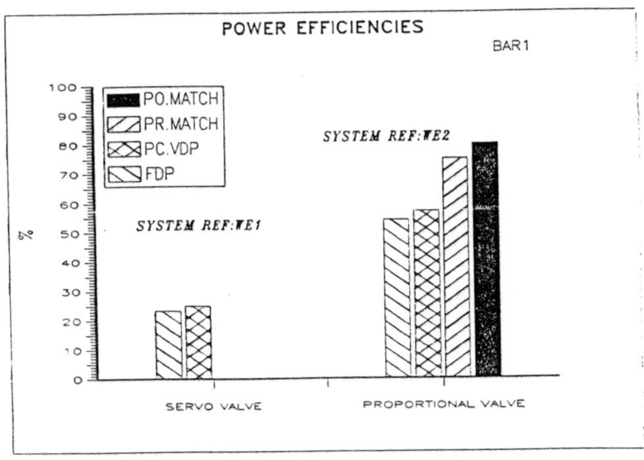

**Question:** **ND Vaughan**
**Fluid Power Centre, Bath, UK**

1. What valve non-linearity data are included in the package and are these taken into account in the determination of frequency response and steady state error?

2. You have explained the improvement in efficiency with proportional valves and fixed capacity pump as being possible because of a reduced valve pressure drop. Does this imply inferior controllability?

**Answer:**

1. Non-linearities such as hysteresis are included in the package and taken into account in the determination of steady-state errors.

2. It can be seen from the data given in the paper that both steady-state and dynamic performance are inferior with proportional valves. This is due to two factors:

   (i) the lower dynamic response of proportional valves
   (ii) the higher flow gain of proportional valves resulting in a correspondingly lower electrical gain

By applying Hydrasoft's loop gain optimisation feature, the highest system performance for the given operating conditions can be achieved.

**Question:** **F Conrad**
**Technical University of Denmark**

Can the program deal with:

(i) MIMO cases?

(ii) digital controllers?

**Answer:**

(i) Yes. MIMO cases can be dealt with by making use of the 'Customised Option' feature which allows for customised sub-routines to be incorporated in the program.

(ii) A special optional program covering digital controllers is available.

# 11. Adaptive Control of a Pilot Operated Solenoid Valve

J B Gamble


## Abstract

This paper describes the development of an adaptive controller for a pilot operated proportional solenoid valve. In such valves, the dynamic response of the main stage is dependent on the supply pressure to the pilot stage. A model of the valve is developed, which is used to demonstrate that the effects of oil compressibility can be ignored. This allows the main stage to be accurately modelled as a simple integrator, with the velocity gain dependant on the pilot supply pressure.

A novel nonlinear adaptive controller is described which employs a recursive least squares estimator to identify the velocity gain of the valve from the pilot and main stage spool positions. This estimated value is then used to modify a nonlinear controller gain to obtain consistent closed loop response despite wide variations in the pilot supply pressure.

Experimental step response results are presented which demonstrate that invariant performance is obtained for supply pressures between 40 and 200 bar.


## Notation

| | |
|---|---|
| $A$ | Main spool cross sectional area |
| $a_i$ | $i^{th}$ Filter polynomial denominator coefficient |
| $b_i$ | $i^{th}$ Filter polynomial numerator coefficient |
| $c_p$ | Pilot valve pressure gain |
| $c_x$ | Pilot valve flow gain |
| $K$ | Kalman gain |
| $k$ | Main spool return spring stiffness |
| $k_q$ | Velocity gain |

| | |
|---|---|
| $m$ | Main spool mass |
| $n$ | Sample number |
| $P$ | Estimator convariance |
| $P_A$ | Pilot valve port A pressure |
| $P_B$ | Pilot valve port B pressure |
| $Q$ | Flow rate |
| $x$ | Pilot spool displacement |
| $y$ | Main spool displacement |
| $\beta$ | Bulk modulus (Section 2) |
| $\beta$ | Forgetting factor (Section 3) |
| $\theta_i$ | Filter input |
| $\theta_o$ | Filter output |
| $\upsilon$ | Main spool friction |
| $\omega_n$ | Natural Frequency |

# 1 Introduction

## 1.1 Pilot Operated Proportional Valves

In pilot operated proportional valves, a small proportional solenoid valve is used to control the movement of a larger main stage valve. They are used to control high flow rates, where the larger spool mass and higher flow reaction forces prevent the spool from being directly driven by the solenoid. The main disadvantage of this arrangement is that the dynamic response of the main stage spool is dependent on the supply pressure to the pilot stage; higher supply pressures giving faster response. This dependency arises because the flow gain of the pilot valve is directly dependent on the supply pressure.

These valves are normally operated under closed loop control, with both pilot stage and main stage spool position feedback. Both loops are typically closed using Proportional plus Integral plus Derivative (PID) controllers. However, this scheme is unable to eliminate the effects of variations in the pilot supply pressure. Consistent operation has traditionally been only possible by employing a separately regulated pilot supply.

The valve used in this study is illustrated in Figure 1. This valve had a rated flow of 150$l$/min and differed from commercially available models only in that a higher flow pilot valve was used, as described in section 4.

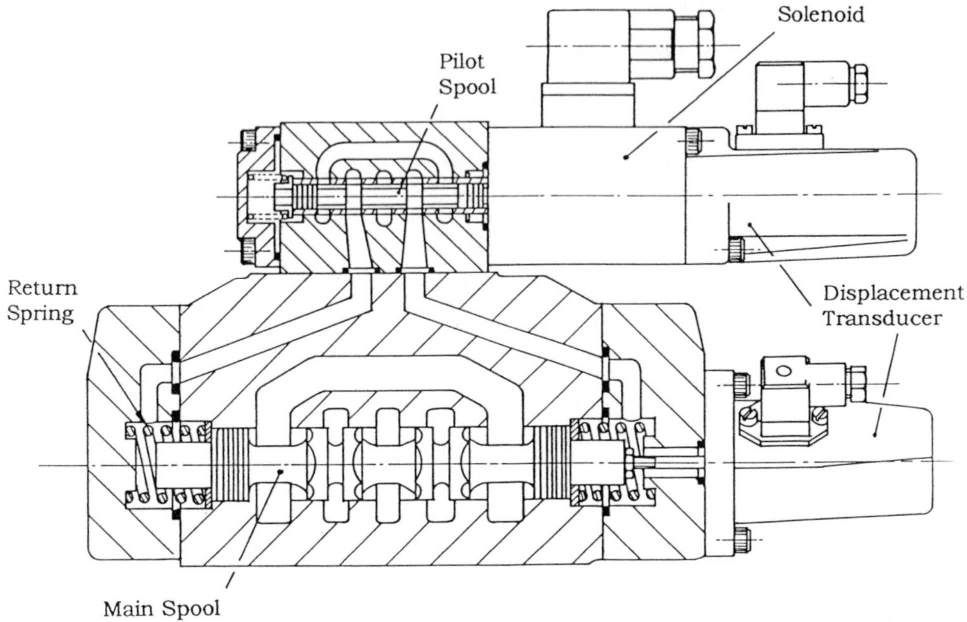


Fig 1. Pilot Operated Solenoid Valve

1.2 Self Tuning Adaptive Control

Adaptive control schemes can be divided into two groups; Model Reference Adaptive Control (MRAC), and Self Tuning Control (STC). In the MRAC method, the output of the plant is compared to the output from a reference model, and the controller is modified to eliminate differences between the two. In contrast, the STC technique uses plant input/output data to estimate the plant parameters. The controller is then redesigned based on these estimates. Despite their differences however, both methods require that the control signal has a sufficiently rich frequency content to excite all the dynamic modes within the plant.

The nonlinear, time varying nature of hydraulic systems makes them attractive candidates for adaptive control, and a number of successful adaptive controllers for hydraulic systems have been reported. Figuredo [1] describes the application of MRAC to a linear actuator servomechanism, and Plummer [2] demonstrates the application of STC to a similar system. Despite the success of these studies however, the requirement for a sufficiently rich plant input has prevented widespread commercial application of this technology.

The dynamics of the valve are analysed in the next section. In Section 3, describes the development of a self tuning controller, and experimental results are presented in Section 4.

## 2 Valve Model

Figure 2 shows a schematic diagram of the two-stage valve used in this study. A transfer function is required which describes the dynamics from pilot spool displacement to main stage spool displacement.

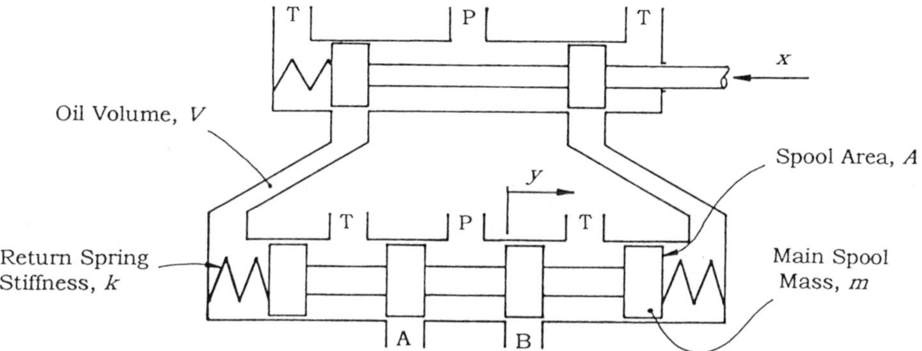

Fig 2. Schematic Diagram of Pilot Operated Solenoid Valve

Equating forces on the main stage spool:

$$(P_A - P_B)A = m\frac{d^2y}{dt^2} + \upsilon\frac{dy}{dt} + 2ky \tag{1}$$

The flow relationship for the pilot valve is very nonlinear. However, for small perturbations about an initial operating point, the flow can be expressed:

$$\delta Q = c_x \delta x - \frac{c_p}{2}(\delta P_A - \delta P_B) \tag{2}$$

In this equation, $c_p$ and $c_x$ are respectively the pilot valve pressure and flow gains:

$$\left.\begin{aligned} c_p &= \frac{\partial Q}{\partial P} \\ c_x &= \frac{\partial Q}{\partial x} \end{aligned}\right\} \tag{3}$$

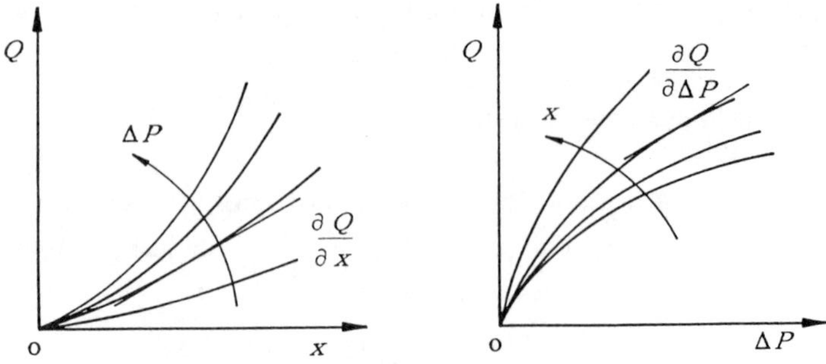


Fig 3. Flow Characteristics of Pilot Valve

Values for $c_p$ and $c_x$ can be obtained by linearising the pressure flow characteristics of the pilot valve as shown in Fig 3 [4]. If it is assumed that the main stage spool in its centre position (ie the oil volumes on either side of the spool are identical), then equating the flow through the pilot valve with the flow into the main stage, and taking the effects of oil compressibility into account gives:

$$c_x\,\delta x - \frac{c_p}{2}(\delta P_A - \delta P_B) = A\frac{\mathrm{d}}{\mathrm{d}t}\delta y + \frac{V}{4\beta}\frac{\mathrm{d}}{\mathrm{d}t}(\delta P_A - \delta P_B) \tag{4}$$

Substituting for $P_A$-$P_B$ from equation (1):

$$c_x\,\delta x = \frac{Vm}{4\beta A}\frac{\mathrm{d}^3}{\mathrm{d}t^3}\delta y + \left(\frac{V\upsilon}{4\beta A} + \frac{c_p m}{2A}\right)\frac{\mathrm{d}^2}{\mathrm{d}t^2}\delta y + \left(A + \frac{Vk}{2\beta A} + \frac{c_p \upsilon}{2A}\right)\frac{\mathrm{d}}{\mathrm{d}t}\delta y + \frac{c_p 2k}{2A}\delta y \tag{5}$$

Taking Laplace transforms, the small perturbation transfer function is:

$$\frac{\delta y}{\delta x}(\mathrm{s}) = \frac{c_x/A}{\dfrac{Vm}{4\beta A^2}\mathrm{s}^3 + \left(\dfrac{V\upsilon}{4\beta A^2} + \dfrac{c_p m}{2A^2}\right)\mathrm{s}^2 + \left(1 + \dfrac{Vk}{2\beta A^2} + \dfrac{c_p \upsilon}{2A^2}\right)\mathrm{s} + \dfrac{c_p 2k}{2A^2}} \tag{6}$$

If the pilot spool is assumed to operate about its centre position, and to be perfectly "zero lapped", then the pressure gain can be taken as zero. Under these conditions, equation (6) reduces to:

$$\frac{\delta y}{\delta x}(\mathrm{s}) = \frac{c_x/A}{\mathrm{s}\left(\dfrac{Vm}{4\beta A^2}\mathrm{s}^2 + \dfrac{V\upsilon}{4\beta A^2}\mathrm{s} + 1 + \dfrac{V2k}{4\beta A^2}\right)} \tag{7}$$

The natural frequency of the second order term is:

$$\omega_n = \sqrt{\frac{4\beta A^2}{Vm} + \frac{2k}{m}} \qquad (8)$$

For the particular valve under investigation:

$A = 4.34\text{x}10^{-4}\ \text{m}^2$ $\qquad$ $m = 0.265$ kg

$k = 8680$ N/m $\qquad$ $V = 26\text{x}10^{-6}\ \text{m}^3$

Taking $\beta = 120\text{x}10^9\ \text{N/m}^2$ gives:

$$\omega_n = \sqrt{\frac{4\ \text{x}120\text{x}10^9\ \ (4.34\text{x}10^{-4})^2}{26\text{x}10^{-6}\ \ 0.265} + \frac{2\text{x}8680}{0.265}}$$

$$= 18.2\ \text{kHz}$$

Earlier work on the control of the pilot valve indicated that a bandwidth of between 120 and 180Hz was possible [3]. Thus, the natural frequency of the main stage is at least a hundred times the bandwidth of the pilot valve, and it would therefore be unrealistic to attempt to control this mode. For controller design therefore, the effects of oil compressibility can be neglected. Under these conditions, the second order term in equation (8) can be ignored, and the transfer function of the main stage reduced to a simple integrator:

$$\frac{\delta y}{\delta x}(s) = \frac{c_x}{As} = \frac{k_q}{s} \qquad (9)$$

Although $A$ will remain constant for any given valve design, the pilot flow gain $c_x$ (and hence the velocity gain $k_q$) changes with both supply pressure and pilot spool displacement, as shown in Fig 3.

## 3 Adaptive Controller

Self tuning adaptive control has the advantage that the parameter estimation and controller design stages are separate. Thus the controller design method cannot influence the convergence of the estimator, and can be considered in isolation.

3.1 Parameter Estimator

There are a considerable number of parameter estimation strategies that could have been used [5]. Of these, good results have been reported using Recursive Least Squares (RLS) estimators [2]. This method has the advantages of simplicity and ease of implementation.

The analysis presented in the previous section shows that valve can be modelled as an integrator with a variable gain:

$$\frac{y}{x}(z) = \frac{k_q}{(1-z^{-1})} \tag{10}$$

The task is to estimate the velocity gain $k_q$ based on a number of measurements of $x$ and $y$. For this simple first order case, the RLS estimator will have the form:

$$\hat{k}_q(n) = \hat{k}_q(n-1) + K(n)\left[(1-z^{-1})y(n) - \hat{k}_q(n-1)x(n)\right] \tag{11}$$

Where $\hat{k}_q(n)$ is the estimated value of $k_q$ at sample $n$, and $K(n)$ is the Kalman gain. For an RLS estimator, the Kalman gain will have the form:

$$\left.\begin{aligned} K(n) &= P(n)x(n) \\ P(n) &= \frac{P(n-1)}{\beta + P(n-1)x^2(n)} \qquad 0<\beta<1 \end{aligned}\right\} \tag{12}$$

In this equation, $P(n)$ is the (normalised) covariance, and $\beta$ is a "forgetting factor" [5]. Since all the terms in equations (11) and (12) are scalar, convergence of this estimator is easily determined by inspection. For a sufficiently large value of $x(n)$, the term $P(n)$ (and hence the Kalman gain) diminishes. The parameter estimate converges (ie adaptation ceases) when the Kalman gain reduces to zero. This implies that even a constant value of $x(n)$ can be "sufficiently exciting" to produce a good estimate. The speed of the convergence is controlled by the forgetting factor $\beta$. Smaller values slow the rate of convergence and enable the estimator to track faster changes in $k_q$. A problem ocurrs however, when there are either offset errors in the spool displacement measurements, or friction in the main spool. Both result in zero main stage velocity with a non-zero spool displacement. This causes the velocity gain estimate to drift from the correct value. To prevent this happening, the estimator was turned off when the measurement of $x$ reduced below a predetermined threshold.

The above analysis has not allowed for noise on the measurements $y(n)$ and $x(n)$. Any noise correlated with the measurements will produce biased results. Greatly reduced susceptibility can be obtained by pre-filtering the measurements. Plummer [2] has shown that the filter need not be exactly matched to the noise characteristics in order to obtain improved estimates. For the valve controller, bandpass filters were used, having the form:

$$\frac{\theta_o}{\theta_i}(z) = (1-z^{-1})\frac{b_o + b_1 z^{-1} + b_2 z^{-2}}{1 + a_1 z^{-1} + a_2 z^{-2}} \tag{13}$$

The "backwards difference" term $(1\text{-}z^{-1})$ was used to remove offset errors, and the polynomial $b(z)/a(z)$ was designed to have a low pass characteristic to remove high frequency noise. Both the pilot displacement measurement $x(n)$ and the mains stage displacement measurement $y(n)$ were filtered using identical filters.

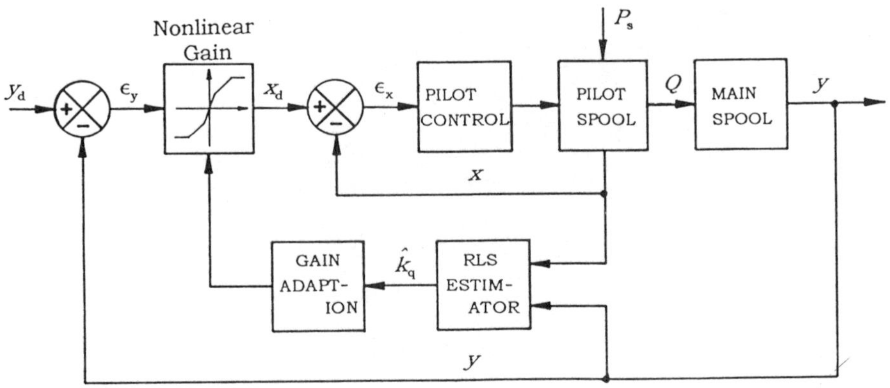


Fig 4. Adaptive Controller

3.2 Nonlinear Controller Design

A block diagram of the full adaptive controller is presented in Fig 4. The small perturbation analysis, presented in Section 2, implies that a simple proportional controller is sufficient to control the position of the main stage spool. In practice however, problems were encountered due to the nonlinear flow characteristics, and limited bandwidth of the pilot valve. Nonlinear flow characteristics in the pilot spool resulted in the main stage exhibiting different response times for different amplitude step inputs. The limited bandwidth of the pilot valve had a tendency to destabilise the main stage loop at low supply pressures, since the pilot valve was required to open further in an attempt to maintain a constant main stage response.

Both these problems were solved by using a nonlinear proportional gain in the main stage controller. Optimum gain settings were obtained by manually tuning the controller for a range of different pressures. During tuning, the adaptation mechanism was disabled, but the estimator was allowed to run so that the estimated gain could be noted. This procedure enabled a map to be constructed of control effort (ie desired pilot spool displacement) as a function of estimated pressure and main stage spool position error. The resultant map is shown in Fig 5.

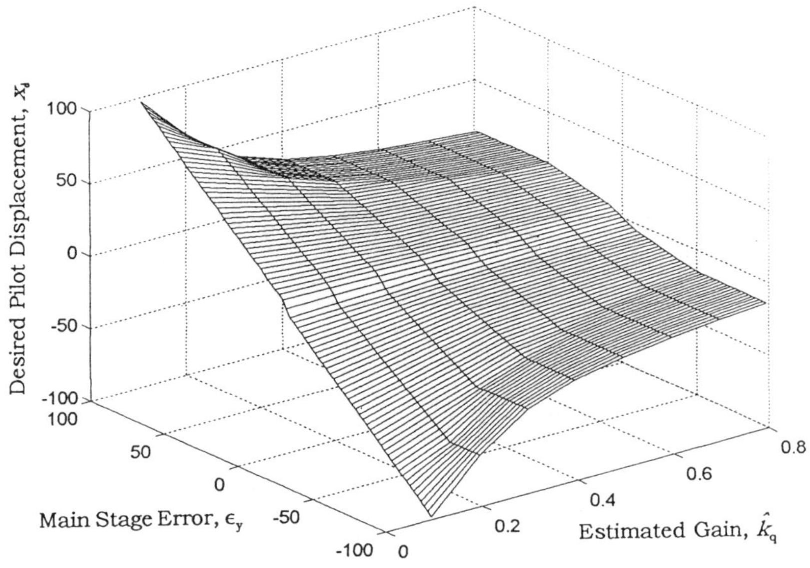


Fig 5. Nonlinear Control Map

The controller was first tuned to give the best possible response with a pilot pressure of 30bar. This determined the minimum achievable response time over all pressures. The controller was then re-tuned at successively higher pressures with the aim of maintaining a consistent response time.

It is easy to see that the response of the main spool can be made faster by increasing the flow rating of the pilot valve. In practice, the bandwidth and accuracy of the pilot loop limit the response that can be achieved. Instability will result if the pilot valve is too slow, and any steady state errors or oscillations in the pilot valve control loop will be amplified by the higher supply pressure. It was discovered by experiment that the highest available flow rate version of the pilot valve could be used without loss of control at high pressures. The pilot valve used in the following experiments had a rated flow approximately 8 times that of the standard valve.

## 4 Experimental Results

The adaptive controller was implemented on a Texas Instruments TMS320C30 Digital Signal Processor (DSP) mounted inside a personal computer. The system provided many times the processing power necessary to implement the controllers. This enabled all programs to be written using the C programming language, with floating point calculations used throughout; greatly reducing the programming effort required to evaluate new control algorithms. The DSP also controlled the pilot valve, and the sample rate was chosen to meet this requirement rather than the requirements of the adaptive controller. The sample period was 0.25ms, and $\beta$=0.95.

Figure 6 shows the pilot and main stage response to a step input for a range of different pressures from between 30 and 200 bar. It can be seen that the main stage response remains virtually constant over the entire pressure range. A noticeable difference occurs at a supply pressure of 30bar where a slight undershoot is just discernible. This was attributed to friction in the main spool. The graph of pilot displacement vs. time shows how the controller reduces the spool displacement as the pressure increases. Note that the nonlinear gain was tuned to just allow the pilot valve to reach full displacement during a 100% step demand with a supply pressure of 30bar.

The pressure estimates are shown in Fig 7. The sudden drop in estimated velocity gain at the start of the step may either have been a tuning transient in the estimator, or be an accurate indication of pilot supply pump limitation effects at high pressure. The large changes in estimated gain during the 140 and 200bar step responses were attributed to the greater curvature of the pilot flow characteristic at high pressure (see Fig 3). Although the estimator was designed to switch off when the main spool was stationary, it appears that the velocity gain estimate drifted during the 200bar test.

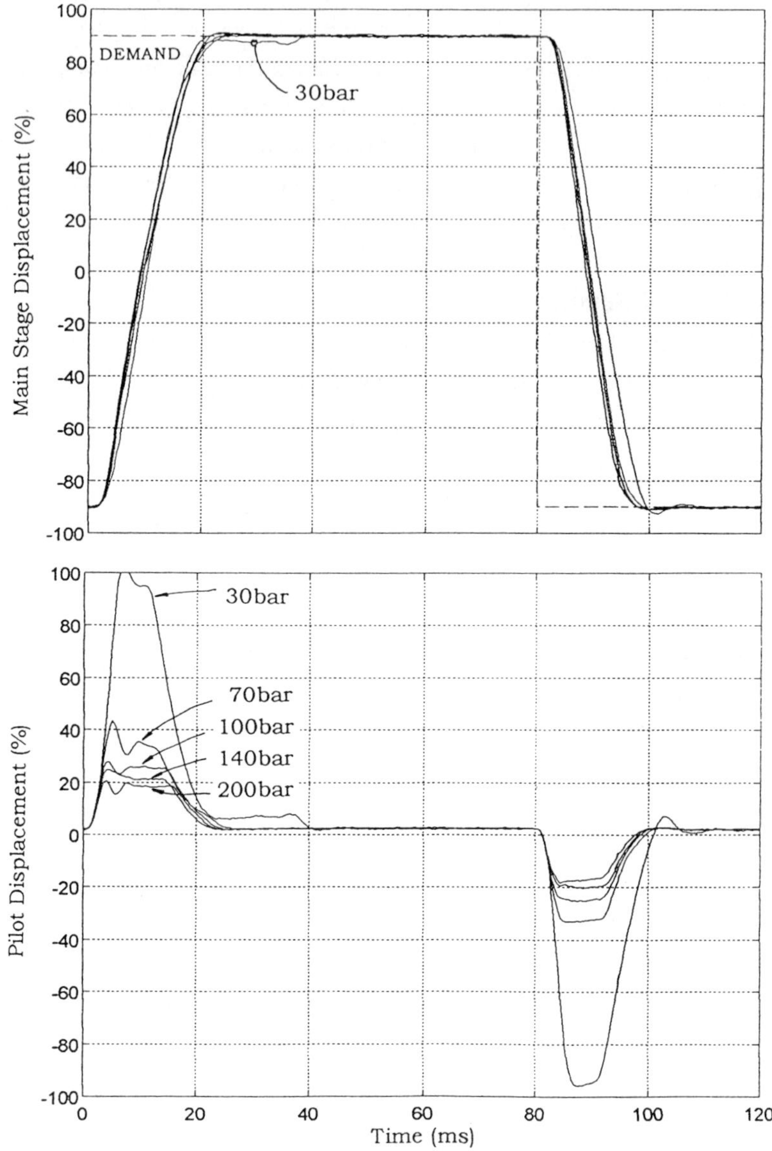


Fig 6. Step Response - Adaptive Control

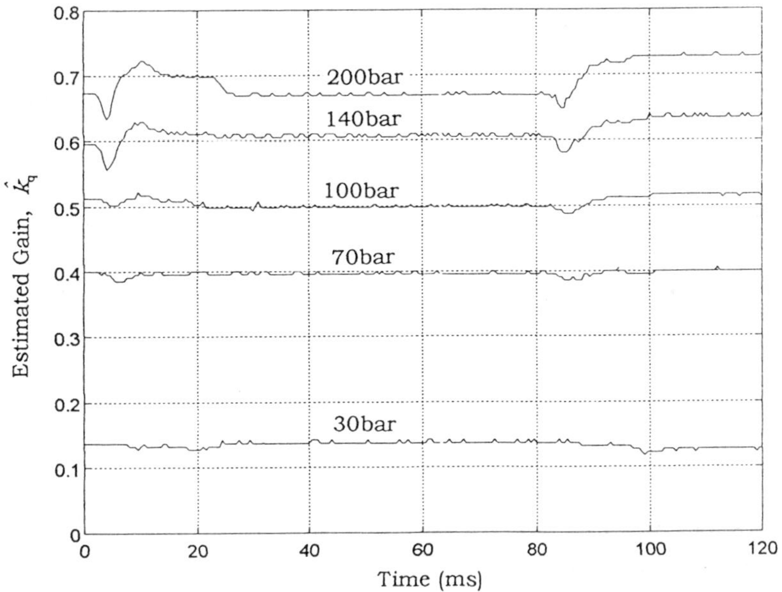


Fig 7. Parameter Estimate During Step Response

A similar step response test was conducted using an equivalent commercially available two-stage valve. These results are presented in Fig 8, and show that the main stage response changes radically with pressure. The graph of pilot displacement vs. time shows that the pilot is opened to the same position for all pressures. Note that the flow rating of the pilot valve is lower than that used with the adaptive controller.

## 5 Conclusions

An adaptive controller for a two-stage proportional valve has been demonstrated. The controller allows the pilot supply pressure to be varied from 30 to 200 bar without significantly changing the valve response.

A theoretical analysis of the valve dynamics indicates that the effects of oil compressibility can be ignored. The valve can thus be accurately modelled as a simple integrator. A major benefit of this simplification is that only a single parameter estimate is required in order to isolate the effects of changes in the supply pressure. This in turn implies that the supply pressure can be determined from any non-zero pilot spool displacement, and thus "persistency of excitation" within the RLS estimator ceases to be a problem.

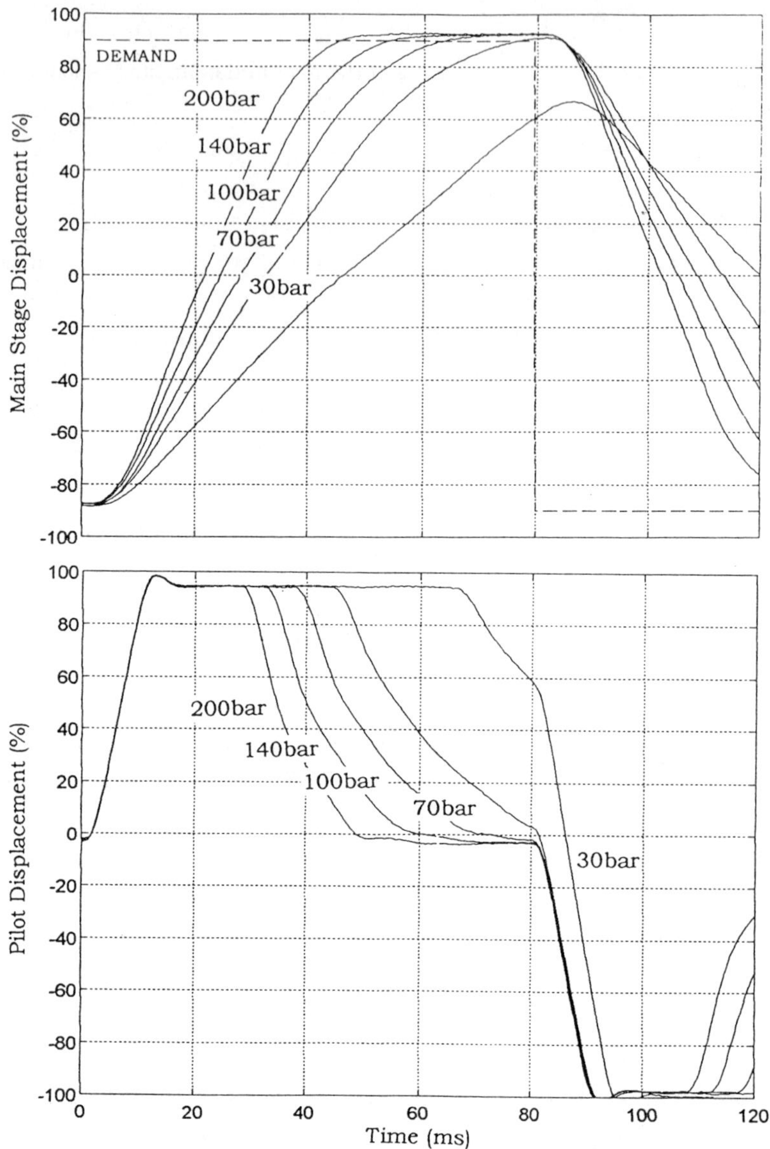


Fig 8. Step Response - Conventional Valve

The experimental results confirm that it is possible to accurately estimate the flow gain of the pilot valve from measurements of the pilot and main stage spool positions. The use of a nonlinear proportional gain in the main stage control loop has enabled consistent response to be obtained over the full pressure range using a single parameter estimate. It is possible however that response at lower pressures could be improved by incorporating a phase advance compensator in the main stage control loop.

The experimental results shown in Fig 6. indicate that the velocity gain estimate may have drifted whilst the main spool was stationary. This could have been caused by friction stopping the main spool before the pilot spool displacement had fallen below the threshold at which the estimator was switched off. Drift within the velocity gain estimate could be detected by examining the size of the covariance *P*. This parameter could also be used to prevent drift by switching off the estimator when *P* exceeds a predetermined value. Additional jacketing techniques that could be implemented include disabling the estimator when the main stage velocity reduces below a predetermined value.

## References


[1] **Figueredo K R A**. *Application of microprocessor based model reference adaptive control to servomechanisms*. PhD Thesis, University of Bath, 1988

[2] **Plummer A R**. *Digital control techniques for electro-hydraulic servosystems*. PhD Thesis, University of Bath, 1991

[3] **Gamble J B**. *Sliding mode control of a proportional solenoid valve*. PhD Thesis, University of Bath, 1992

[4] **Clarke D W**. "Introduction to self-tuning controllers", in *Self tuning* and adaptive control, ed Harris C J and Billings S A, Peter Peregrenus, London, 1985

[5] **M^c^Cloy D and Martin H R**. *Control of Fluid Power* 2nd Ed, Ellis Horwood, Chichester, 1980


## Acknowledgements


I would like to thank my colleagues at Vickers Systems for their help and assistance during the course of this study. In particular I would like to thank Dr B Riedle for the conversation which led directly to the concept of an adaptive two stage valve.

**WRITTEN DISCUSSION**
**Adaptive control of a pilot operated solenoid valve**
**JB Gamble (Vickers Systems Division, Trinova Ltd, UK)**

**Question:** **W Backé**
**IHP, RWTH Aachen, Germany**

When you drive a closed-loop position control having different overall gains with an input ramp, you get nearly the same curves as in your Figure 7. Is it possible to actuate your valve by a ramp instead of a step input and get the same result as with adaptive control?

**Answer:**

As I see it, there are two problems with using ramps. Firstly, in order to keep the valve stable at the maximum expected supply pressure, the gain would be so low that the valve would be very slow at low supply pressures. This would give rise to large following errors at low pressure. Although the errors would reduce with increasing pressure, the overall effect would be something akin to a pressure dependent delay. Secondly, in order for the valve to follow the ramp, the ramp rate would have to be less than the maximum speed of the valve. Thus a ramp driven valve would always be slower than an adaptive valve. One of the aims of this study was to try to increase the speed of the valve.

**Question:** **W Rampen**
**University of Edinburgh, UK**

1. Can analogue multipliers be used in place of the digital controller for setting the gain?

2. Given that your valve works well over a broad range of supply pressures and that valves are primarily dissipative devices what promise is there in using variable pressure compensation such that overall system power requirements and noise are reduced to minimal levels?

**Answer:**

1. Yes, I suppose that analogue multipliers could be used for setting the gain, but since some form of microprocessor is needed to estimate the required gain, it seems logical to also use this to implement the gain as well. I do not know if anyone has implemented an entirely analogue parameter estimator.

2. I see great promise for this type of valve in load sensing applications. Currently, any load sensing system employing pilot operated valves must maintain a high pressure pilot supply circuit. With the adaptive valve controller this is no longer required, and the valve can be run directly from the main supply circuit, even when the pressure is varied in accordance with the requirements of the load. This would obviate the need for a separate high pressure pilot supply pump, thereby reducing noise, power losses, and installed cost.

**Question:** **P Krus**
**University of Linköping, Sweden**

What is the reason that you have parameter drift when the valve is in neutral position, although you switch the parameter updates off in that case?

**Answer:**

Parameter drift occurs in the null position for two reasons. Firstly, temperature drift in the pilot spool position transducer means that the pilot valve may be in the centre null position when the transducer is indicating otherwise. Secondly, nonlinear effects due to slightly different lap conditions on each of the metering edges in the pilot valve spool mean that port A may close at a different position to port B.

Both these problems can be cured by detecting parameter drift directly rather than trying to predict when it will occur. Probably the most promising method is to detect an increase in the covariance gain the estimator, and switching off the parameter updates when this reaches a predetermined value. Thus the gain estimate will not be updated unless the input signals are sufficiently exciting to reduce the covariance gain below the limiting value.

**Question:** **A Breugelmans**
**British Steel Technical Teeside Laboratories, UK**

Is the intention to run the valve from a single supply; has it been tested in this way?
How does drift in the pilot position transducer effect the adaption? How is it taken into account?
Have you considered measuring the pressure and use this signal to schedule gain?

**Answer:**

Yes, the intention is that the valve can be run from a single supply, though this has yet to be tested. The tests described in the paper were conducted without flow through the main spool.

Drift of the sensor does not affect the adaptation. The readings of pilot (and main stage) position are "backwards differenced", in order to avoid offset errors. In essence the estimator works on the relative rather than absolute spool positions.

The use of a pressure transducer would not be a cost effective solution. The cost of the additional transducer would increase the cost of the valve significantly, whereas the solution adopted here can be achieved at little or no additional cost since both position transducers are already fitted to the valve as standard.

# 12. Application of Fuzzy Reasoning to Stabilization of an Adaptive Control System with Unmodelled Dynamics

K Sanada *and* A Kitigawa

## Abstract

This paper demonstrates that an adaptive control system with unmodelled dynamics can be stabilized by limiting an adjustable gain. Due to unmodelled dynamics, a feedback loop through an adjustable gain can be unstable. If the adaptive gain is limited, the feedback loop may be stabilized. If unmodelled dynamics are excited, output error contains high frequency content. Checking the high frequency content, the limitation is invoked to suppress unmodelled dynamics. The upper and lower limit may not be determined precisely, because of unknown plant parameters. Here Fuzzy reasoning is successfully applied to give automatic adjustment of the limits. The adjustment is done by Fuzzy reasoning on high-pass filtered output error. It is demonstrated that an adaptive control system with unmodelled dynamics is stabilized by adjustable limitation through simulation of an electro-hydraulic servo motor system.

## 1 Introduction

Adaptive control theory has been established on ideal assumptions. As K. J. Åström pointed out[1], the most restrictive assumption is that a mathematical model is at least of the same order as the real plant. This implies that the mathematical model must be at least as complex as the true system which may be nonlinear. Control systems are often designed on reduced order models. High frequency dynamics are neglected. When designing an adaptive control system with unmodelled dynamics, it is very important that control laws can cope with unmodelled dynamics. Several studies have been made on the unmodelled dynamics, in which adaptive control laws were modified. R. Ortega and T. Yu[2] demonstrated that internal signals of an adaptive control system become relatively bounded with a normalized signal. A $\sigma$-modification approach proposed by P. A. Ioannou and P. V. Kokotovic[3][4] guarantees local stability when relative degree of a reduced order model is one. G. Kreisselmeier and K. S. Narendra[5][6] showed that a dead band based on a norm of adjustable parameters is effective when unmodelled dynamics are relatively small. B. D. O. Anderson[7] pointed out that local stability is guaranteed if signals are persistently excited. R. L. Koust and B. Friedlander[8] derived conditions for global stability of a system which has relative degree of one with stable unmodelled dynamics. A technique of cutting out high frequency components due to unmodelled dynamics by low pass filters was proposed[9]. A technique in which a normalized signal and a projection algorithm of estimated parameters were combined was proposed[10]. P. A. Ioannou and K. S. Tsakalis[11] proposed a robust adaptive control law using normalized signals which guarantees global stability. Although adaptive control theories for unmodelled dynamics have been investigated by many researchers, a practical robust technique for a real system has not been well developed.

K. A. Edge and F. Hu[12] demonstrated an application of a model reference adaptive control scheme to the control of delivered pressure of a swash plate piston pump. Yamahashi et. al[13] applied $\delta$-transform to an electro-hydraulic servo system in order to improve frequency responses. The transient behavior of a continuous-time adaptive control system was investigated by the authors[14] and a technique of tuning an adaptive controller was proposed. However, only a few studies have been done on robust adaptive control of an electro-hydraulic servo system[15]. The purpose of this paper is to investigate a robust adaptive control of an electro-hydraulic servo system which copes with instability due to unmodelled dynamics.

In this study, limitation of an adjustable gain is examined to stabilize an adaptive control system with unmodelled dynamics. The limitation is automatically adjusted by Fuzzy reasoning based on a high-pass filtered output error signal. Adjustable limitation on Fuzzy reasoning is examined through simulation work. An electro-hydraulic servo motor system is concerned. Servo valve dynamics and oil compressibility are regarded as unmodelled dynamics in this paper. In a section 3, a nonlinear model of the system is described and a linearized model is presented. Performance of an adaptive control for the linearized model under ideal conditions is presented. In a section 4, instability due to unmodelled dynamics is demonstrated. A technique of limiting an adjustable gain is introduced. In a section 5, Fuzzy reasoning is applied to adjust the limitation and it is presented that the adjustable limitation improves adaptive control performance. Finally, results are summarized.

## 2 Nomenclature

| | | |
|---|---|---|
| $a$ | parameter of a linearized model | 1/s |
| $b$ | parameter of a linearized model | $1/(\mathrm{V{\cdot}s^2})$ |
| $C_d$ | coefficient of viscous friction | N·m/(rad/s) |
| $c$ | a coefficient of discharge | - |
| $D$ | motor displacement | $\mathrm{m^3/rad}$ |
| $d$ | width of a port | m |
| $i$ | drive current | A |
| $J$ | inertia | $\mathrm{kg{\cdot}m^2}$ |
| $K_a$ | servo amplifier gain | A/V |
| $K_f$ | feedback gain | V/rad |
| $K_v$ | current-spool displacement gain | m/A |
| $K_1$ | adjustable gain | V/(rad/s) |
| $K_2$ | adjustable gain | - |
| $K_{10}, K_{20}$ | initial values of $K_1$ and $K_2$ | |

| | | |
|---|---|---|
| $p_l$ | load pressure | Pa |
| $p_0$ | supply pressure | Pa |
| $\Delta p$ | pressure drop | Pa |
| $q$ | flow rate | $m^3/s$ |
| $r$ | demand signal | V |
| $s$ | Laplace variable | rad/s |
| $t$ | time | s |
| $u$ | control input | V |
| $V$ | oil volume | $m^3$ |
| $x$ | spool displacement | m |
| $\beta$ | oil compressibility | $Pa^{-1}$ |
| $\gamma_1$ | adaptive gain | $V \cdot s^2/rad^3$ |
| $\gamma_2$ | adaptive gain | $1/(V \cdot rad)$ |
| $\zeta$ | damping coefficient | - |
| $\varepsilon$ | output error $= \omega - \omega_M$ | rad/s |
| $\theta$ | load angle | rad |
| $\rho$ | oil density | $kg/m^3$ |
| $\omega_n$ | natural angular frequency | rad/s |
| $\omega$ | rotating speed | rad/s |

superscript

| | |
|---|---|
| * | operating point |

subscript

| | |
|---|---|
| $M$ | reference model |
| $s$ | specified values |

# 3 Mathematical models

This paper is concerned with an electro-hydraulic servo motor system which drives an inertial load shown in figure 1. In this section, a nonlinear model and a linearized model are discussed. The considered system has a servo valve, a servo motor, an inertial load, a rotary encoder, a servo amplifier and an adaptive controller. The rotary encoder is used to measure rotating speed of an inertial load. Specification of the system is listed in table 1.

## 3.1 Nonlinear model

A servo amplifier had a constant gain $K_a$, but output current was saturated at 60 mA. Flow rate through a servo valve is determined by drive current and load pressure :

$$q = cdK_v i \sqrt{\frac{p_0 - sgn(x)p_l}{\rho}} \tag{1}$$

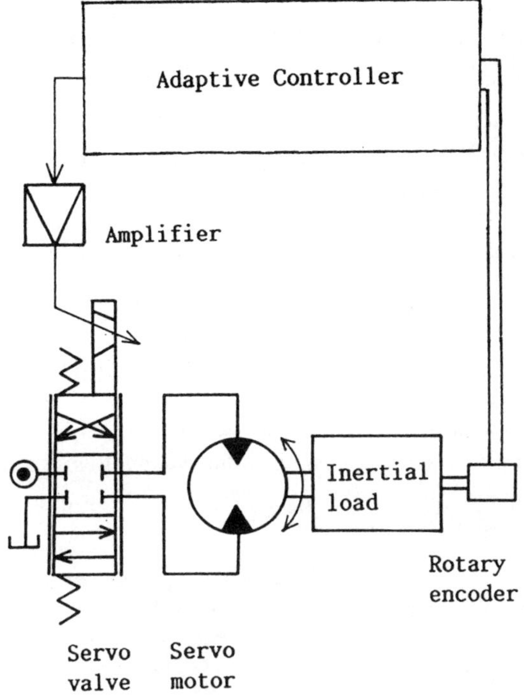


Figure 1 A schematic diagram of an electro-hydraulic servo motor system

Table 1 Specification of an electro-hydraulic servo motor system

| | | | |
|---|---|---|---|
| servo valve | flow rate | $q_S = 3.2 \times 10^{-4}$ | $m^3/s$ |
| | pressure drop | $\Delta p_S = 6.8$ | MPa |
| | drive current | $i_S = 15$ | mA |
| servo motor | displacement | $D = 1.6 \times 10^{-6}$ | $m^3/rad$ |
| load | inertia | $J = 6.2 \times 10^{-3}$ | $kgm^2$ |
| supply pressure | | $p_0 = 13.7$ | MPa |
| servo amplifier | gain | $K_a = 9.1 \times 10^{-3}$ | A/V |

where $c$ is a coefficient of discharge and $d$ is width of a port. Using specification of a servo value, $cdK_v$ was calculated from the following relationship and it was assumed to be constant :

$$q_s = cdK_v i_s \sqrt{\frac{\Delta p_s}{\rho}} \tag{2}$$

where $q_s$, $\Delta p_s$ and $i_s$ are specified flow rate, specified pressure drop and specified drive current respectively, which are constant.

In equation (1), $x$ is displacement of a servo value spool. A servo valve has dynamics between drive current $i$ and displacement $x$. Here the dynamics were approximated as a second order lag element:

$$\frac{X(s)}{I(s)} = \frac{K_v \omega_n^2}{s^2 + 2\zeta\omega_n s + \omega_n^2} \tag{3}$$

Parameters $\omega_n$ and $\zeta$ were estimated from frequency characteristics presented in a data sheet.

When a servo valve is opened, a motor rotates according to oil flow rate $q$ from a servo valve. Considering oil compressibility, a continuity condition is written as follows:

$$D\omega + \frac{\beta V}{4}\frac{dp_l}{dt} = q \tag{4}$$

An inertial load is driven by torque generated by load pressure $p_l$. The inertial load was supported by two ball-bearings. A shaft of the load was connected with a motor shaft by a coupling. The motor was an axial piston motor. Considering inertia and viscous friction, an equation of motion is written as follows :

$$Dp_l = J\frac{d\omega}{dt} + C_d\omega \tag{5}$$

where $C_d$ is a coefficient of viscous friction which was estimated through experiment. First, rotating speed $\omega$ was measured under a steady condition. Figure 2 shows the results for various supply pressures. As seen from an equation (1), flow rate $q$ depends on drive current $i$, supply pressure $p_0$ and load pressure $p_l$. Drive current $i$ and supply pressure $p_0$ were directly measured. Secondary, calculating a ratio of flow rate $q$ to specified flow rate $q_s$, load pressure $p_l$ was derived:

$$p_l = p_0 - (\frac{qi_s}{q_s i})^2 \Delta p_s \tag{6}$$

Assuming that the load pressure was generated by viscous friction, a coefficient of viscous friction $C_d$ was calculated:

$$C_d = \frac{Dp_l}{\omega} \tag{7}$$

The results are plotted in figure 3. Negative sign corresponds to opposite direction of rotation. The coefficient $C_d$ depends on rotating speed $\omega$. When rotating speed is large, $C_d$ is approximately equal to 0.05 N·m/(rad/s). Equation (1), (3), (4) and (5) are basic equations of a nonlinear model.

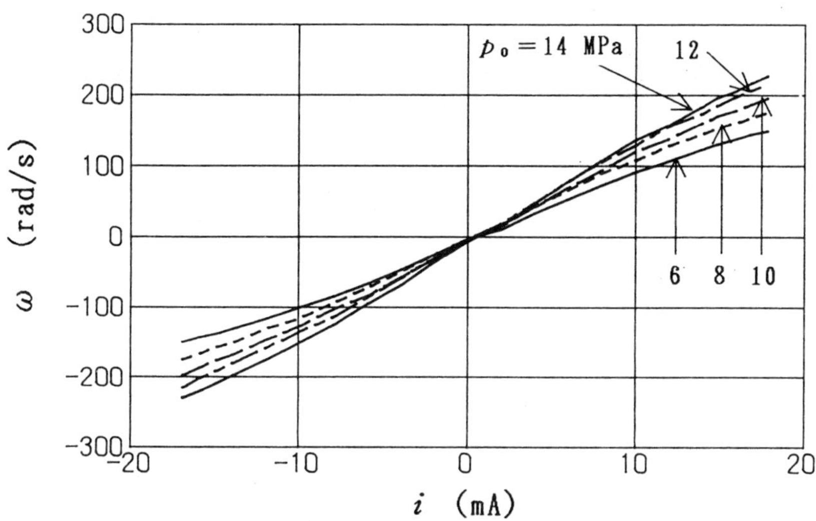


Figure 2 Relationship between drive current $i$ and rotating speed $\omega$

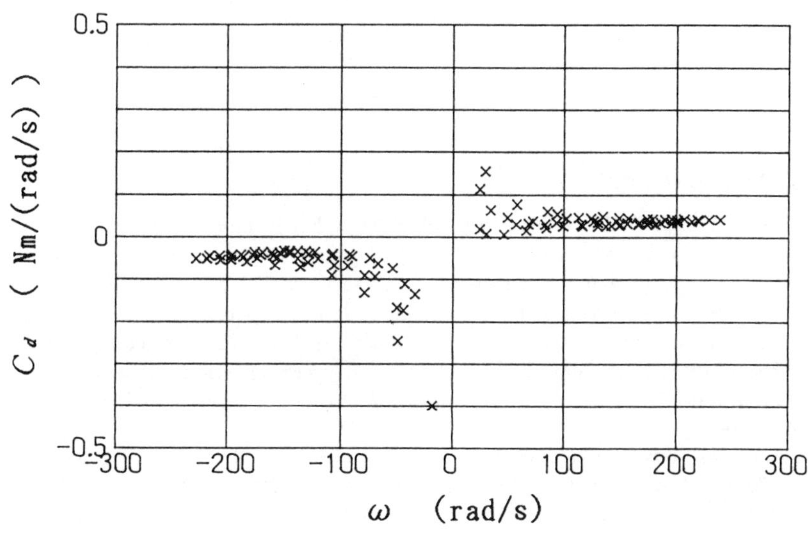


Figure 3 Coefficient of viscous friction $C_d$

### 3.2 Reduced order model

When designing adaptive control system, a plant model is often approximated as a reduced order model. Linearization of the equation (1) yields:

$$\Delta q = k_1 \Delta i - k_2 \Delta p_l \tag{8}$$

$$k_1 = \frac{q^*}{i^*} \tag{9}$$

$$k_2 = \frac{q^*}{2(p_0 - p_l^*)} \tag{10}$$

where * denotes a variable at an operating point. Neglecting servo valve dynamics and oil compressibility, Laplace transforming (4), (5) and (8) and rearranging :

$$\frac{\Omega_{(s)}}{U_{(s)}} = \frac{b}{s+a} \tag{11}$$

where

$$a = \frac{D^2 + k_2 C_d}{J k_2} \tag{12}$$

$$b = \frac{K_a k_1 D}{J k_2} \tag{13}$$

Equation (11) is a linearized reduced order model. Parameters $a$ and $b$ depend on an operating point where a nonlinear model is linearized. For example, assuming that $u_s = 0.2$ V is an operating point, from experimental results shown in figure 1, the following parameters were calculated: $k_1 = 0.022$ (m$^3$/s)/A, $k_2 = 4.5 \times 10^{-12}$ (m$^3$/s)/Pa, $a = 67$ 1/s, and $b = 11700$ 1/(Vs$^2$).

### 3.3 Adaptive control under ideal conditions

If a considered plant is perfectly modelled by a linear model (11), an ideal adaptive control system can be constructed. In this paper, model reference adaptive control system[14] shown in figure 4 is concerned. A servo motor drives an inertial load. The final purpose of the system is to control the load angle position. The load angle is fed back through a feedback gain $K_f$. To a block which is a first order lag element (11), adaptive control theory is applied. A reference model is specified by parameters $a_M$ and $b_M$. Gains $K_1$ and $K_2$ are adaptively adjusted through integrators which have adaptive gains $\gamma_1$ and $\gamma_2$.

Simulation results are shown in figure 5 in which the following parameters were used: $a_M = 70$ 1/s, $b_M = 12000$ 1/(Vs$^2$), $a = 67$ 1/s, $b = 8000$ 1/(Vs$^2$), $\gamma_1 = 1$ V·s$^2$/rad$^3$, $\gamma_2 = 10^3$ 1/(V·rad), $K_f = 0.1$ V/rad, $K_{10} = 0$ V/(rad/s), $K_{20} = 1$. A step change of a demand signal caused transient responses. Rotating speed $\omega$ of the load followed reference model output $\omega_M$.

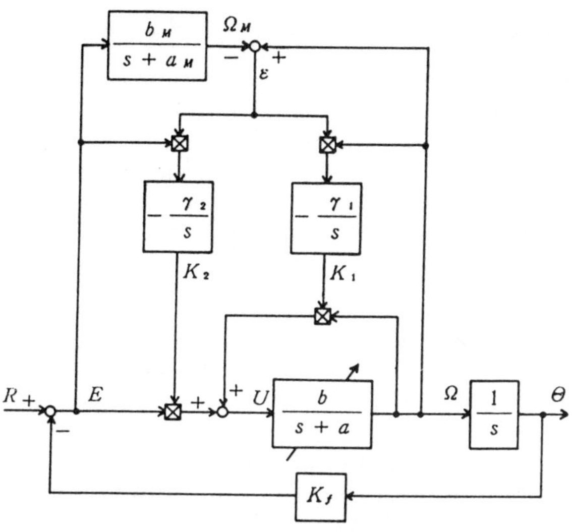


Figure 4 A block diagram of an adaptive control system under ideal conditions

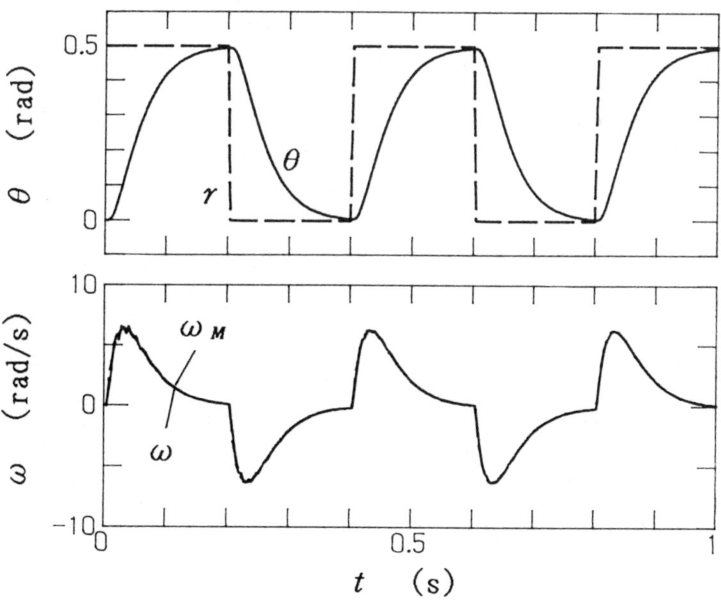


Figure 5 Simulation results of an adaptive control system under ideal conditions

# 4 Instability due to unmodelled dynamics

In the previous section, for a linearized model (11) performance of the adaptive control system under ideal conditions was demonstrated. However, the controlled plant has more complex dynamics, that is, nonlinear flow characteristics, servo valve dynamics and oil compressibility, as described in 3.1. Here as a simulation model of a servo system, the nonlinear model is examined. A linearized model of a plant shown in figure 4 which is a reduced model is replaced by the nonlinear model. The following parameters were used: $\zeta = 0.7$, $\omega_n = 628$ rad/s, $\beta = 7 \times 10^{-10}$ Pa$^{-1}$, $V = 4.3 \times 10^{-6}$ m$^3$ and $\gamma_2 = 5$ 1/(V·rad). Other parameters were the same as those in the previous section.

An example of simulation results is presented in figure 6. Rotating speed $\omega$ showed unstable responses. An adjustable gain $K_1$ changed rapidly and had periods of being positive. Multiplied by the gain $K_1$, the rotating speed is fed back to make a feedback loop as shown in figure 4. The positive feedback gain caused instability.

To overcome the instability due to positive feedback, a technique of limiting the feedback gain $K_1$ was examined. Considering a linearized model (11) with servo valve dynamics(3), a transfer function between a control input and rotating speed is :

$$\frac{\Omega(s)}{U(s)} = \frac{{\omega_n}^2}{s^2 + 2\zeta\omega_n s + {\omega_n}^2} \frac{b}{s+a} \tag{14}$$

Using Routh-Hurwitz stability criterion, a condition for stability of a feedback loop is obtained:

$$\frac{a}{b} > K_1 > -\frac{2\zeta({\omega_n}^2 + 2\zeta\omega_n a + a^2)}{\omega_n b} \tag{15}$$

In order to make the system stable, the feedback gain $K_1$ must be limited to satisfy equation (15). When all parameters are known, upper limit and lower limit can be precisely determined. An example of simulation results for such an ideal case is presented in figure 7. The upper limit was 0.0083 and the lower limit was –0.127. By the limitation of $K_1$, the system was stabilized.

# 5 Adjustable limitation on Fuzzy reasoning

In the previous section, instability due to unmodelled dynamics and stabilization by limiting a feedback gain $K_1$ are demonstrated. In the simulation, oil compressibility was not considered. Parameters of the controlled plant may change depending upon an operating point. Hence, the limitation should be more restrictive and adjustable according to instability. The more unstable the system is, the more the gain $K_1$ must be restricted. When the system behaves calmly, the restriction can be relaxed. The degree of instability can not be determined a priori. In this paper, as a measure of instability, high frequency contents $\varepsilon_h$ of output error $\varepsilon = \omega - \omega_M$ is used. Absolute value of the high-pass filtered output error was integrated every 0.05 second. Resultant signal of the integration was referred as the high frequency contents $\varepsilon_h$. When the high frequency contents are detected, it can be said that the system is going to be unstable and a gain $K_1$ should be restricted. When the high frequency contents become small and when output error $\varepsilon$ still remains, the restriction can be relaxed. In order to do the restriction

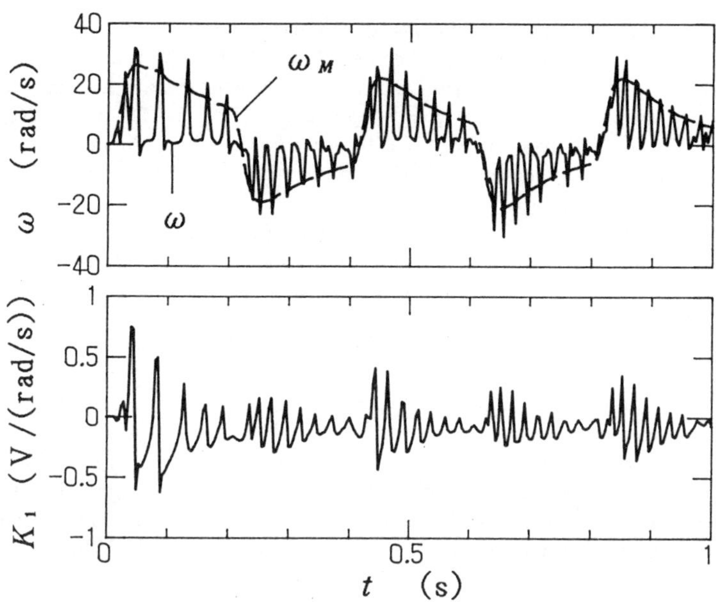


Figure 6 Simulation results of adaptive control of a nonlinear model

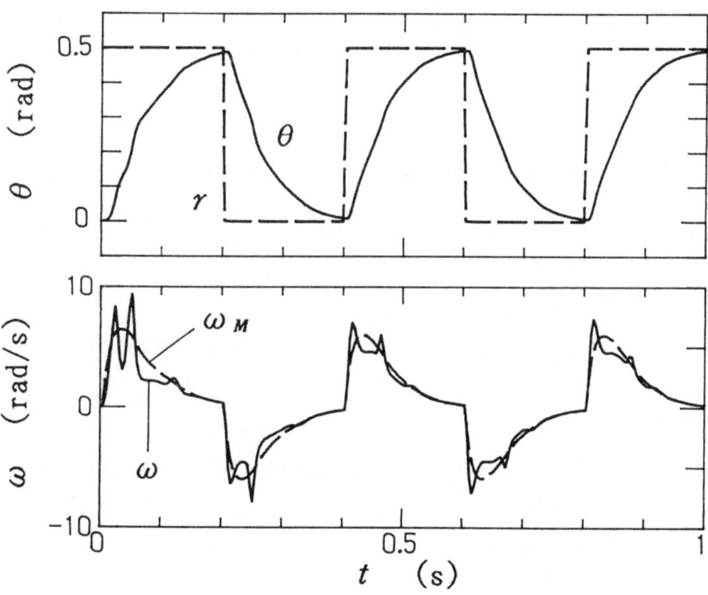


Figure 7 Stabilization by limitation of a gain $K_1$
(a linearized model with servo valve dynamics)

and the relaxation, Fuzzy reasoning was applied. In table 2, a rule of Fuzzy reasoning used for simulation is presented. High frequency contents $\varepsilon_h$ and absolute value of output error $\varepsilon$ were used as input information for the rule. For each signal, three labels PS, PM, and PL were assigned. There were four output labels. DM and DL mean restriction of limitation. EL and EM mean relaxation. HO means no change. Membership functions are shown in figure 8. Upper limit and lower limit were multiplied by the result of Fuzzy reasoning every 0.05 second.

Simulation results using the Fuzzy reasoning is presented in figure 9. A plant model was a linearized model combined with servo valve dynamics which was the same as figure 7. A gain $K_1$ and rotating speed $\omega$ showed rapid changes in the early stage of responses. High frequency content contained in the rapid change of rotating speed were detected. Upper limit and lower limit were restricted as a result of Fuzzy reasoning. $K_1$ was limited mainly by the lower limit. The rotating speed $\omega$ followed the reference model output $\omega_M$. The system was made stable and the adjustment of limits based on Fuzzy reasoning improved control performance.

A nonlinear model considering servo valve dynamics, flow characteristics, and oil compressibility was examined. Simulation results are shown in figure 10. Compared with figure 6, the system was well stabilized. Output error $\varepsilon = \omega - \omega_M$ became larger than that of figure 9 in which a linearized model was examined. When controlling a nonlinear plant, fast adaption is essential for good performance. In this simulation, to cope with instability due to unmodelled dynamics, adaptive gains were reduced ($\gamma_1 = 1$, $\gamma_2 = 5$), and it caused slow adaptation.

Through these simulation works, it was demonstrated that the adjustable limitation of $K_1$ based on Fuzzy reasoning may cope with instability due to unmodelled dynamics.

## 6 Conclusions

A robust technique using adjustable limitation is effective to stabilize an adaptive control system of an electro-hydraulic servo motor. The technique must be validated with experimental works.

An electro-hydraulic system is essentially nonlinear system. When designing an adaptive control system based on a linearized reduced model, it may cause instability. Adjustable limitation of a feedback gain is effective to cope with the instability due to unmodelled dynamics. Fuzzy reasoning can be applied to adjust the limitation.

The authors thank Mr. A. Oota who contributed his simulation skills.

## References

[1] **Åstoröm K.J.** "*Adaptive Feedback Control*", Proc. of the IEEE, Vol.75, No.2, February 1987, pp.185-217

[2] **Ortega R.** and **YU T.** "*Theoretical Results of Robustness of Direct Adaptive Controllers*", Proc. of IFAC World Congress, München(1987)

[3] **Ioannou P.A.** and **Kokotovic P.V.** "*Robust Redesign of Adaptive Control*", IEEE Trans. AC-29-3, pp.202-211, 1984

Table 2 A rule of Fuzzy reasoning

| | | $\varepsilon_h$ | | |
|---|---|---|---|---|
| | | PS | PM | PL |
| $\lvert\varepsilon\rvert$ | PS | HO | DM | DL |
| | PM | EM | HO | DM |
| | PL | EL | EM | HO |

PS positive small
PM positive medium
PL positive large
HO no change
DM restrict medium
DL restrict much
EM relax medium
EL relax much

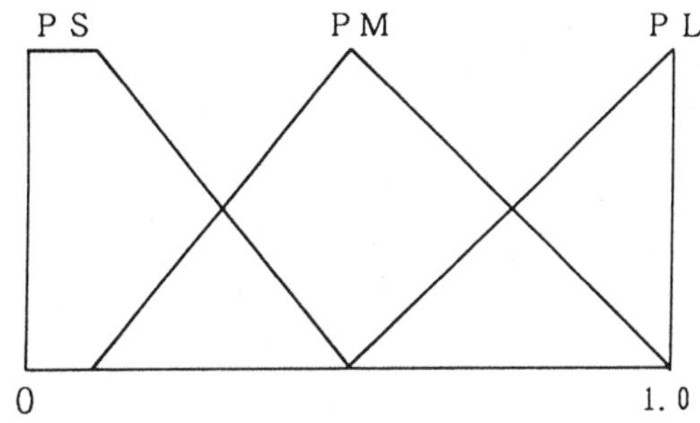


(a) membership functions for $\varepsilon_h$ and $\lvert\varepsilon\rvert$

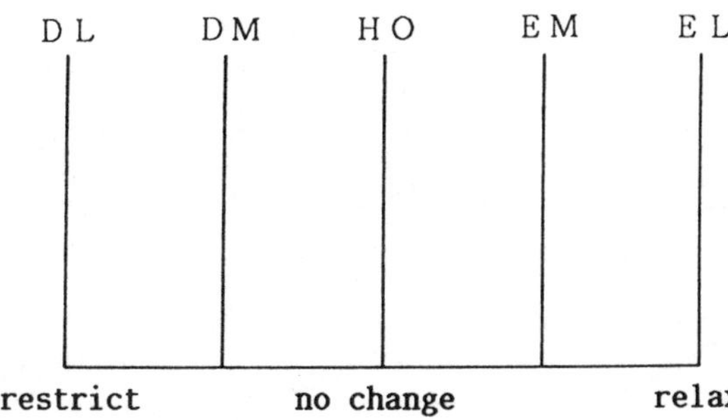


(b) membership functions for output of Fuzzy reasoning

Figure 8 Membership functions

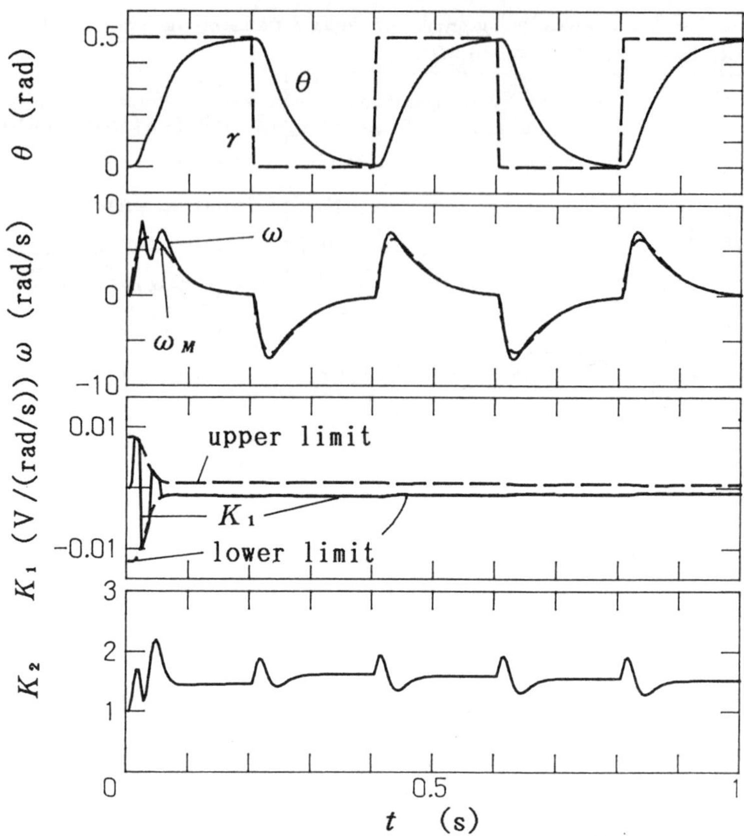


Figure 9 Improvement of performance by adjusting the limitation on Fuzzy reasoning (a linearized model with servo valve dynamics)

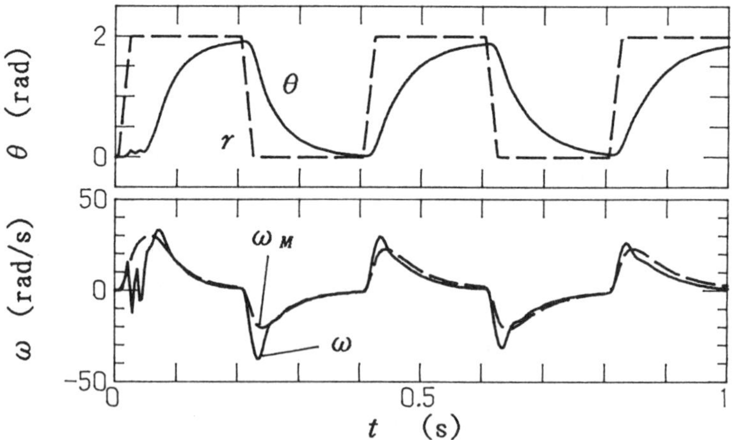


Figure 10 Adaptive control of a nonlinear model with adjustment by Fuzzy reasoning

[4] **Ioannou P.A.** "*Robust Adaptive Controller with Zero Residual Tracking Errors*", IEEE Trans AC-31-8, pp.773-776, 1986

[5] **Kreisselmeier G.** and **Narendra K.S.** " *Stable Model Reference Adaptive Control in the Presence of Bounded Disturbances*", IEEE Trans. AC-27-6, pp.1169-1175, 1982

[6] **Narendra K.S.** *Adaptive and Learning System-Theory and Applications*, Plenum Press,1986

[7] **Anderson B.D.O.** " *Exponential Convergence and Persistent Excitation*", Proc. 21th IEEE CDC, pp.12-17, 1982

[8] **Kosut R.L.** and **Friedlander B.** " *Robust Adaptive Control,Conditions for Global Stability*", IEEE Trans. AC-30-7, pp.610-624, 1985

[9] **Elliott H.**,**Das M.** and **Ruiz G.** " *A Frequency Domain Analysis of Direct Adaptive Pole Placement Algorithms in the Presence of Unmodelled Dynamics*", IFAC Workshop, June, San Francisco, 1983

[10] **Praly L.** " *Robustness of Model Reference Adaptive Control*", Proc. of 3rd Yale Workshop on Adaptive Control System Theory, June, New Haven, 1983

[11] **Ioannou P.A.** and **Tsakalis K.S.** " *A Robust Direct Adaptive Controller*", IEEE Trans. AC-31-11, pp.1033-1043, 1986

[12] **Edge K.A.** and **Hu F.** " *Constant Pressure Control of an Axial Piston Pump - a Simulation Study of an Adaptive Control Scheme*", Fluid Power Systems Modelling and Control Fourth Bath International Fluid Power Workshop, pp.77-94, 1991

[13] **Yamahishi K.**,**Takahashi K.**,**Tamoto Y.** and **Ikeo S.** " *The Improvement of the Frequency Response of an Electrohydraulic Servo System Using Model Reference Adaptive Control Techniques*", Fluid Power Systems Modeling and Control,Fourth Bath International Fluid Power Workshop, pp.309-320, 1991

[14] **Sanada K.** and **Kitagawa A.** " *A Method of Tuning a Continuous-Time Adaptive Controller for an Electrohydraulic Servomechanism*", Fluid Power Systems Modelling and Control Fourth Bath International Fluid Power Workshop, pp.321-335, 1991

[15] **Yun J.S.** and **Cho H.S.** " *Adaptive Model Following Control of Electrohydraulic Velocity Control Systems Subjected to Unknown Disturbances*", IEE Proceedings, Vol.135, Pt.D, No.2, March 1988, pp.149-156

**WRITTEN DISCUSSION**

**Application of fuzzy reasoning to stabilization of adaptive control system with unmodelled dynamics**

**K Sanada & A Kitigawa (Tokyo Institute of Technology, Japan)**

**Question:** **PS Keogh**
**Fluid Power Centre, Bath UK**

An alternative approach towards achieving stable closed-loop control in the presence of unmodelled dynamics is the use of H-infinity controller design. Is this a possibility for future work?

**Answer:**

We agree that the use of H-infinity controller design is a possibility for future work. Model reference adaptive control systems can achieve exact model matching of a plant through the use of a reference model. However the stability of the system in the presence of unmodelled dynamics must be guaranteed utilizing appropriate techniques. An H-infinity controller does not have an adaptation function and hence may guarantee robust stability and robust performance against unmodelled dynamics. The modelling of a plant or evaluation of unmodelled dynamics seems to be a key point to design a good H-infinity controller.

**Question:** **RE Koski**
**Sun Hydraulics Corpn., USA**

Your model showed an inertial load driven by a servo-motor and controlled by a servo valve. Then you demonstrated improved stability for an unmodelled load (presumably the same inertial load). How did your system "adapt" to loads with varying inertia? Was this tested?

**Answer:**

We have tested loads with varying inertia. The system can adapt to some variations. However, when the load shows a marked variation in inertia (say 10 times as heavy as the original inertia), our system cannot adapt to the change. The effect of the variation in inertia may not be accounted for by the stabilization mechanism. Re-design of a rule of Fuzzy reasoning offers the potential to add an adaptation function to the system for such marked variations in inertia.

# 13. A Novel Adaptive Control Scheme for Hydraulic Actuator Motion Systems

T O Andersen, F Conrad, P E Hansen *and* J J Zhou

**ABSTRACT**

An adaptive controller for hydraulic actuators is developed by using hyperstability theory and evaluated by simulation and experimental tests carried out on a fast hydraulic test robot.
The control law consists of an adaptive feedforward and an adaptive auxiliary signal. Only position feedback is required. For the adaptive controller the load is considered as unknown, and the controller requires a minimum knowledge of the actuator dynamics. The results obtained are compared with PD and model-based schemes.

## 1. INTRODUCTION

In many hydraulic applications operated by a servo system external load disturbances and variation in system parameters usually influence the system. To achieve satisfactory process control the controller has to be designed based upon an exact process dynamic model which has to include all the physical disturbances and uncertainties. Lack of complete knowledge of the dynamic model results in degradation of the performance and even instability can occur.
When an exact dynamic model is known model-based controllers outperform most conventional controllers, particularly when the actuators are required to move fast and precisely. When a reasonably accurate process-model is not available, the performance of an adaptive scheme is better than for model-based controllers.
In this paper we combine the advances of the above mentioned controllers into a new adaptive control scheme called AMAC (Adaptive Model-based Actuator Control).
The hyperstability theory will be used to derive a new control law for a hydraulic servo system subjected to unknown disturbances. The control law consists of two parts, namely, a model-based feedforward part and a part that originates from a direct adaptation scheme.

## 2. THE FAST HYDRAULIC TEST ROBOT

The fast hydraulic test robot shown in Fig. 1 has been developed, designed and implemented as a test facility by the Control Engineering Institute, The Technical University of Denmark.

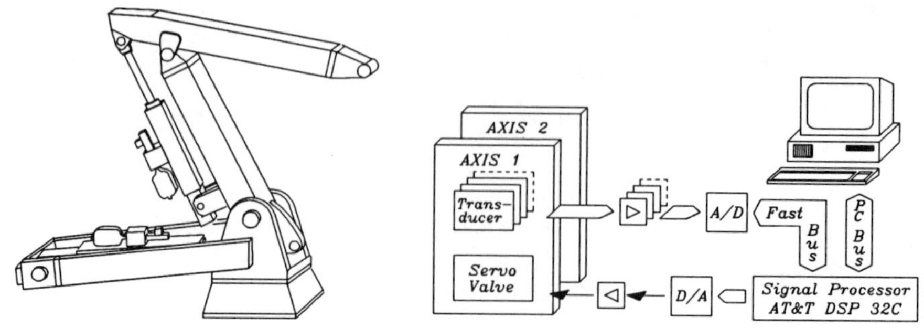


Fig. 1 *The hydraulic test robot facility*

Today the robot is implemented as a two-link rotary arm manipulator with two linear high performance electro-hydraulic servo actuators to drive and control the motion of the two robot links. The advantage of employing linear actuators is that the robot arms can be controlled directly without gears, which are necessary in robot systems applying rotational motors. The hydraulic cylinders are specially designed and manufactured in order to obtain high performance. The piston rods have hydrostatic bearings and low friction seals for minimum friction.
A specially designed membrane accumulator on the pressure port and the tank port help to obtain an approximate constant supply pressure and a low constant return pressure.
Two two-stage high frequency servo valves control the piston motions. The volumes of the cylinder chambers and the piston masses are minimized in order to obtain a high response speed.
Both actuators can achieve a static force of 20 kN and the robot has a load capacity of 50 kg at the tool center with a maximum velocity around 3.5 m/s. In this case the centrifugal and the Coriolis forces influence the dynamics and lead to dynamic coupling between the two links.

## 3. DYNAMIC MODEL

We shall restrict overselves to the class of actuator models which can be described by the following equation

$$\Gamma(t)\dot{x} + \eta(t) = u , \qquad \text{with } \Gamma(t) > 0 \tag{1}$$

where u is the process input, x is the process output and $\Gamma(t)$, $\eta(t)$ are load dependent time-functions to be adapted. Eq.(1) is general for many hydraulic drive systems. Examples of drive systems which belong to this class of actuator models are a servo valve combined either with a hydraulic cylinder or a hydraulic motor.

The experimental results presented in this paper are carried out with drive systems consisting of double rod equal area hydraulic cylinders controlled by servo valves. For such drive systems the relation between an input voltage u to the load flow rate Q, for a servo valve, is governed by the well known orifice law and is given by

$$Q = ku\sqrt{P_S - \mathrm{sign}(x_v)P_L} \tag{2}$$

where $P_S$ is the supply pressure, $P_L$ is the load pressure, $x_V$ is the spool position of the valve and k is a static valve constant. The linearized continuity equation of the cylinder operating around the mid-stroke position yields

$$Q = A\dot{x} + cP_L + \frac{V}{4\beta}\dot{P}_L \tag{3}$$

where A is the cylinder piston area, c is the total leakage coefficient, $\dot{x}$ is the velocity of the piston, V is the total volume of the valve and cylinder chamber and $\beta$ is bulk modulus of oil. Combining Eq.(2) and Eq.(3) yields

$$\frac{1}{k\sqrt{P_S - \mathrm{sign}(x_v)P_L}}(A\dot{x} + cP_L + \frac{V}{4\beta}\dot{P}_L) = u \tag{4}$$

This equation belongs to the actuator model described by Eq.(1) when $\Gamma$ and $\eta$ are chosen as

$$\Gamma(t) = \frac{A}{k\sqrt{P_S - \mathrm{sign}(x_v)P_L}} \quad , \quad \eta(t) = \frac{\Gamma(t)}{A}(cP_L + \frac{V}{4\beta}\dot{P}_L) \tag{5}$$

The load pressure is normally generated by viscous damping, coulomb friction, stiction, inertia and a load subjected to an external disturbance.

In the experimental study the hydraulic robot manipulator is considered as an external disturbance in order to realize a reaction force of a real process. In this system the main disturbance is caused by the manipulator dynamics. In the absence of friction and other disturbances, the dynamics of the two link rigid manipulator (with the load considered as part of the last link) can be derived from well known results of classical mechanics, thus

$$\tau = H(q)\ddot{q} + C(q,\dot{q}) + G(q) \tag{6}$$

where q and $\tau$ are $2\times1$ vectors of joint displacements and applied torques, respectively. H(q) is the $2\times2$ manipulator inertia matrix, $C(q,\dot{q})$ is the $2\times1$ vector of centripetal and Coriolis torques and G(q) is the $2\times1$ vector of gravitational torques.

## 4. CONTROL LAW

Motivated by the computed torque control scheme a model-based control law related to Eq.(1) is given by the following structure

$$u = \Gamma(t)[\dot{x}_r + k_p(x_r - x)] + \eta(t) \tag{7}$$

It is assumed that $\Gamma$ and $\eta$ can be accurately computed. $k_P$ is a constant feedback gain, $x_r$ is the reference piston position.
Combining Eq.(1) and Eq.(7) the resulting system error equation will be

$$\dot{e} + k_p e = 0 \tag{8}$$

where e is defined as: $e = x_r - x$. Eq.(8) states that the closed loop system becomes asymptotically stable by proper choice of $k_P$.
In Zhou (1989) it is shown that the compressibility and leakage effects of the fluid are negligible for the robot system. Under these assumptions the continuity equation for the servo valve and the cylinder chamber yields $Q = A\dot{x}$. If further the pressure ratio $P_L/P_S$ is minimized the control law Eq.(7) can be put in the form

$$u = \frac{A}{k\sqrt{P_S}}(\dot{x}_r + k_p e) \tag{9}$$

which is the CVI-controller proposed by [Conrad, *et al*, 1992]. When $\Gamma(t)$ and $\eta(t)$ are unknown or are subject to variations then estimates $\hat{\Gamma}(t)$ and $\hat{\eta}(t)$ should be used in Eq.(7). Then the closed loop error equation becomes

$$\dot{e} + k_p e = \Gamma^{-1}[(\Gamma - \hat{\Gamma})(\dot{x}_r + k_p e) + \eta - \hat{\eta}] \tag{10}$$

The control problem is now to find an adaptation law to adjust $\hat{\Gamma}(t)$ and $\hat{\eta}(t)$ such that

$$\lim_{t \to \infty} e(t) = 0$$

To find a general adaptation law to assure global asymptotically stability the hyperstability approach in combination with the properties of positive dynamic systems is used. In order to use the hyperstability theory Eq.(10) can be written in the following structural way.

$$\begin{aligned} \dot{e} + k_p e &= \Gamma^{-1}[(\Gamma - \hat{\Gamma})(\dot{x}_r + k_p e) + \eta - \hat{\eta}] = \omega_1 \\ \omega_1 &= -\omega = -\psi_1(v, \dot{x}_r, e, t)(\dot{x}_r + k_p e) - \psi_2(v, t) \\ v &= D(p)e \end{aligned} \tag{11}$$

Fig. 2 shows the block diagram of the equations.

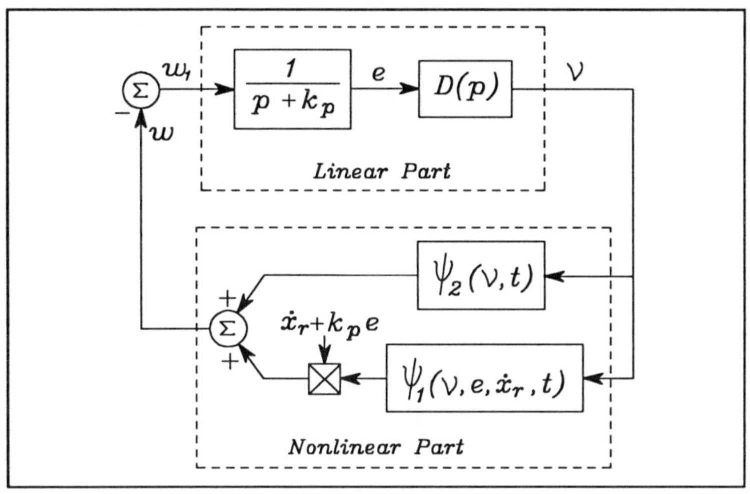


Fig. 2 *Equivalent feedback system for Eq.(11)*

Recalling the hyperstability theory [Landau 1979] the necessary and sufficient conditions assuring the asymptotic stability of the equivalent representation of Eq.(11) shown in Fig. 2 are the following.

i) The transfer function of the linear time-invariant part should be strictly positive real (SPR).
ii) The non-linear time-varying block in the feedback part should satisfy the integral inequality

$$\int_0^t \nu \, \omega \, d\tau \geq -\delta^2 \qquad \forall t \geq 0$$

where $\delta$ is an arbitrary constant.

The first condition can be satisfied by choosing the linear compensator as

$$D(p) = \frac{1 + a_0 p}{1 + a_1 p} \qquad \text{where } p = \frac{d}{dt} \tag{12}$$

with $k_p > 0$ , $a_0 \geq a_1/(1 + a_1 k_p)$ , $a_1 > 0$
Note that the first order error dynamics itself are SPR for $k_p > 0$, but using the linear compensator one can influence the adaptation transients in a convenient way, by proper choice of the linear compensator constants $a_0$ and $a_1$.

The second condition ( Appendix A ) can be satisfied by choosing

$$\begin{aligned}\psi_1(e,v,\dot{x}_r,t) &= k_1\frac{1}{\sqrt{\Gamma}}\int_0^t\frac{1}{\sqrt{\Gamma}}v(\dot{x}_r+k_pe)d\tau+k_2\frac{1}{\Gamma}v(\dot{x}_r+k_pe)\\ \psi_2(v,t) &= k_3\frac{1}{\sqrt{\Gamma}}\int_0^t\frac{1}{\sqrt{\Gamma}}v\,d\tau+k_4\frac{1}{\Gamma}v\end{aligned}\tag{13}$$

In the above analysis we have assumed that $\Gamma$ is bounded and positive, and $k_1,k_3 > 0$ , $k_2,k_4 \geq 0$. Combining Eq.(11) and Eq.(13) gives

$$\begin{aligned}\hat{\Gamma}(t) &= \hat{\Gamma}(0)+k_1\int_0^t v(\dot{x}_r+k_pe)d\tau+k_2v(\dot{x}_r+k_pe)\\ \hat{\eta}(t) &= \hat{\eta}(0)+k_3\int_0^t v\,d\tau+k_4v\end{aligned}\tag{14}$$

In deriving Eq.(14) we have treated the actuator parameters $\Gamma$ and $\eta$ as unknown and slowly time-varying in comparison with the adaptation scheme. The overall control scheme including the actuator is shown in Fig. 3.

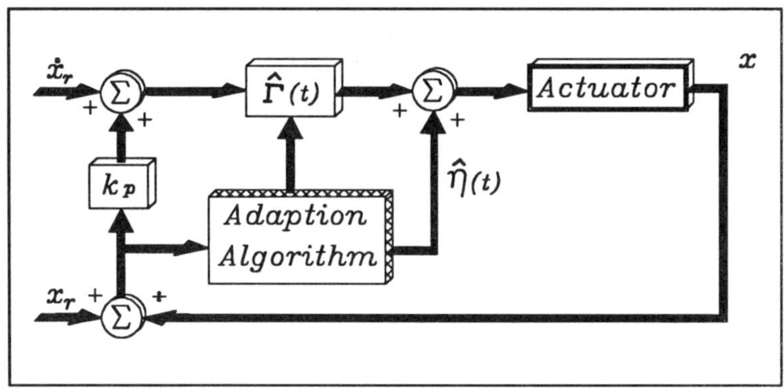


Fig. 3 *Complete Control Scheme*

## 5. SIMULATION AND EXPERIMENTAL RESULTS

A simulation and experimental study have been performed to investigate the effectiveness of the adaptive control design. The results were compared with those of the conventional PD-controller and the CVI-controller given in [Conrad, *et al,* 1992] and reviewed in this paper. The same reference trajectory, a square, was chosen for both the simulation and experimental study and is placed in the work-space of the robot as shown in Fig. 4.

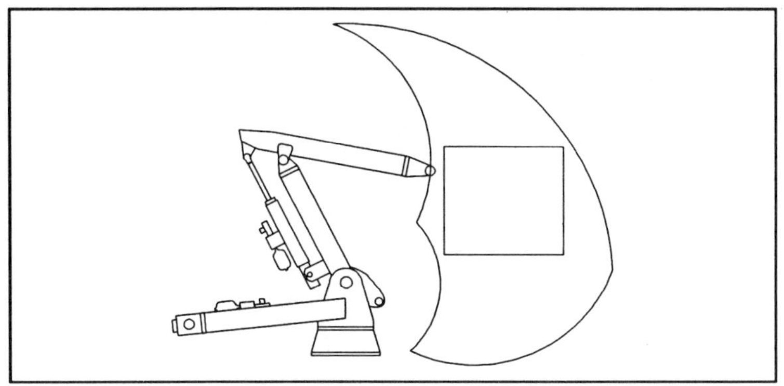

Fig. 4 *Reference trajectory in the robot work-space*

The sides in the square have a length of 0.8 meters and the time for one pass is 4 sec. Each side has the same acceleration profile. The profile in Cartesian coordinates is given in Fig. 5.

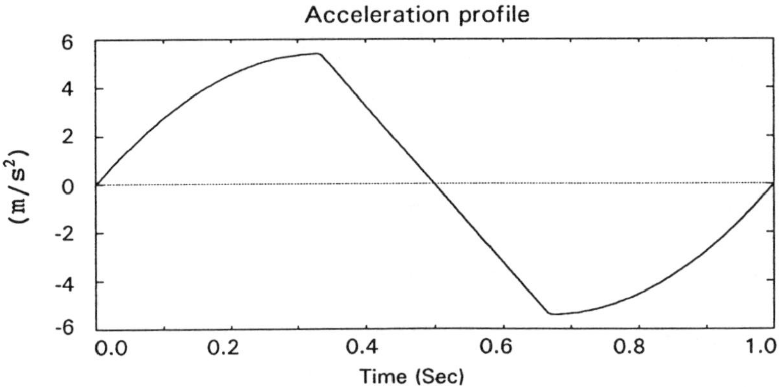


Fig 5. *Acceleration profile for each segment in square*

The system and control parameters used are given in Table 1.

**Table 1. Numerical Values of System Parameters**

---

*COMMON ACTUATOR CONSTANTS*

$P_S$ = 180 bar
$A = 9.46(10)^{-4}$ m$^2$
$\beta = 7.0(10)^8$ N/m$^2$
$c = 1.667(10)^{-12}$ (m$^3$/s)/Pa
$V = 3.2(10)^{-4}$ m$^3$

*SERVO VALVE GAIN*

Actuator 1 (Lower link) $k = 3.8(10)^{-8} \dfrac{m^3/s}{V\sqrt{Pa}}$

Actuator 2 (Upper link) $k = 1.19(10)^{-8} \dfrac{m^3/s}{V\sqrt{Pa}}$

*CONTROL PARAMETERS*

| LOWER LINK | UPPER LINK |
|---|---|
| PD-controller: $[\ u = K_p(e + \tau_D \dot{e})\ ]$ | |
| $K_P$ = 160 V | $K_P$ = 550 V |
| $\tau_D$ = 0.027 sec | $\tau_D$ = 0.0054 sec |
| CVI-controller: | |
| $k_P$ = 22 m/s | $k_P$ = 24.5 m/s |
| AMAC-controller: | |
| $\hat{\Gamma}(0) = A/(k\sqrt{P_S})$ V/(m/s) | $\hat{\Gamma}(0) = A/(k\sqrt{P_S})$ V/(m/s) |
| $\hat{\eta}(0)$ = 0 V | $\hat{\eta}(0)$ = 0 V |
| $a_0$ = 1/37 sec | $a_0$ = 1/185 sec |
| $a_1$ = 1/600 sec | $a_1$ = 1/600 sec |
| $k_1$ = 7000 V/(m/s)$^2$ | $k_1$ = 10000 V/(m/s)$^2$ |
| $k_2$ = 2000 V/(m/s)$^2$ | $k_2$ = 3000 V/(m/s)$^2$ |
| $k_3$ = 0.5 V | $k_3$ = 0.5 V |
| $k_4$ = 110 V | $k_4$ = 550 V |
| $k_P$ = 10 m/s | $k_P$ = 10 m/s |

---

## 5.1 SIMULATION RESULTS

A computer simulation model of the two degree of freedom robot manipulator was used for initial testing and evaluation of the adaptive control algorithm. The simulation program takes into account most nonlinearities in the manipulator and actuator dynamics.

In order to compare simulation and experimental data the same sampling time of 1.5 milliseconds was used. The integrated values required for adaptation were approximated by the trapezoidal integration rule.

Fig. 6(a,b) and Fig.7(a,b) shows the joint tracking error.

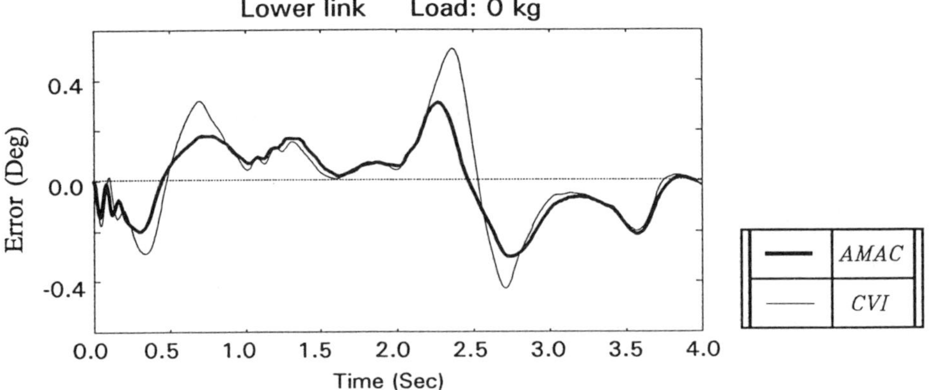


Fig. 6(a) *Lower link tracking accuracy with no load*

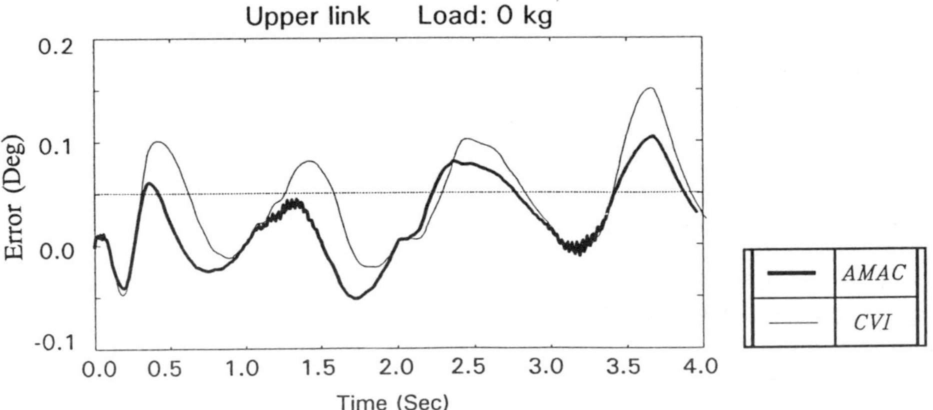


Fig. 6(b) *Upper link tracking accuracy with no load*

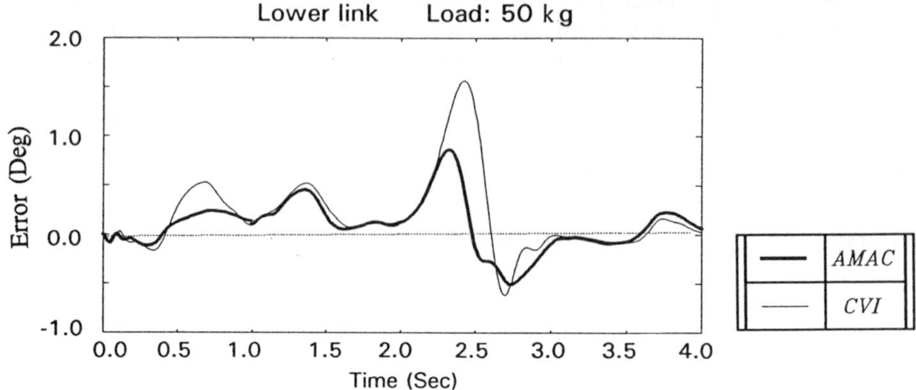


Fig. 7(a) *Lower link tracking accuracy with load*

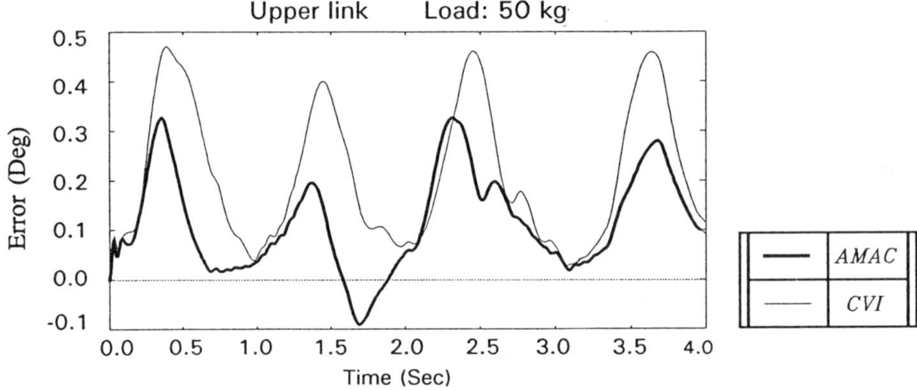


Fig. 7(b) *Upper link tracking accuracy with load*

Simulation results revealed that the adaptive controller was effective over a wide range of payload masses.

The CVI-controller contained a larger tracking error, especially with a payload of 50 kg. This was expected since the feedforward part of the CVI-controller neglects the load pressure $P_L$ which is not insignificant with a large payload.

It should be recalled that both the AMAC and the CVI-controller do not require a priori knowledge of the properties of the payload.

The study also showed that the adaptive controller is able to adapt quickly as the manoeuvre begins.

## 5.2 EXPERIMENTAL RESULTS

The schematic diagram of the overall system shown in Fig. 1 was used for the experimental study. The experiments were mainly performed to verify the simulation results discussed in the previous section.

In order to test the "adaptability" of the AMAC-scheme the initial conditions $\hat{\Gamma}(0)$, $\hat{\eta}(0)$ were set equal to zero. In this case the AMAC-scheme generates the control action without a priori knowledge of the actuator model. The results showing the joint tracking error are given in Fig. 8(a,b).

It is seen that for both links the position errors converge, to a value similar to those obtained in the simulation study, in about 2 seconds.

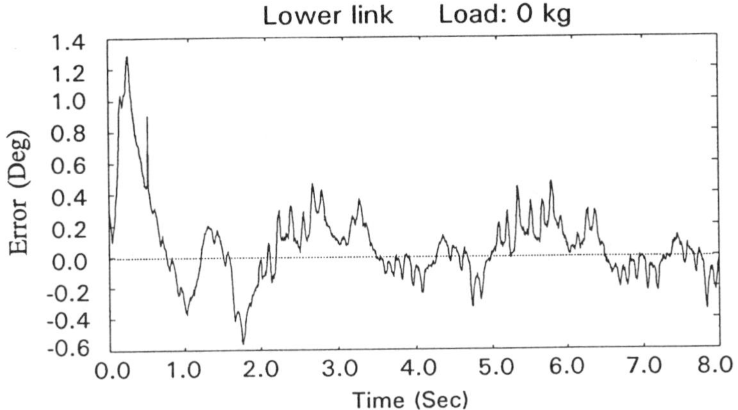


Fig. 8(a) *"Adaptability" of lower link*

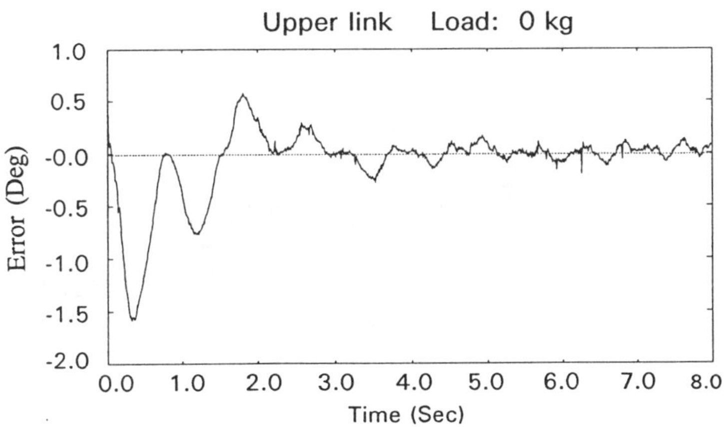


Fig. 8(b) *"Adaptability" of upper link*

Next, the AMAC-scheme was compared with the CVI-controller and a simple PD-controller. In this case $\hat{\Gamma}(0)$ and $\hat{\eta}(0)$ are given by the values shown in Table 1. The results are shown in Fig. 9(a,b) and Fig. 10(a,b). The gains used are the same ones as in the simulation study (Table 1).
The results coincide very well with the simulated results. The maximum real errors equals the maximum simulated errors, but the error curve is somewhat different and may be due to unmodelled dynamics in the simulation study.
In conclusion the experimental study revealed that the AMAC-scheme exhibits a much better performance as compared with the CVI- and specially the PD-controller.

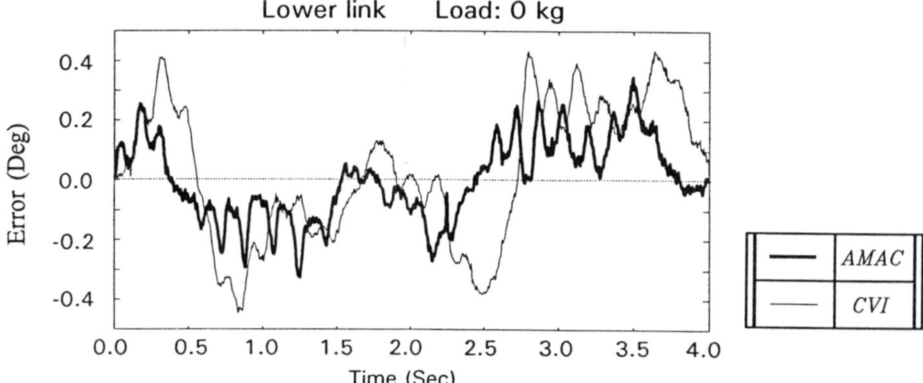


Fig. 9(a) *Lower link tracking accuracy*

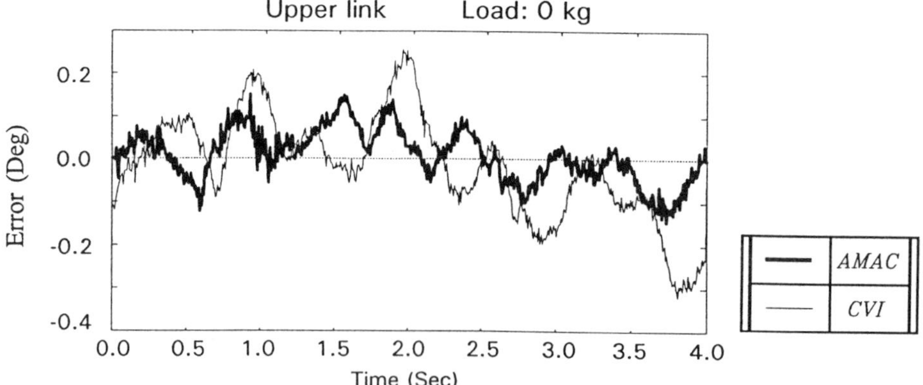


Fig. 9(b) *Upper link tracking accuracy*

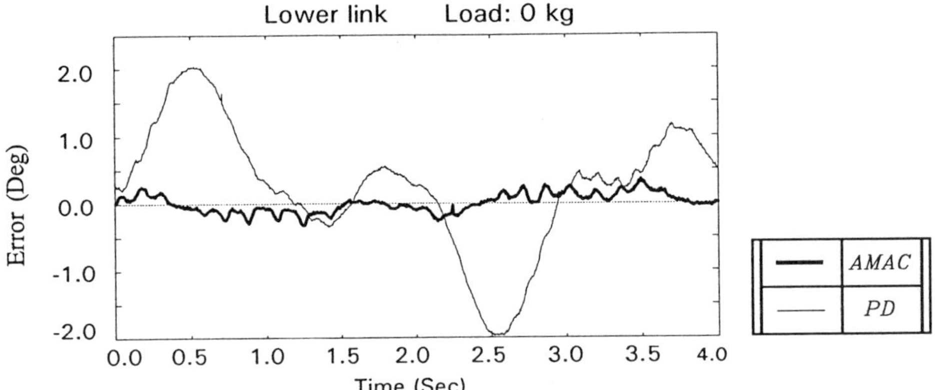


Fig. 10(a) *Lower link tracking accuracy*

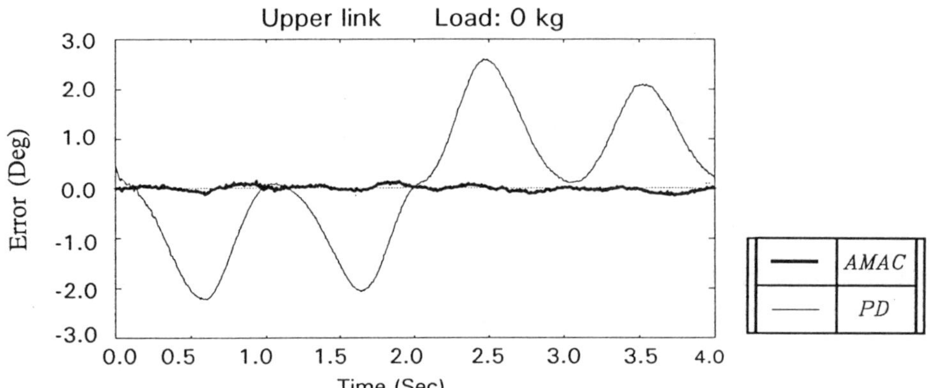


Fig. 10(b) *Upper link tracking accuracy*

## 6. CONCLUSION

The theory and implementation results of an adaptive controller for hydraulic actuators, called AMAC, have been presented. The control scheme is computationally efficient and requires no exact knowledge of the system parameters.

No velocity or acceleration feedback of any kind are required - only position feedback. The algorithm has been implemented for the control of an hydraulic direct drive robot. The adaptive controller was compared with a model-based constant gain controller (the CVI) and a simple PD feedback loop. The adaptive scheme outperformed these controllers by a significant margin,specially with the robot carrying a payload of 50 kg. The experimental results demonstrate that the AMAC is capable of controlling a robot

by independent joint control, in spite of the high degree of static and dynamic couplings between the links.
In summary, the AMAC simplicity, ease of implementation and robustness to variations in payload parameters make the proposed scheme a simple practical solution to the actuator control problem.

## References


[1] **Conrad F., Sørensen P.H.,Trostmann E., Zhou J.J.**,(1990), "*Design and Digital Control of a Fast Hydraulic Test Robot Manipulator*", Proc. of the 9th BHRA International Conference On Fluid Power, Cambridge, England, 25-27 April.

[2] **Conrad F., Zhou J.J., Siggaard M.**,(1992), "*Experimental evaluation of two feedback linearisation control schemes*", Proc. of ASME Winter Annual Meeting in Fluid Power systems and Technology Symposia, Anaheim, California, Nov. 8-113.

[3] **Craig J.J., Hsu P., Sastry S.S.**,(1986), "*Adaptive Control of Mechanical Manipulators*", Proc. IEEE Robotics and Automation.

[4] **Landau Y.D.**,(1979), "Adaptive Control, the Model Reference Approach", Marcel Dekker, N.Y.

[5] **Tarokh M.**(1992), "*Combined Adaptive and Computed Torque Control of Robot Manipulators*", ASME 1992, DSC-Vol. 42, Advances in Robotics.

[6] **Zhou J.J.**,(1989) "*Adaptive Control and Applications to Hydraulic Robot Manipulators*", Ph.D. Thesis, The Control Engineering Institute, Technical University of Denmark.


## Appendix A

In Eq.(11) the problem is to choose $\psi_1$ and $\psi_2$ in order to satisfy Popov´s integral inequality

$$\int_0^t \nu\,\omega\,d\tau \geq -\delta^2 \quad ; \quad \forall t \geq 0 \tag{A1}$$

While the procedures are the same for both $\psi_1$ and $\psi_2$ it will be shown for only $\psi_1$. If $\psi_1$ is given by an expression having the structure of a proportional and integral adaptation law, it can be written as

$$\psi_1 = \gamma_1(t)\int_0^t \gamma_1(t)\phi_1(\nu,\dot{x}_r,e,t)d\tau + \gamma_2(t)\phi_2(\nu,\dot{x}_r,e,t) \tag{A2}$$

where $\gamma_1, \gamma_2, \phi_1$ and $\phi_2$ are time functions to be determined. Inequality (A1) can then be split into two inequalities, i.e.

$$I_1 = \int_0^{t_1} \nu(\dot{x}_r+k_pe)[\gamma_1(t)\int_0^t \gamma_1(t)\phi_1(\nu,\dot{x}_r,e,t)d\tau]dt \geq -\delta_1^2 \tag{A3}$$

$$I_2 = \int_0^{t_1} \nu(\dot{x}_r+k_pe)[\gamma_2(t)\phi_2(\nu,\dot{x}_r,e,t)]dt \geq -\delta_2^2 \tag{A4}$$

Consider Eq.(A4) first. This inequality will be satisfied for $\delta_2^2 \geq 0$ if the integrand is non-negative, i.e.

$$\nu(\dot{x}_r+k_p e)\phi_2(\nu,\dot{x}_r,e,t)\gamma_2(t) \geq 0 \tag{A5}$$

Let $\phi_2$ be given by

$$\phi_2(\nu,\dot{x}_r,e,t) = k_2\nu(\dot{x}_r+k_p e) \tag{A6}$$

then Eq.(A5) will simplify to

$$k_2\gamma_2(t)\nu^2(\dot{x}_r+k_p e)^2 \geq 0 \tag{A7}$$

which is true for $k_2$ being a positive constant and $\gamma_2(t)$ being a positive function. Next, consider the inequality of Eq.(A3). Using the relation

$$\int_0^{t_1}\dot{f}(t)k_1 f(t)dt = \frac{k_1}{2}[f^2(t_1)-f^2(0)] \geq -\frac{1}{2}k_1 f^2(0) \quad ; \quad k_1 \geq 0 \tag{A8}$$

one can from Eq.(A3) identify that

$$\dot{f}(t) = \nu(\dot{x}_r+k_p e)\gamma_1(t) \tag{A9}$$

$$k_1 f(t) = \int_0^t \gamma_1(t)\phi_1(\nu,\dot{x}_r,e,t)d\tau \tag{A10}$$

By differentiating Eq.(A10) with respect to time and using Eq.(A9) , one gets

$$\phi_1(\nu,\dot{x}_r,e,t) = k_1\nu(\dot{x}_r+k_p e) \tag{A11}$$

We now need to determine $\gamma_1(t)$ and $\gamma_2(t)$ in order to specify the adaptation law explicitly. By choosing $\gamma_1(t) = 1/\sqrt{\Gamma(t)}$ and $\gamma_2(t) = 1/\Gamma(t)$ one gets from Eq.(11)

$$\frac{\Gamma-\hat{\Gamma}}{\Gamma} = -k_1\frac{1}{\sqrt{\Gamma}}\int_0^t\frac{1}{\sqrt{\Gamma}}\nu(\dot{x}_r+k_p e)d\tau - k_2\frac{1}{\Gamma}\nu(\dot{x}_r+k_p e) = -\psi_1 \tag{A12}$$

By defining $s = \nu(\dot{x}_r+k_p e)$ and $H = \Gamma-\hat{\Gamma}$ Eq.(A12) can be written as

$$\frac{H}{\Gamma} = -k_1\frac{1}{\sqrt{\Gamma}}\int_0^t\frac{1}{\sqrt{\Gamma}}s\,d\tau - k_2\frac{1}{\Gamma}s \tag{A13}$$

Multiplying both sides by $\sqrt{\Gamma}$ and differentiating Eq.(A13) with respect to time give

$$\frac{\dot{H}}{\sqrt{\Gamma}} - \frac{H\dot{\Gamma}}{2\sqrt{\Gamma}\Gamma} = -k_1\frac{1}{\sqrt{\Gamma}}s + k_2\frac{\dot{\Gamma}}{2\sqrt{\Gamma}\Gamma}s - k_2\frac{1}{\sqrt{\Gamma}}\dot{s} \tag{A14}$$

Eq.(A14) can be reduced to

$$2\Gamma\dot{H} - H\dot{\Gamma} = -k_1 2\Gamma s + k_2(\dot{\Gamma}s - 2\Gamma\dot{s}) \tag{A15}$$

Now, if the adaptation process is fast, then $\Gamma$ can be treated as an unknown and slowly time varying quantity. This is obtained by updating $\hat{\Gamma}$ at a high sample rate. Under this condition $\Gamma$ can be considered as a constant quantity, and then Eq.(A15) can be reduced to

$$\dot{H} = -k_1 s - k_2\dot{s} \tag{A16}$$

Eq.(14) can now be obtained from Eq.(A16) as

$$\hat{\Gamma} = \Gamma + k_1\int_0^t\nu(\dot{x}_r+k_p e)d\tau + k_2\nu(\dot{x}_r+k_p e) \tag{A17}$$

In Eq.(A17) $\Gamma$ can, under the above conditions, be considered representing the initial value of the integral, as done in Eq.(14).

**WRITTEN DISCUSSION**
**A novel adaptive control scheme for hydraulic actuator motion systems**
**TO Andersen, F Conrad, PE Hansen & JJ Zhou (Technical University of Denmark)**

**Question:** **RE Koski**
**Sun Hydraulics Corpn., USA**

Have you included any allowance for structural flexibility in defining the position of the robot?

**Answer:**

No, the position of the tool centre point is calculated based on knowledge of the geometry of the robot and measurement of the piston positions. The reason is, that the developed control scheme is mainly for controlling actuators. It might have been a good idea to show the piston position error, but showing the robot position also indicates the performance of the scheme when controlling a hydraulic robot, and the ability to cope with load disturbances.

**Question:** **P Krus**
**University of Linköping, Sweden**

What is the *a priori* information needed to initialise the AMAC-controller?

**Answer:**

In the adaptation laws the initial conditions are meant to include *a priori* information, but the *a priori* information is not needed to ensure stability of the proposed control scheme. However, having *a priori* information included in the adaptation laws, such as that used in this paper, the speed of convergence is significantly increased.

# 14. A Study of Gain-Scheduling Method for Controlling the Motion of Pneumatic Servos

J Pu, P R Moore, R Harrison *and* R H Weston

**Abstract**

Pneumatic drives are characterised by low damping and high compressibility which leads to significant variation in load stiffness with position. The transient behaviour of pneumatic servos also varies with the initial conditions of the system. Optimum tuning is difficult to achieve. For example, a set of control gain terms may be selected which work well for one particular move but may fail to operate satisfactorily for others.

This paper studies the approach of "gain-scheduling" to improve the performance consistency of pneumatic servos. The scheduling scheme is based upon a simplified model of pneumatic servos. An experimental evaluation is conducted, comparing the effectiveness of the chosen gain-scheduling scheme against the typical response characteristics of the drive using fixed gain constants for control.

**Nomenclature**

| | |
|---|---|
| A | cylinder diameter |
| $C_o$ | valve spool coefficients |
| G(s) | open-loop transfer function of the cylinder, valve and load system |
| L | half of total cylinder stroke length |
| k | ratio of specific heat |
| $K_a$ | acceleration feedback gain |
| $K_p$ | proportional gain |
| $K_v$ | velocity feedback gain |
| M | total moving mass |
| $p_q$ | quiescent pressure |

$r_m$ amplifier gain
R gas constant
$V_o$ half of total control volume of the cylinder
$T_s$ supply temperature
$Y_i$ initial displacement from mid-position of cylinder stroke
Y(s) actual displacement
$Y_d(s)$ demand displacement

## 1. Introduction

PID control algorithms have been successfully implemented for a wide range of applications and drive systems. An extension and modification of this control algorithm leads to a three loop control method: namely proportional, velocity and acceleration feedback control. This control method has proven sufficient to stabilise pneumatic servo drive systems [1,2,3].

The experimental set-up of this study involves a rodless cylinder with encoder feedback device and a five port servo valve (See Figure 1). Details and operating conditions of the system are listed in Table 1.

Table 1 Details and Operational Conditions of the Drive System

| | |
|---|---|
| Rodless Cylinder | Diameter: 40 mm<br>Stroke length: 1000mm |
| Control Valve | Torque motor driven; two stage;<br>Input current: - 800 to + 800 mA;<br>Rated flow: 680 l/min at supply pressure 6 bar with pressure drop of 1 bar |
| Servo Amplifier | Input voltage: - 10 to + 10 volts;<br>Output current: - 790 to + 790 mA; |
| Digital to Analogue Converter | 12 bits |
| Controller Sampling Rate | 4 ms |
| Position Measurement | 0.02 mm/count |
| Total Moving Mass | 11.5 kg |

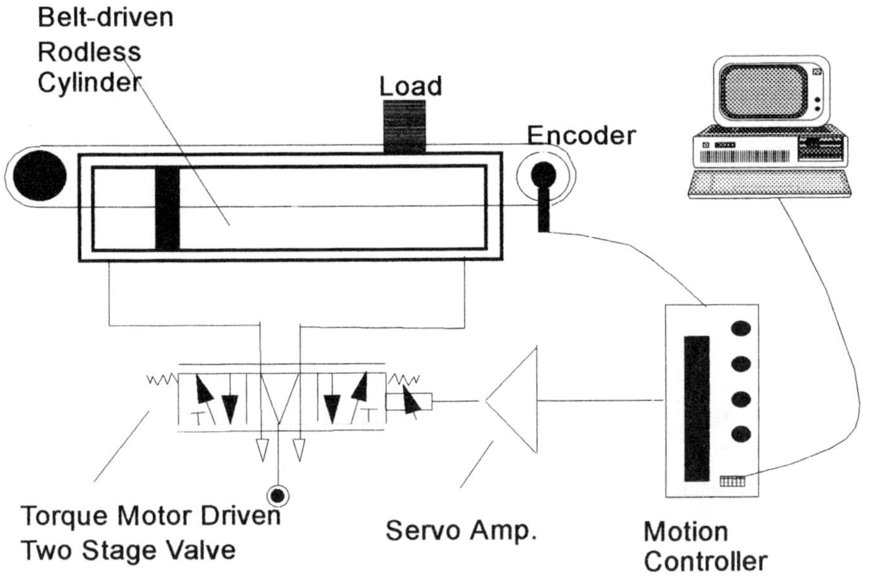


**Figure 1 Experimental System**

**2. Limitations with Constant-Gain Control**

It is generally found that the load stiffness of pneumatic drives is dependent upon the load position, with the minimum stiffness typically occurring at the mid-stroke position when using symmetric or rodless cylinders [4,5]. The damping in pneumatic systems is inherently low and is generally related to valve opening conditions [3,5].

Figure 2 illustrates typical system responses for moves; (a) where the direction of the move is reversed, (b) where the length of move is varied.[6] and (c) where the demand velocity is changed [7]. A higher load velocity may also result in system oscillation and overshoot [8]. The results presented were obtained using a conventional fixed gain controller with proportional, velocity and acceleration feedback.

These experimental results show that inconsistency in the system response of the drive can be significant when the initial state of the system and the parameters of a required move (e.g. speed, acceleration and distance of moves) are changed. This situation is exaggerated when the control gains are poorly tuned. This inconsistency in the system response of the drive can be surpressed to a large extent if high gain values are used and the right balance is found between the control terms used to achieve stability [1]

Response inconsistency can pose a number of problems when tuning the drive: a set of control parameters may work well for one particular move but fail to operate

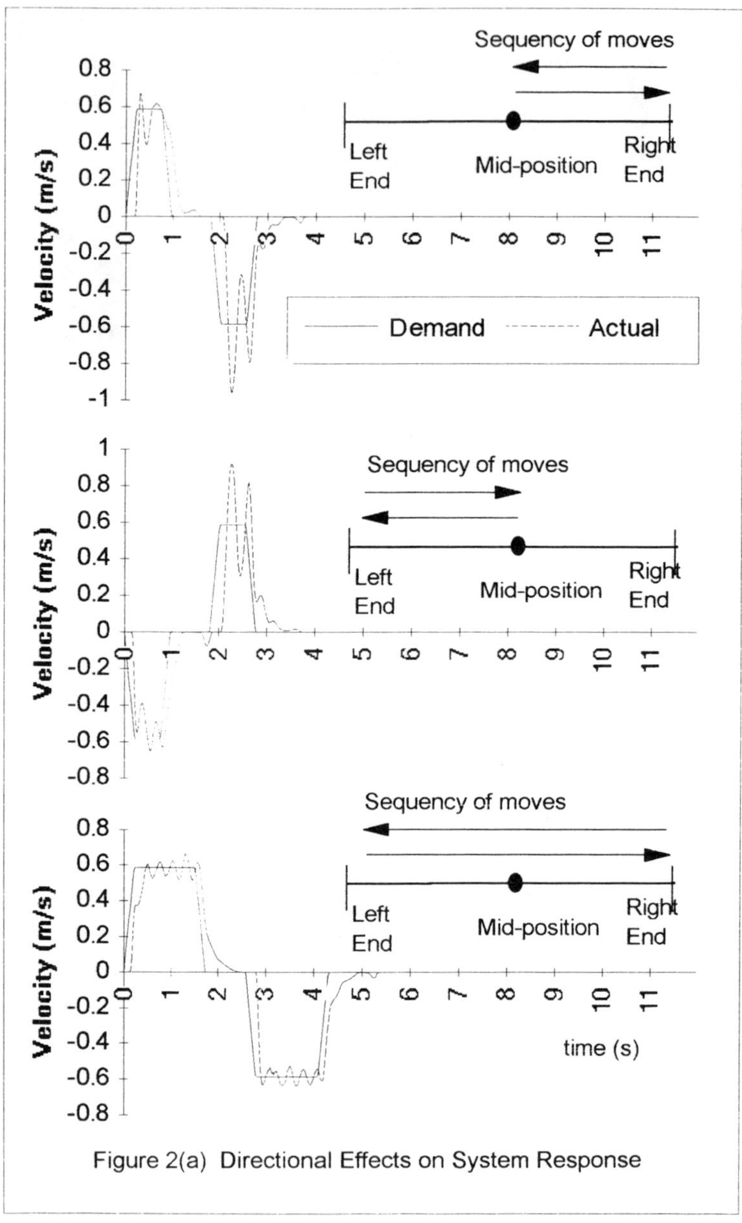


Figure 2(a) Directional Effects on System Response

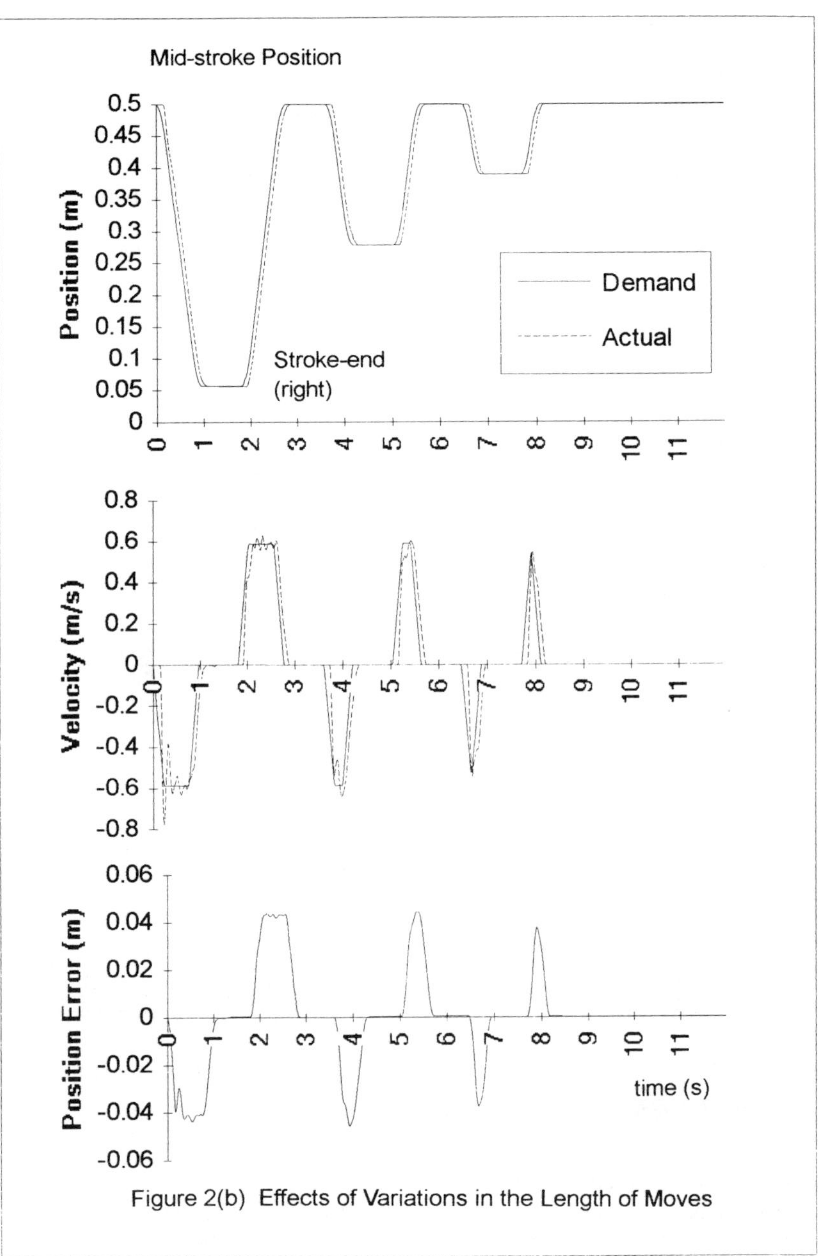

Figure 2(b) Effects of Variations in the Length of Moves

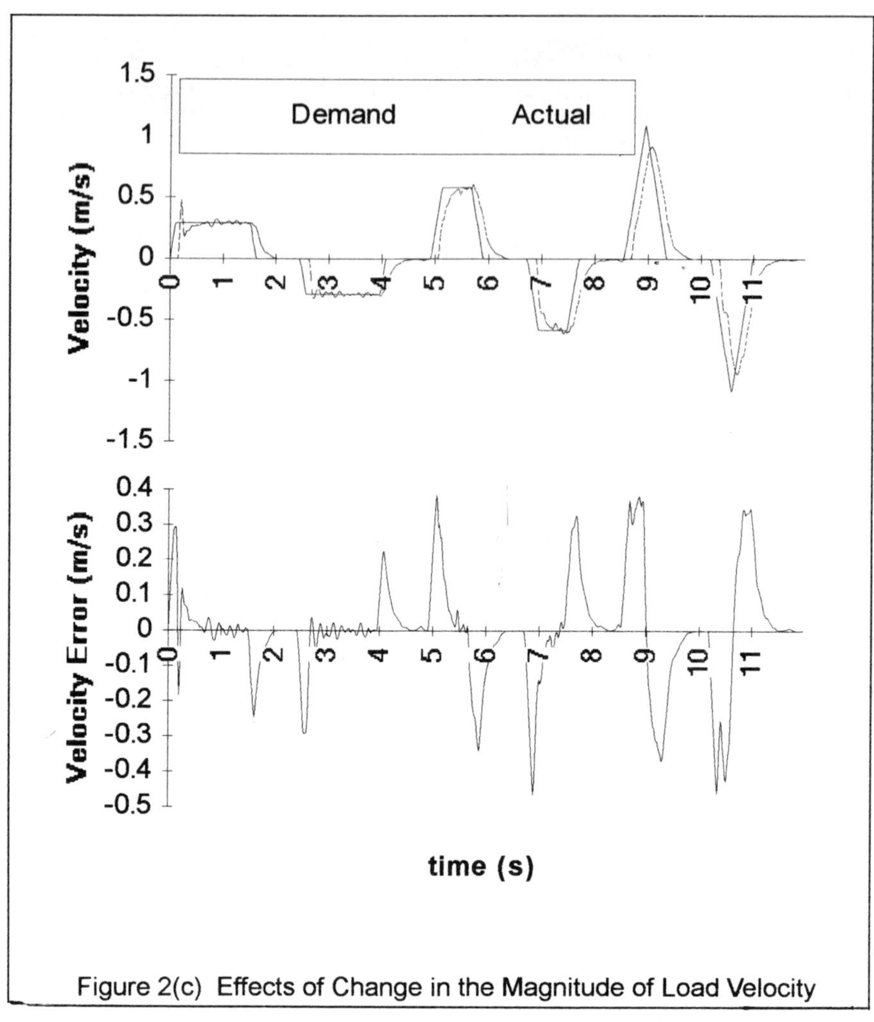

Figure 2(c) Effects of Change in the Magnitude of Load Velocity

satisfactorily for another move. It is highly desirable to achieve consistent system response for a defined range of expected load movements.

### 3. Gain-Scheduling Scheme

Gain scheduling is a method which can be used to improve the performance consistency of non-linear systems. Various types of gain scheduling scheme can be derived, targeted at specific non-linear elements in the system. In this paper, a simplified model derived previously [1] is used as a theoretical guideline for devising the scheduling scheme. The model assumes that the valve is operated near its null position and the damping ratio can be approximated as zero. Note that the model refers to a system with proportional, velocity and acceleration feedback. For reasons of convenience, the model is quoted as follows

$$\frac{Y(s)}{Y_d(s)} = \frac{b_m K_p}{s^3 + b_m K_a s^2 + (a^2 + b_m K_v)s + b_m K_p} \qquad (1)$$

where

$$b_m = \frac{2kRT_s r_m A C_o}{M[V_o^2 - (AY_i)^2]} \qquad (2)$$

$$a2 = \frac{2P_q V_o k A^2}{M[V_o^2 - (AY_i)^2]} \qquad (3)$$

Applying Routh's stability criteria, the following relationship needs to be satisfied to achieve stability

$$\frac{K_p}{K_a} < w^2(1 + h_v K_v) \qquad (4)$$

with

$$w^2 = \frac{2kAP_q}{ML} f(Y_i/L) \qquad (5)$$

$$h_v = \frac{RT_s C_o}{AP_q} \qquad (6)$$

and

$$f(Y_i/L) = \frac{1}{1 - (Y_i/L)^2} \qquad (7)$$

Equation (4) indicates that

(i) The incorporation of acceleration feedback is necessary to stabilise the system ; in practice, leakage and frictional forces also introduce damping into the system (These factors however are not taken into account in the model described above).

(ii) Although the presence of velocity feedback is not essential to achieving system stability, it will significantly improve the stability margin of the system.

A scheduling scheme can be derived as a function of several associated variables, e.g. valve flow-displacement relationship, load velocity, load position etc. In this study, only load position is considered where Equation (7) is used to characterise the position dependence of load stability of pneumatic drives. It can be seen that $f(Y_i/L)$ approaches infinity when $Y_i=L$. However, a practical system can never reach this state. Thus, Equation (7) needs to be modified for gain scheduling. Figure 3 illustrates a comparison between actual and theoretical estimations of the critical proportional gain which is the magnitude of the "proportional gain" that will start to induce instability.

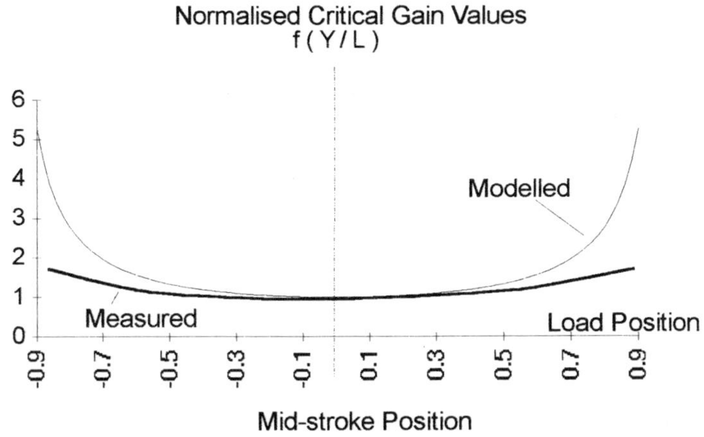


Figure 3 A Comparison between Measured and Modelled Critical Gain Values (when setting velocity feedback gain value to zero)

For industrial applications, the load is rarely positioned at the two extreme ends of a pneumatic cylinder stroke. The home position is often some distance away the end stroke positions. The function $f(Y_i/L)$ is thus modified as follows for implementing gain scheduling.

$$f_m(Y_i/L) = \begin{cases} f(Y_i/L) & \text{when } |Y_i| < Y_a \\ f(Y_a/L) & \text{when } |Y_i| > Y_a \end{cases} \quad (8)$$

where $Y_a$ is the distance between the home position and the mid-stroke position.

There are three loop gains in the control method chosen for this study which can be scheduled. The stability constraint of Equation (4) indicates that scheduling $K_p$ and $K_a$ independently may not guarantee stability. The ratio of proportional gain to acceleration should be scheduled to maintain stability.

Pneumatic servo drive systems typically only incorporate a position feedback device (e.g. a rotary encoder). Velocity and acceleration information is subsequently derived in software from the position data obtained from the sensor. The signal to noise ratio of the acceleration information is lower than that of velocity signal (which is the first derivative of the position signal). The derivation of velocity and acceleration signals in this study uses a simple backward difference algorithm.

## 4. Experimental Results and Discussions

A range of experimental work has been conducted as a proof-of-concept study of gain scheduling. The gain-scheduling scheme is schematically shown in Figure 4. The effects of gain-scheduling are individually examined rather than in combination.

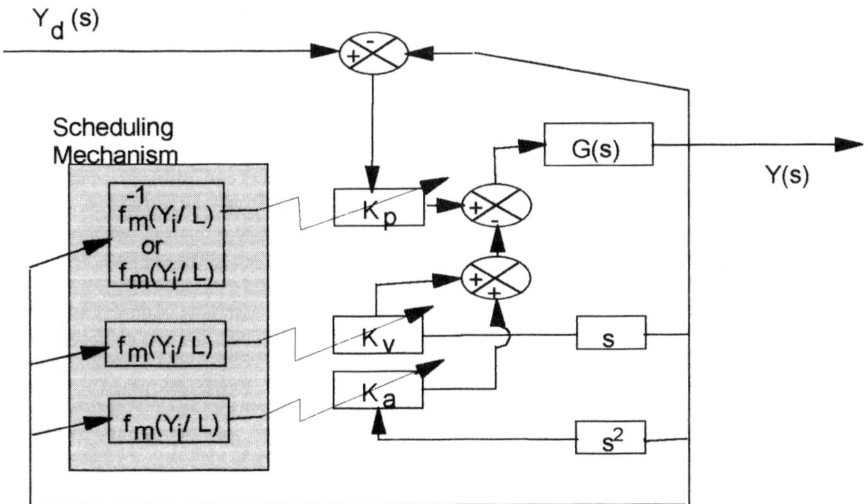


Figure 4 Schematic Diagram of the Chosen Gain Sheduling Method

(1) Proportional Gain $K_p$

Two schemes have been implemented for scheduling the proportional gain as a function of load position

(i) for uniform stability margin

According to the stability constraint dictated by Equation (4), a scheduling scheme is derived as follows

$$K_{p(s)} = f_m(Y_i/L)\, K_p \qquad (9)$$

It should be noted that as $f_m$ is greater than 1 the scheduled proportional gain $K_{p(s)}$ at the given position $Y_i$ will be always greater than the original fixed gain $K_p$. In other words, the overall stability margin of the system is reduced as a result of applying gain scheduling dictated by Equation (9).

(ii) for uniform system stiffness

The natural frequency of a system is dependent upon the stiffness of the system and the mass of the load. Equivalent to a mass-spring system, the relationship can be described as follows

$$w2 = \frac{K}{M} \qquad (10)$$

where K represents the stiffness of the system. Comparing Equation (10) with Equation (5), it can be found that

$$K = \frac{2kAPq}{L} f(Y_i/L) \qquad (11)$$

To achieve uniform stiffness, the following relationship can be adopted for scheduling the gain of the closed-loop system.

$$K_{p(s)} = \frac{1}{f_m(Y_i/L)} K_p \qquad (12)$$

In Figure 5, it is shown that increased oscillation can result when the proportional gain is scheduled according to Equation (9). In this case, $K_{p(s)}$ alters with load position, permitting a reduced stability margin. Conversely, $K_p$ can be scheduled to achieve consistency in system stiffness. This arrangement will result in excessive stability margin, resulting in less oscillation but larger following errors.

(2) Velocity Feedback Gain $K_v$

To achieve consistent stability conditions (referring to Equation (4)), the scheduling scheme adopted for velocity feedback loop gains should be more complex in nature than that used to schedule proportional and acceleration feedback loop gains. However, as a matter of simplicity, Equation (13) is selected for scheduling $K_v$ in this study. Figure 6 indicates that the resulting system response lies somewhere between those obtained when using fixed gain constants of the two extreme values, i.e. the minimum and maximum gain values of $K_{v(s)}$.

$$K_{v(s)}= f_m(Y_i/L)\, K_v \tag{13}$$

The scheduling scheme of $K_v$ should be a function of not only load position but also the valve displacement or alternatively the load velocity. Obviously, more studies are necessary to reveal the full significance of scheduling the velocity feedback gain to improve performance consistency when controlling pneumatic servos.

(3) Acceleration Feedback Gain $K_a$

Equation (14) is used for scheduling the acceleration feedback gain

$$K_{a(s)}= f_m(Y_i/L)\, K_a \tag{14}$$

In Figure 7, two set of moves are presented. It is shown that the impact of gain scheduling on the system response is marginal when the move is initiated from one

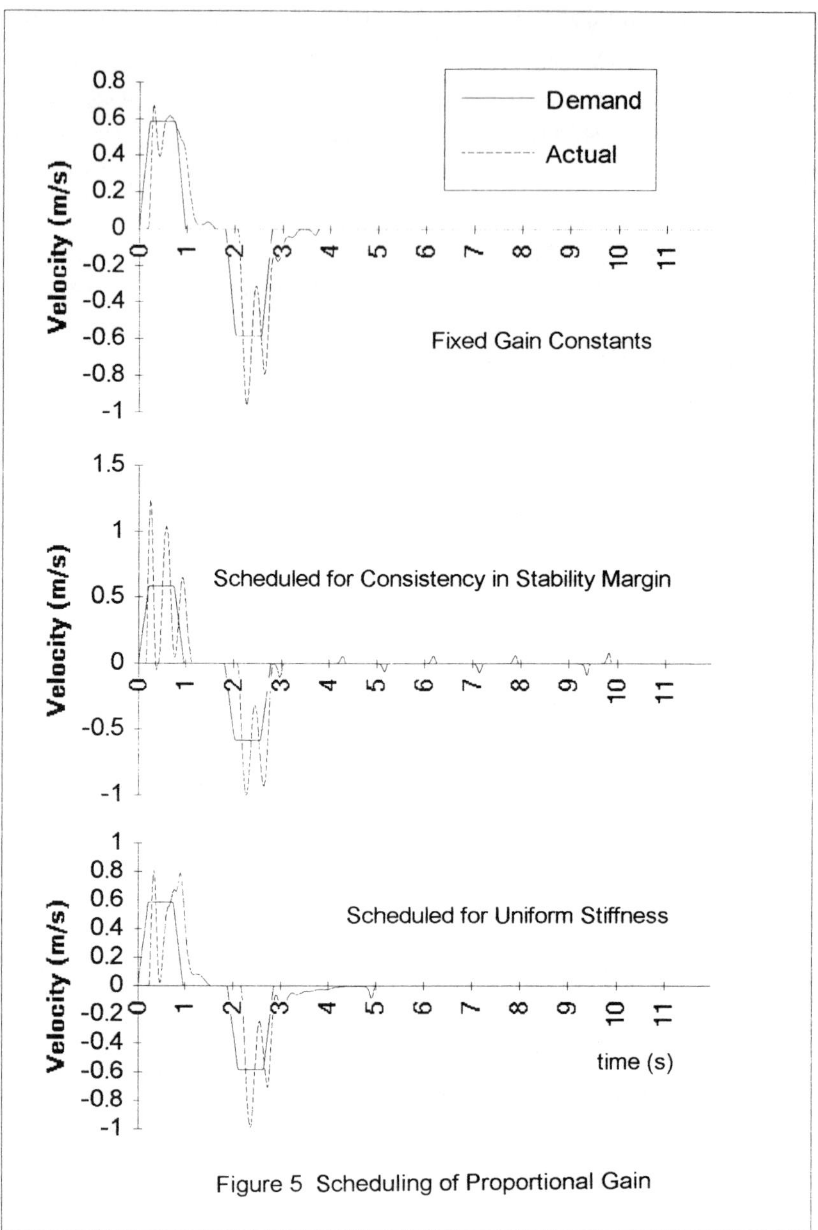


Figure 5 Scheduling of Proportional Gain

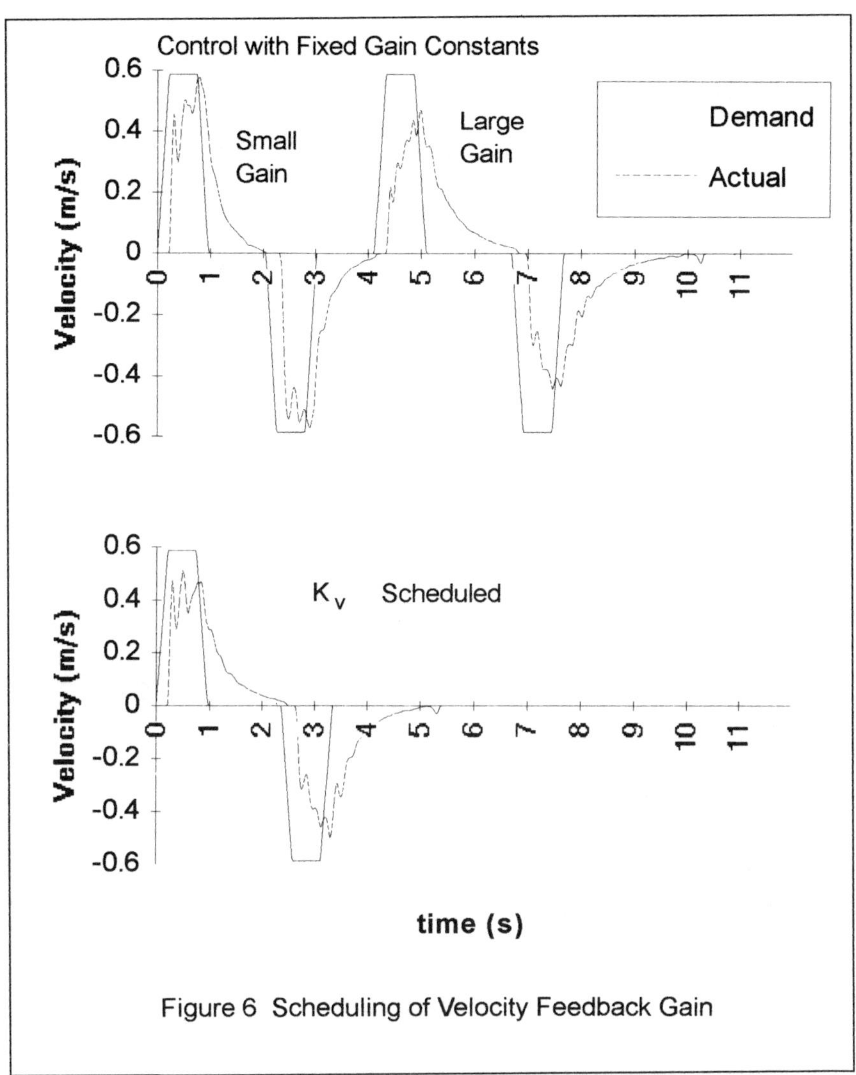


Figure 6 Scheduling of Velocity Feedback Gain

cylinder end to the other. However, significant improvement can be obtained when the move is initiated from the mid-stroke position which exhibits minimum stiffness or the least stable position.

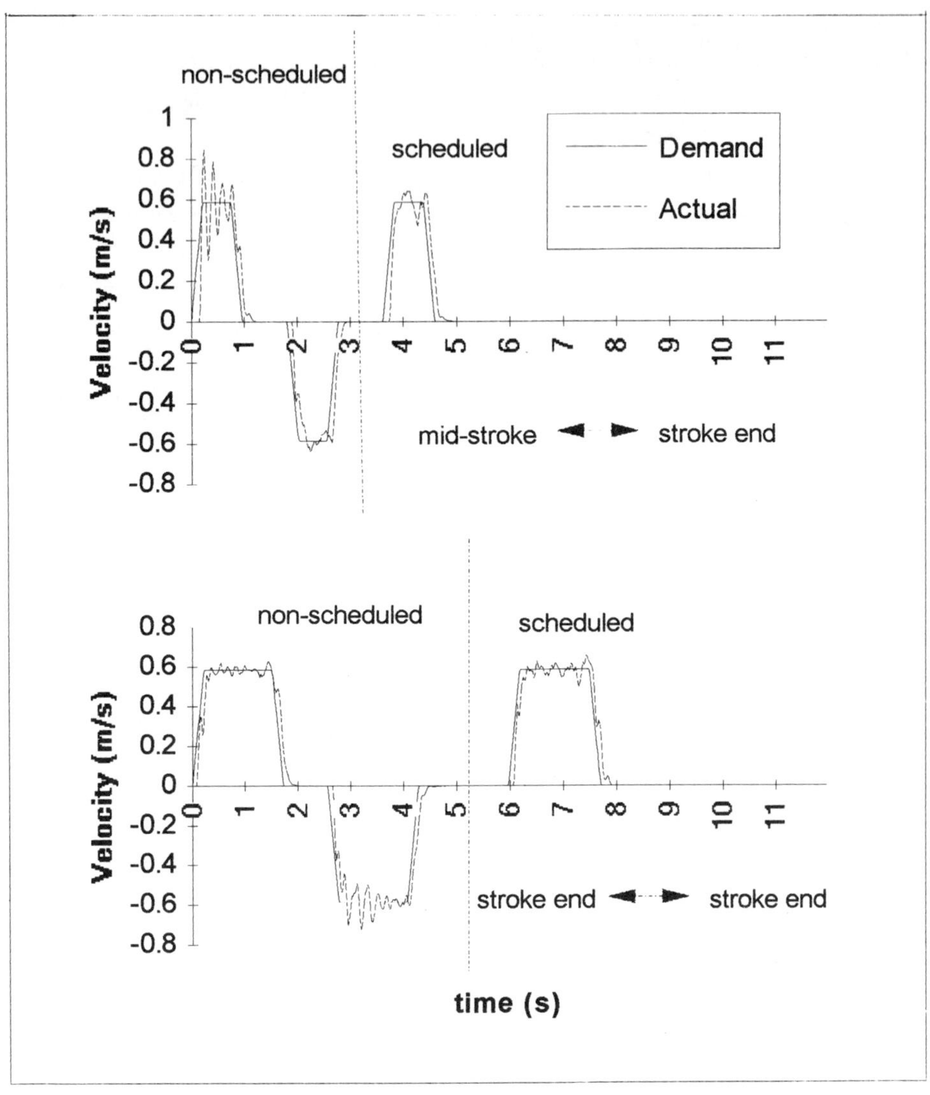


Figure 7 Scheduling of Acceleration Feedback Gain

**5. Conclusions and Future Work**

This report presents the preliminary results of an investigation into the use of "gain scheduling" methods applied to pneumatic servos. The authors wish to emphasise that the results reported in this paper are not yet conclusive in nature, as:

- The scheduling schemes are based on a simplified model involving many assumptions. The model assumes small perturbations for linear analysis; the conclusions thus derived may not be entirely applicable to pneumatic servos which exhibit such highly non-linear system dynamics.

- The gain scheduling scheme should ideally be continuous. Due to technical constraints, the scheduling is currently implemented as a relatively small number of discrete steps (typically 10 steps over the actuator stroke).

However, the results gained do indicate the potential significance of gain-scheduling as a method for improving the performance consistency when controlling pneumatic servo drives. The results obtained from this study indicate that the acceleration and velocity feedback gain constants are the most appropriate subjects for scheduling. The scheduling of the proportional gain certainly produces a significant effect on the system response but may give rise to other more complex problems. The following work is recommended in order to establish a systematic approach for implementing gain-scheduling methods with a sound theoretical and experimental foundation.

(i) A simulation based study of the actual system is considered to be important to enable the effects of various scheduling schemes to be thoroughly investigated, which is difficult to achieve by pure experimental means. For example, it would be desirable to see how characteristic equation root locations are affected by load position and how various schedules accommodate these variations.

(ii) Appropriate software tools need to be devised to aid in the implementation of gain scheduling methods and to assist in drive system tuning.

Current research work is progressing at Loughborough in these associated areas.

**Acknowledgement**

The authors would like to express their sincere thanks to Mr. C B Wong for his help in formatting the experimental results.

**6. References**

[1] **Pu, J., Moore, P.R., and Weston, R.H.**, "High-gain Control and Tuning Strategy for Vane-type Reciprocating Pneumatic Servo Drives", International Journal of Production Research, Vol.29, No.8, pp1587-1601, 1991

[2] **Weston, R.H., Moore, P.R., Thatcher, T.W., and Morgan, G.**, "Computer Controlled Pneumatic Servo Drives", I.Mech.E, 198B(14), pp 175-281, 1984

[3] **Backe W**, 1986, "The Application of Servo-Pneumatic Drives for Flexible Mechanical Handling Techniques", Robotics 2, North Holland, pp45-56.

[4] **Burrows, C.R.**, 1969, "Effect of Position in the Stability of Pneumatic Servo Mechanisms", Research Notes in Mechanical Engineering Science, II, 615-616

[5] **Pu, J., and Weston, R.H.**, 1988, "Motion Control of Pneumatic Drives", Journal of Microprocessors and Microsystems, Volume 12, No 7, pp373-382

[6] **Nagarajan, R., and Weston, R.H.**, "Front End Control Schemes for Pneumatic Servo Driven Modules", IMechE Part B 271-277 (1985)

[7] **Pu, J., and Weston, R.H.**, 1990, "Steady State Analysis of Pneumatic Servo Drives", Proc Instn Mech Engrs Vol 204, Part C, Journal of Mechanical Engineering Science, pp377-387

[8] **Pu, J., Weston, R.H., and Moore, P.R.**, "Digital Motion Control and Profile Planning for Pneumatic Servos", Journal of Dynamic Systems, Measurement and Control, ASME, December Issue, Vol.114 pp634-640 , 1992

**WRITTEN DISCUSSION**
**A study of gain-scheduling method for controlling the motion of pneumatic servos**
**J Pu, PR Moore, R Harrison & RH Weston (Loughborough University of Technology, UK)**

**Question:** **W Backé**
**IHP, RWTH Aachen, Germany**

Do you think it is appropriate to change the proportional gain $K_p$? The proportional gain $K_p$ determines the overall amplification and by this the accuracy reached in the end position. Thus it seems to be better to let $K_p$ remain constant and adapt $K_v$ and $K_a$ to produce a fast and well-damped dynamic performance.

**Answer:**

The effects of scheduling $K_p$ are investigated in this paper for comparison with those of scheduling $K_v$ and $K_a$. The paper does suggest in the conclusion that it would be more appropriate to schedule $K_v$ and $K_a$ rather than $K_p$ due to unpredictable effects which can thus be produced.

However, the authosr would like to point out that the accuracy attained will be dependent upon several factors, e.g. hysteresis and lap condition of the controlling valve, friction and leakage characteristics of the actuator and the type of the payload to be positioned (i.e. horizontal or vertical position). A detailed discussion of such factors can be found in reference [5] and [7]. In other words, the accuracy achieved at different locations along the working stroke can vary.

A constant $K_p$ may generate a constant accuracy along the working range of the stroke if (i) an over-lapped control valve is used (so that no compressed air will be allowed to exit the actuator chambers when the payload has reached its final position); (ii) a symmetric or rodless cylinder is used; and (iii) only pure inertia load is applied.

Nevertheless, if the use of constant $K_p$ at the target location is desirable to achieve good accuracy, the proportional gain can still be scheduled in such a way to meet this requirement.

**Question:** **ND Vaughan**
**Fluid Power Centre, Bath, UK**

Please explain what you mean by "uniform stiffness".

**Answer:**

'Uniform stiffness' is a borrowed term, referring to the forward path gain of the closed-loop system or the ability of the closed-loop system to react to errors caused by external disturbances.

**Question:** **JB Gamble**

**Vickers Systems Division, Trinova Ltd, UK**

Please could you comment further on how you chose the control gains $K_p$, $K_v$ and $K_a$, and how you decide on the range over which they are scheduled.

Also, how did you obtain the velocity and acceleration in the experimental system?

**Answer:**

As mentioned in the paper, the velocity and acceleration signal is derived in software, using a simple backward difference algorithm.

The choice of $K_p$, $K_v$ and $K_a$ can be established via theoretical or empirical means, with reference to Equation (8) and Figure 3. The range over which they are scheduled can be application dependent.

The results shown in this paper are a preliminary nature. Further investigation is required to provide a definitive method for scheduling the control gains to achieve improved or optimised results.

# 15. Frequency Characteristics Analysis of a Pneumatic Servo System using M-Sequence and Sinusoidal Input Signals

K Araki, X Liu *and* T Ohnishi


## Abstract

This paper describes the comparison of the pneumatic system frequency characteristics obtained with the M–sequence and the sinusoidal input methods. The digital simulation of the system frequency characteristics with sinusoidal input signal is very time consuming, so it has been proposed to use the M–sequence instead of the sinusoidal signal for shortening the calculation time. However, it is important to be aware to what extent the analysis results obtained from the M–sequence method agree with those obtained from the sinusoidal input method for a system including nonlinear components. This paper discusses the different effects of the two signals on the system frequency characteristics through the analysis of a pneumatic valve–controlled cylinder system.

The system includes two parts, namely the valve and the cylinder. The valve has nonlinear characteristics between the flow versus valve displacement and the flow versus valve port pressures. The former nonlinearity is mainly studied here. The cylinder is assumed to be linear and the system frequency characteristics are determined. From the analysis, it is found that the bias error of the frequency characteristics obtained from the M–sequence method can be predicted and compensated for when the nonlinear component is placed at first in the loop of the system, and as long as the component property does not depend on time. Comparisons between the two methods show that the results agree well when the valve port pressures are high and differ a little when the valve port pressures are low.

Finally, as an application of the M–sequence method, a parasitic oscillation phenomenon of a pneumatic servo system with the valve–controlled cylinder is simulated and explained using the system frequency characteristics obtained from the M–sequence method.


## 1 Introduction

The describing function method is often used for the frequency characteristics analysis of a system which includes nonlinear components. In many cases, the describing function is difficult to obtain in a function form, especially when the system includes multiple nonlinear components, e g the pneumatic system. In such cases, computer analysis is necessary. However, the ordinary sinusoidal signal input method to get the frequency characteristics is too time consuming for practical use. To shorten the calculation time, it has been proposed that the M–sequence be used instead of the sinusoidal input [1], [2], [3]. Since the nonlinear component shows different characteristics for different input signals, the substitution of the M–sequence will also introduce some errors. If the error can be estimated, then the necessary compensation can be included in the analysis.

This paper describes some features of the M–sequence method through the analysis of a pneumatic valve–controlled cylinder system using both the M–sequence and the sinusoidal input methods. The cylinder is assumed to be a linear second order element, and the valve is simplified to a typical nonlinear model. The different effects of the valve nonlinearity on the system characteristics to the M–sequence and the sinusoidal inputs are discussed.

A program for the analysis of the frequency characteristics of the valve–controlled cylinder system has been previously developed based on the sinusoidal input method [5]. Here the analysis results obtained from the program and from the M–sequence method are compared for various system parameters and signal amplitudes.

It has been found that in some cases, a parasitic oscillation phenomenon may occur according to the valve underlap condition when the valve–controlled cylinder system is connected into a closed–loop system. This phenomenon will be simulated and explained from the system open–loop frequency characteristics obtained from the M–sequence method.

## 2 Nomenclature

| | |
|---|---|
| $A_m$ | sectional area of cylinder piston 12.1cm$^2$ |
| $b_p$ | port width of control orifice 1.61cm |
| $c_d$ | discharge coefficient 0.6 |
| $c_p$ | number of shift registers |
| $D_m$ | damping factor of piston 0.4Ns/m |
| $e_d$ | feedback signal V |
| $e_i$ | input signal V |
| $E_{av}$ | average error |
| $E_{rr}(\ )$ | errors of frequency characteristics |
| $F_c$ | Coulomb friction force |
| $f_0$ | fundamental frequency of m(t) |
| $G_L(\ )$ | shown in figure 4 |
| $G_{nm}(\ )$ | decaying function of components for M–sequence input |
| $G_{ns}(\ )$ | describing function |
| $G_m(\ )$ | system frequency characteristics for M–sequence input |
| $G_r$ | system stability redundancy |
| $G_s(\ )$ | frequency characteristics for sinusoidal input |
| i | unevenness of valve underlap $(\Delta_1/\Delta)-1$ |
| $K_{ci}$ | amplifier gain of servo–valve 0.03cm/V |
| $K_m$ | forward gain |
| $K_{fd}$ | feedback gain 31.6V/cm |
| m( ) | M–sequence |
| $M_m$ | piston mass with load 0.39kg |
| $p_e$ | atmospheric pressure 0.1013MPa |
| $p_L$, $p_L'$ | pressures in two chambers of cylinder MPa |
| $p_s$ | supply pressure 1.0MPa |
| $T_d$ | minimum time interval of m(t) |
| $T_p$ | period of m(t) |
| $T_s$ | time constant of valve 0.015s |
| $V_m$ | half internal volume of cylinder 50cm$^3$ |
| w, w' | mass flows into cylinder kg/s |
| v | velocity of piston |
| $y_i$ | spool displacement cm |
| $y_0$ | normalized $\lvert y_i \rvert$ $(=\lvert y_i \rvert/\Delta)$ |
| z | piston displacement cm |
| $\Delta$ | mean underlap 37μm |
| $\Delta_1$, $\Delta_2$, $\Delta_1'$, $\Delta_2'$ | underlaps of valve control orifices |
| $\kappa$ | ratio of specific heats of air 1.4 |
| $\xi_m$ | damping factor 0.3 |
| $\rho_e$ | density of air at standard condition |
| $\omega_m$ | resonance frequency 80rad/s |

## 3 Pneumatic valve–controlled cylinder and the analysis method

Figure 1 shows the structure of a valve–controlled cylinder system. The mass flows through the control orifices are determined by

$$\left.\begin{aligned} w_1 &= Ff(\Delta_1 + y_i)h_1, \quad w_2 = Ff(\Delta_2 - y_i)h_2 \\ w'_1 &= Ff(\Delta'_1 - y_i)h'_1, \quad w'_2 = Ff(\Delta'_2 + y_i)h'_2 \end{aligned}\right\} \tag{1}$$

$$w = w_1 - w_2, \quad w' = -w'_1 + w'_2$$

where

$$F = c_d b_p \sqrt{\frac{2\kappa}{\kappa - 1}\frac{\rho_e}{p_e}} p_s H$$

$$H = \sqrt{\frac{\kappa - 1}{\kappa + 1}\left(\frac{2}{\kappa + 1}\right)^{\frac{2}{\kappa - 1}}}$$

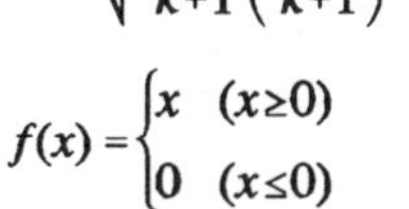

$$f(x) = \begin{cases} x & (x \geq 0) \\ 0 & (x \leq 0) \end{cases}$$

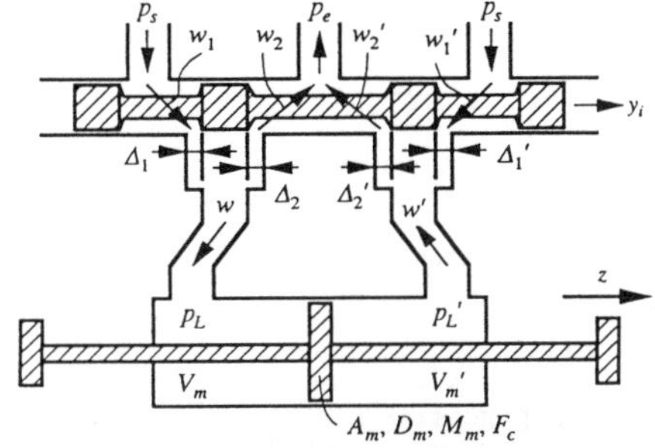


Figure 1 Physical model of the pneumatic valve-controlled cylinder

$$h_1 = h_1(p_L) = \begin{cases} \dfrac{\sqrt{\left(\dfrac{p_L}{p_s}\right)^{\frac{2}{\kappa}} - \left(\dfrac{p_L}{p_s}\right)^{\frac{\kappa+1}{\kappa}}}}{H} & \left(1 \geq \dfrac{p_L}{p_s} \geq 0.5283\right) \\ 1 & \left(0.5283 \geq \dfrac{p_L}{p_s} \geq \dfrac{p_e}{p_s}\right) \end{cases}$$

$$h_2 = h_2(p_L) = \begin{cases} \left(\dfrac{p_L}{p_s}\right)^{\frac{\kappa+1}{2\kappa}} \dfrac{\sqrt{\left(\dfrac{p_e}{p_L}\right)^{\frac{2}{\kappa}} - \left(\dfrac{p_e}{p_L}\right)^{\frac{\kappa+1}{\kappa}}}}{H} & \left(1 \geq \dfrac{p_e}{p_L} \geq 0.5283\right) \\ \left(\dfrac{p_L}{p_s}\right)^{\frac{\kappa+1}{2\kappa}} & \left(0.5283 \geq \dfrac{p_e}{p_L} \geq \dfrac{p_e}{p_s}\right) \end{cases}$$

$$h'_1 = h_1(p'_L), \quad h'_2 = h_2(p'_L)$$

The flow continuity equations are

$$\frac{dp_L}{dt} = \frac{\kappa p_L}{V_m \rho_L} w - \frac{\kappa p_L A_m}{V_m}\frac{dz}{dt}, \quad \frac{dp'_L}{dt} = -\frac{\kappa p'_L}{V'_m \rho'_L} w' + \frac{\kappa p'_L A_m}{V'_m}\frac{dz}{dt} \tag{2}$$

And the motion equation of the piston is

$$M_m \frac{dv}{dt} + D_m v + F_c \operatorname{sgn}(v) = (p_L - p_L^{\prime}) A_m \tag{3}$$

The frequency characteristics of the system from the spool displacement $y_i$ to the piston velocity v has been analyzed in [5] based on the fundamental components like the describing function method. This method is called here the analytical method. The M–sequence method has been introduced in [1], and its flow chart of the calculation is shown in figure 2.

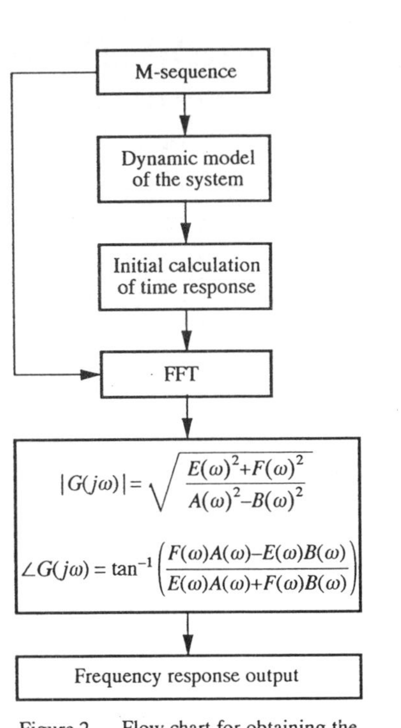


Figure 2 Flow chart for obtaining the frequency response

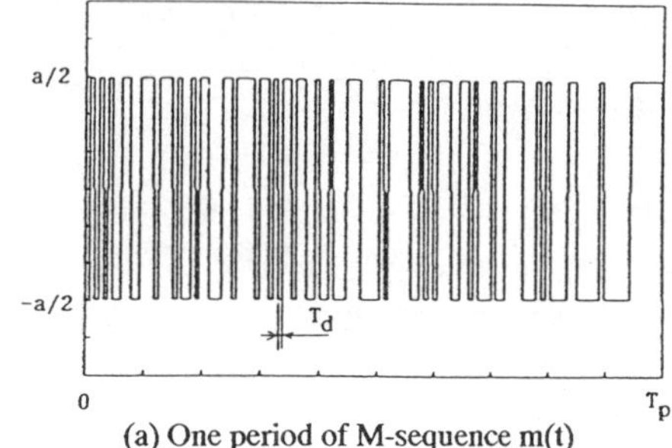


(a) One period of M-sequence m(t)

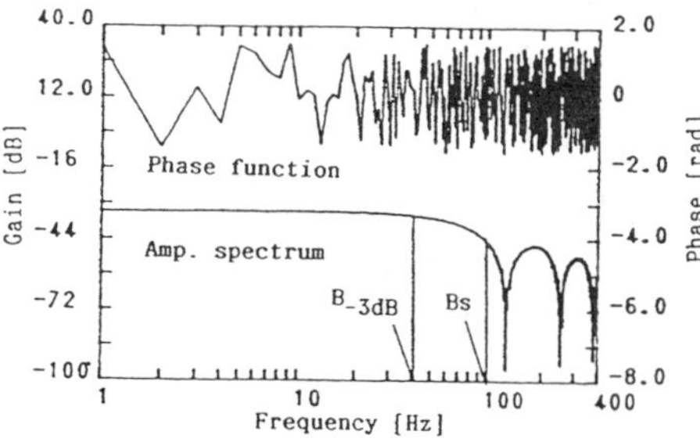


(b) Amplitude spectrum and phase function of m(t)

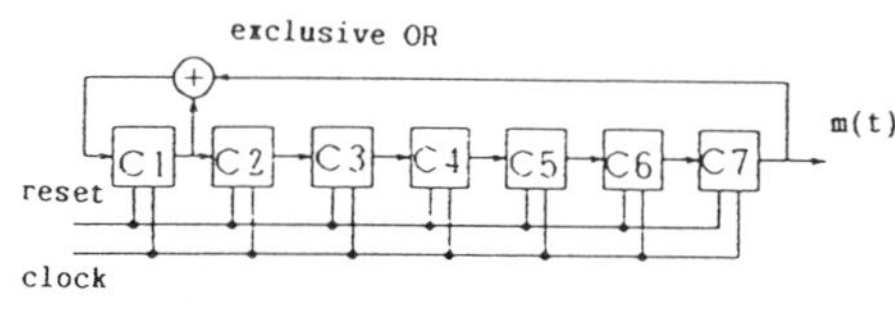


(c) M-sequence generator ($C_p$=7)

Figure 3 M-sequence

The M–sequence is a pseudo random pulse signal, an example of which is shown as a time function m(t) in figure 3(a). This is a two–level M–sequence wave which has the period $T_p$ and the minimum time interval $T_d$ and the constant amplitude (two–levels: +a/2 and –a/2 in this case). Its amplitude spectrum is shown in figure 3(b), from which we can see the M–sequence has the flat amplitude spectrum in the low frequency range and the random phase characteristics. The former characteristics are used for obtaining the frequency response at a time from the M-sequence response. The M-sequence generator of figure 3(a) is shown in figure 3(c), which is composed of 7 shift registers and an exclusive-OR circuit. The parameters of an M-sequence are given by

$$T_p = N_p T_d,\ N_p = 2^{C_p} - 1,\ f_0 = \frac{1}{T_p} \tag{4}$$

# 4 Comparison of analytical results

## 4.1 INFLUENCE OF THE SERVO-VALVE ON THE FREQUENCY CHARACTERISTICS

From the system mathematical model, it can be said that for an operating point the system mainly includes three kinds of nonlinearity, the valve flow versus displacement characteristics, the compressibility of the air in the cylinder, and the Coulomb friction acting on the cylinder piston. As the influence of the Coulomb friction has already been discussed in [8], in this paper $F_c$ is assumed to be 0. For the clear understanding of the influence of each nonlinearity on the system performance, we assume at first the cylinder is a linear second order element, so the system shown in figure 1 can be simplified as shown in figure 4, where $G_n$ indicates the relation of $y_i$ and the mass flow into the cylinder $w_i$. Since $w_i$ versus $y_i$ relation of the valve depends on various factors, the quantitative determination of the relation is difficult, $G_n$ is assumed to have the characteristics as shown in figure 5. Actually, many nonlinear components can be modelled with different $K_1/K_2$. For example, when $K_1/K_2=0$, it becomes a dead zone, and for $K_1/K_2=\infty$, it becomes a saturation.

When the sinusoidal signal passes through the component, the response becomes y(t) as shown in figure 6(a), and the describing function of the component can be derived from it as follows,

$$G_{ns}\left(\frac{\Delta}{y_i}\right) = \frac{2K_1}{\pi}\left[\sin^{-1}\left(\frac{\Delta}{y_i}\right) + \frac{\Delta}{y_i}\sqrt{1-\left(\frac{\Delta}{y_i}\right)^2}\right] + \frac{2K_2}{\pi}\left[\frac{\pi}{2} - \sin^{-1}\left(\frac{\Delta}{y_i}\right) - \frac{\Delta}{y_i}\sqrt{1-\left(\frac{\Delta}{y_i}\right)^2}\right] \tag{5}$$

When the M-sequence passes through the component, the response becomes as shown in figure 6(b). Since the M-sequence is an on-off signal, and $G_n$ is not a function of time, the response of $G_n$ is still an M-sequence which is the same as the input M-sequence but has a different amplitude. A decaying function $G_{nm}$ is introduced to describe the change, which is defined as the ratio of the output to the input signal amplitudes. According to figure 6(b), $G_{nm}$ is given by

$$G_{nm}\left(\frac{\Delta}{y_i}\right) = \frac{w_i}{y_i} = \frac{K_1\Delta + K_2(y_i-\Delta)}{y_i} = (K_1-K_2)\frac{\Delta}{y_i} + K_2 \tag{6}$$

The frequency characteristics of the system in figure 4 obtained from the sinusoidal and the M–sequence signals are expressed with $G_s(j\omega)$ and $G_m(j\omega)$ respectively,

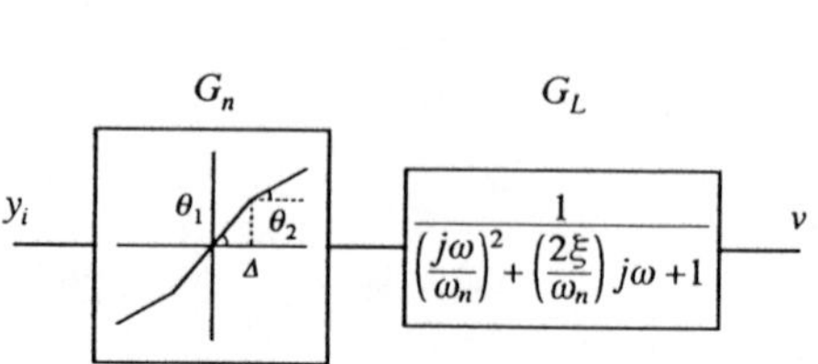


Figure 4 Simplified model of the system in figure 1

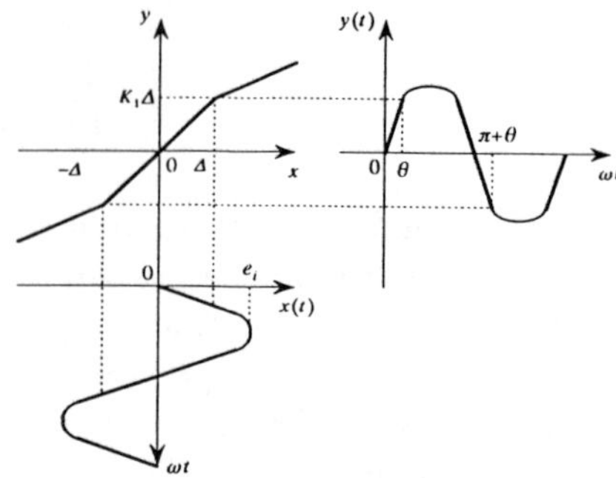


(a) Response to the sinusoidal signal

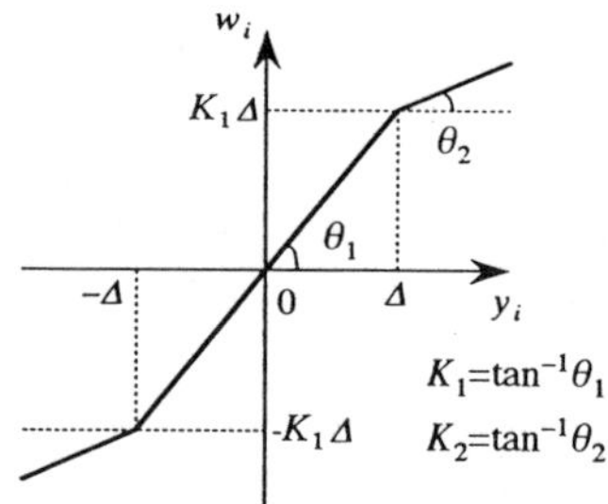


Figure 5 Simplified flow characteristics of the underlap valve

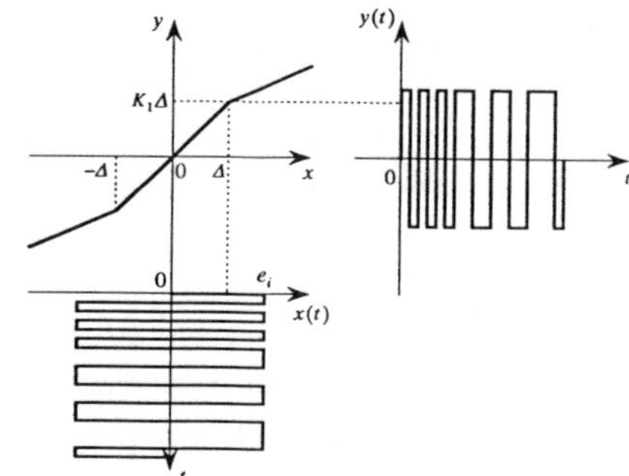


(b) Response to the M-sequence

Figure 6 Responses of the nonlinear component to the sinusoidal and the M-sequence signals

$$G_s(j\omega)=G_{ns}\left(\frac{\Delta}{y_i}\right)G_L(j\omega) \tag{7}$$

$$G_m(j\omega)=G_{nm}\left(\frac{\Delta}{y_i}\right)G_L(j\omega) \tag{8}$$

The error of $G_m(j\omega)$ compared with $G_s(j\omega)$ is

$$E_{rr}(j\omega)=20\log\frac{G_s(j\omega)}{G_m(j\omega)}=20\log\frac{G_{ns}\left(\dfrac{\Delta}{y_i}\right)}{G_{nm}\left(\dfrac{\Delta}{y_i}\right)} \tag{9}$$

$E_{rr}(j\omega)$ versus $\Delta/y_i$ for various $K_1/K_2$ is shown in figure 7. (9) indicates that the error of the two kinds of frequency characteristics only depends on the signal amplitude and does

not depend on the frequency, so the error can be estimated and compensated for.
For example, as the parameters of $G_L(j\omega)$ of the system in figure 4 are $\omega_n$=80[rad/s], $\xi$=0.3, $K_1/K_2$=4, $y_0$=0.1, 0.5 and 0.8, the frequency characteristics obtained from the two methods are shown in figure 8(a), (b), (c). The average error of $G_m(j\omega)$ compared with $G_s(j\omega)$ for all the frequencies is given by

$$E_{av}=\sum [G_s(j\omega)-G_m(j\omega)] \tag{10}$$

The gain errors $E_{av}$ for the three cases are 0.507[dB], 1.05[dB] and 0.683[dB] respectively. According to figure 7, the estimated errors for the relative cases are 0.527[dB], 1.06[dB] and 0.706[dB], which agree well with the practical calculation.
From the above comparison, when the nonlinear component is placed in front of a system, and as long as the component is not a function of time, the bias error caused by the different waveforms of the sinusoidal signal and the M-sequence is independent of the frequency, therefore it can be estimated and compensated for. In practice, as the estimation mentioned above may be difficult to carry out, a simple way is to use a sinusoidal signal as the input to calculate the system frequency characteristics at a frequency, and to use the result to modify the results obtained from the M-sequence method.

## 4.2 FREQUENCY CHARACTERISTICS OF THE VALVE-CONTROLLED CYLINDER SYSTEM

The system shown in figure 1 is analyzed using both the analytical and the M-sequence methods. It has been known that the unevenness of the valve underlap i, which is defined by (11), has a significant influence on the system performance [5], [6].

$$i=\frac{\Delta_1-\Delta}{\Delta},\quad \Delta=\frac{\Delta_1+\Delta_2}{2} \tag{11}$$

As i takes 0.3, 0, −0.3, −0.6 and the peak-to-peak value of the signal is $y_0$=2, the comparison results obtained from the two methods are displayed in figure 9(a), (b), (c), (d). From the figures it can be seen that for i≥0, the two results agree well. But as i decreases, the difference becomes clear, especially when i=−0.6, even though the resonance frequencies show a little difference.
For i=−0.6 and $y_0$=5, 2, 1 and 0.5, the comparison results are shown in figure 10(a), (b), (c), (d). It can be found that among the four cases the bias error of the two gains takes a maximum value when $y_0$ takes a middle value, that is to say $y_0$=2. This is similar to the case shown in figure 8.
The error for the resonance frequency is caused by the compressibility of the air in the cylinder. According to [5], the approximate transfer function of the system is given by

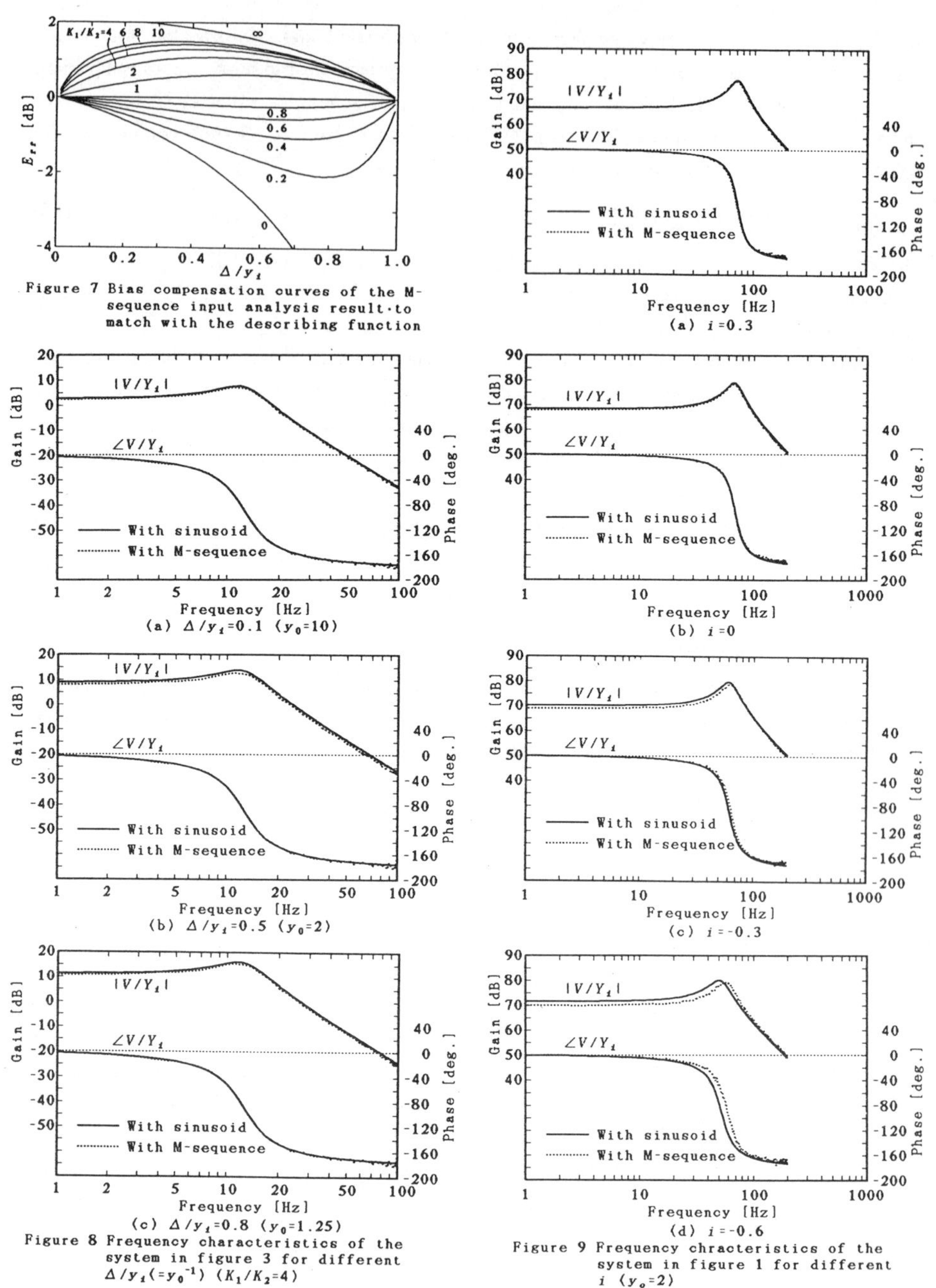


Figure 7 Bias compensation curves of the M-sequence input analysis result·to match with the describing function

Figure 8 Frequency characteristics of the system in figure 3 for different $\Delta/y_i (=y_0^{-1})$ $(K_1/K_2=4)$

Figure 9 Frequency chracteristics of the system in figure 1 for different $i$ $(y_o=2)$

$$G_z(j\omega)=\frac{V(j\omega)}{Y_i(j\omega)}=\frac{K_m}{\left(\frac{j\omega}{\omega_n}\right)^2+\left(\frac{2\xi_m}{\omega_n}\right)j\omega+1} \tag{12}$$

where

$$K_m \propto \frac{w_0}{y_0}\frac{1}{p_L} \tag{13}$$

$$\omega_n \propto \sqrt{p_L} \tag{14}$$

In (13), the change of $w_0/y_0$ is less than that of $1/p_L$. For the M–sequence input signal, the equivalent $K_m$ of $G_m(j\omega)$ is smaller than that for the sinusoidal signal input, so for the same $y_0$, the average pressure in the cylinder for the M–sequence input will be larger. And according to (14), the resonance frequency $\omega_n$ obtained from the M–sequence input would be larger than that obtained from the sinusoidal input.

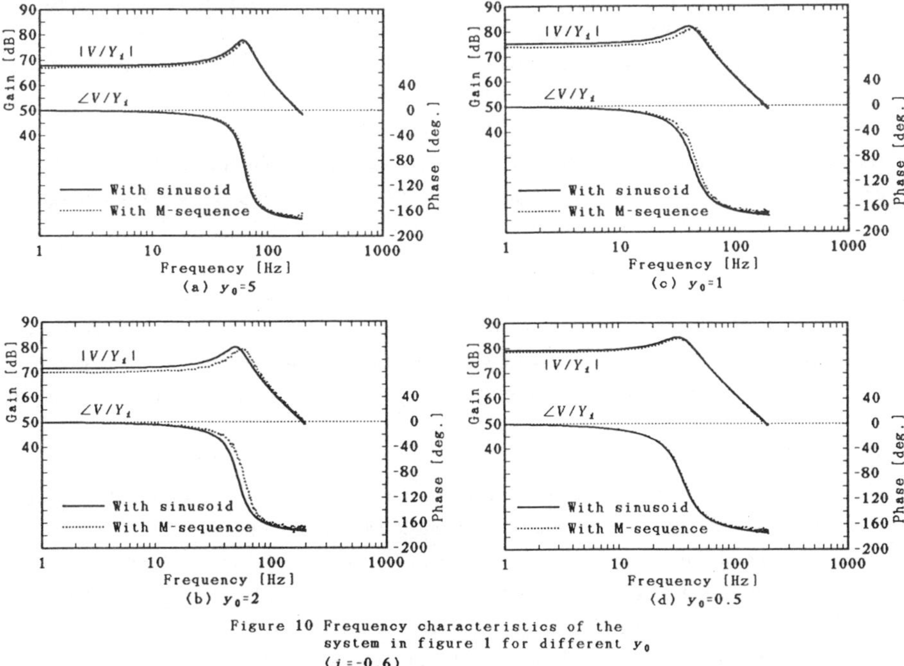


Figure 10 Frequency characteristics of the system in figure 1 for different $y_0$ ($i$=-0.6)

## 5 The analysis of the parasitic oscillation phenomenon with the M–sequence method

When the valve–controlled cylinder in figure 1 is connected into a closed–loop system, we get the block diagram shown in figure 11, where $G_f(j\omega)$ is the equivalent transfer function of the servo amplifier and the first stage of the servo valve which is approximately a first order lag given by;

$$G_f(j\omega)=\frac{K_{ei}}{T_s j\omega+1} \tag{15}$$

$K_{fd}$, shown in figure 11, indicates the total feedback gain of the displacement sensor. It has been found that when i=–0.6, a parasitic oscillation may appear in the system response. Figure 12(a) is a photo taken from the experiment, which shows the step response of the system, and figure 12(b) is the digital simulation results [7]. From the figure it can be seen that following a step change, the system response is oscillatory, and after the oscillation amplitude will gradually increase to a certain value, it will decay and finally settles down. To investigate the reason for the oscillation, the system open–loop frequency characteristic has been analyzed using the M–sequence method. Firstly, for $|e_i|$=0.4[V] (this refers to the beginning of the oscillation), the system frequency characteristics from the input signal $e_i$ to the feedback signal, $E_d(j\omega)/E_i(j\omega)$ is shown by the solid line in figure 13. As the phase diagram passes through the –180[deg] line, the gain diagram is above 0[dB] (the gain margin is negative), so in this case, the system is unstable, which explains why the oscillation at the beginning increases. For $|e_i|$=2[V] (this refers to the case when the oscillation is nearly in the maximum state), the frequency characteristics are shown by the dotted line in figure 13. In this case, the system becomes stable, which explains why the oscillation will subsequently decay. At both the beginning and the maximum oscillating part, the piston velocity is rather large, so the influence of the Coulomb friction is rather small, and it has little effect on the system stability. Friction is ignored in the frequency analysis made above.

From figure 13, it can be said that the system may have a stable limit cycle in ordinary case and the oscillation will keep on with a definite amplitude. But in this special case, the influence of the Coulomb friction (about 10[N]) will become very large as the oscillation decays to rather small amplitude and finally the oscillation will settle down.

## 6 Conclusion

This paper compares the M–sequence and the sinusoidal input methods for the frequency characteristics analysis of a pneumatic system. It is shown that

1) As the nonlinear component is in front of a linear system, and as long as the component is not a function of time, the bias error caused by the different waveforms of the two methods can be estimated and compensated for in the analysis.
2) For the normal cases, that is to say the unevenness of the valve underlap $i \ge 0$, the frequency characteristics obtained from the two methods agree well, but as $i<0$, a difference will occur. The difference on the gains is due to the nonlinear $w_i$ versus $y_i$ characteristics of the valve; And the difference in the resonance frequency is related both the $w_i$ versus $y_i$ characteristics and the compressibility of the air in the cylinder.

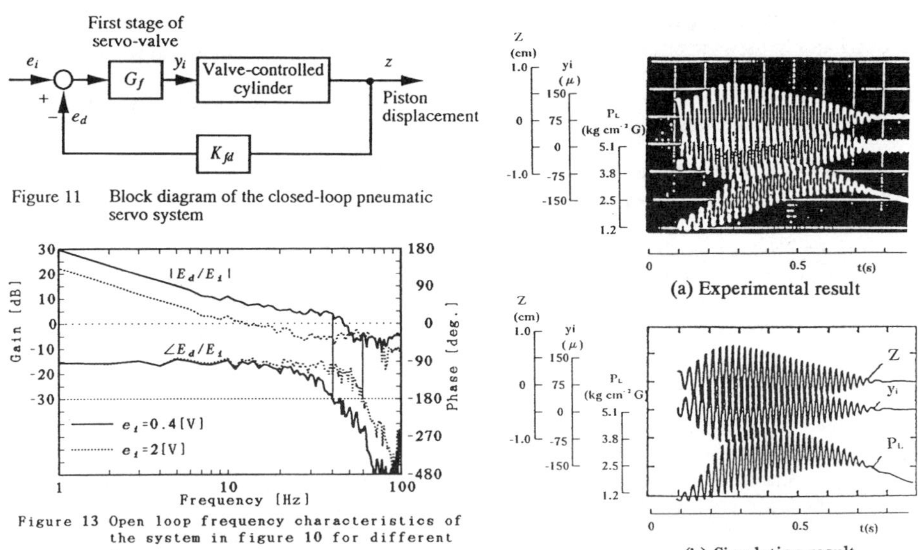


Figure 11 Block diagram of the closed-loop pneumatic servo system

Figure 13 Open loop frequency characteristics of the system in figure 10 for different $e_i$

Figure 12 The experimental and the simulation parasitic oscillation phenomenon of system in figure 11 ($i$=-0.6)

3) The parasitic oscillation of the servo system for i=–0.6 is caused by the system having a stable limit cycle, which makes the system oscillate when the loop gain is high enough.

# References


[1] **Araki K, Liu X–P, Ishino Y.** "Frequency Response Simulation with M–sequence and Its Application in Pneumatics", 3rd Triennial International Symposium on Fluid Control, Measurement, and Visualization (FLUCOME'91), San Francisco, USA, Aug 28–31, 1991 pp 527–534

[2] **Goodwin G C, Payne R L.** *Dynamic System Identification: Experiment Design and Data Analysis*, Academic Press, USA, 1977, ISBN 0–12–289750–1

[3] **Golomb S W.** *Shift Register Sequences*, Holden–day, 1967 pp 85–88

[4] **Ezekiel F D, Shearer J L.** "Pressure–Flow Characteristics of Pneumatic Valves", Trans. ASME, Vol.79, pp 1577–1590, 1957

[5] **Araki K.** "Frequency Response of a Pneumatic Valve–Controlled Cylinder with an Uneven–Underlap Four–Way Valve, Part I and Part II", The Journal of Fluid Control, Vol 15 No 1, pp 47–94, 1986

[6] **Araki K.** "Effect of Valve Configuration on a Pneumatic Servo", 6th BHRA Fluid Power Symposium, Cambridge, UK, pp 271–291, 1981

[7] **Araki K, Liu X–P, Ehana Y, et al.** "The Development of Bond Graph Simulation and Its Application in Pneumatics", 1st International Symposium on Fluid Power Transmission and Control, Beijing, China, 1991, pp 459–463

[8] **Araki K, Liu X–P.** "Frequency Characteristics Analysis of a Pneumatic Fatigue Tester", 3rd International Conference on Fluid Power Transmission and Control ('93 ICFP), Hangzhou, China, Sep, 1993

## WRITTEN DISCUSSION

**Frequency characteristics analysis of pneumatic servo system using M-sequence and sinusoidal input signals**

**K Araki (Saitama University, Japan), X Liu (Hitachi Construction Machinery Co Ltd, Japan) & T Ohnishi (Saitama University, Japan)**

**Question:** **ND Vaughan**
**Fluid Power Centre, Bath, UK**

1. Fig 6b does not take account of valve dynamics. Surely the output will be modified if valve dynamics are included?
2. Please comment on the comparison between M sequence and random noise.
3. Have you considered other transform methods - for example, wavelet analysis?

**Answer:**

1. Fig 6b does not take account of valve dynamics. When the bandwidth of the valve is almost equal to or less than that of the valve-cylinder system, the output may change from sharp-edged pulse train (Fig 6b) into rounded waves and the transfer characteristics, eqn. (6), must be modified. But, when the bandwidth of the valve is about one order higher than that of the valve-cylinder system, the parameters $T_d$, and $T_p$ and $C_p$ of an M-sequence will be properly decided from the dynamics of the valve-cylinder system, and the output is almost the same as Fig 6b. In pneumatic systems, valve responses are mostly very fast compared with valve-cylinders because of the difference in size.

2. Random noise generated by the Random generator block of Vis Sim ver. 1.2 (Visual Simulations) were applied to the input of the system. Here the average of the absolute values of any random noise used is the same value as the amplitude of the M-sequence pulse train and the data number of any random noise used is the same as the number, $2^{12}$=4096, of the time increments in the R.K.G. analysis.

   Fig a-1 shows the simulation results obtained with the sinusoidal, M-sequence and random noise signal inputs to the system. Both the errors of the M-sequence and random noise results are small in this case. But the characteristic curves in the case of the random noise input are not so smooth and have spiky noise. Other filters and windows were used for the random noise, but the simulation results changed little. The spiky noise may be removed by applying the coherence blanking but to achieve smoothness is not so easy.

   Fig a-2 shows a result which was obtained from averaging 10 sets of frequency characteristics and shows less noise. However, the calculation time for this figure is about 10 times that of the M-sequence. If we limit the calculation time to a few minutes, which will be generally acceptable, the M-sequence methods looks better than the random noise method.

3. The wavelet transform method looks very appropriate for analysis of pneumatic systems, whose characteristics change significantly with change of the operating point (which may change over the duration of the M-sequence input). We would like to apply wavelet analysis to the pneumatic system in the future.

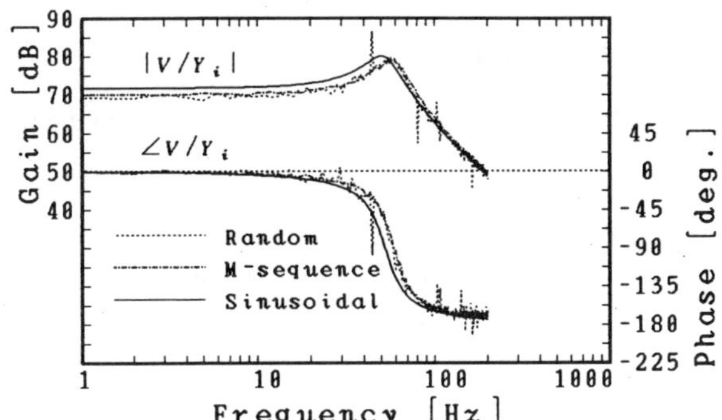


Figure a–1

$$\begin{pmatrix} y_0=2 & i=-0.6 \\ C_p=8 & T_p=1[\text{s}] \end{pmatrix}$$

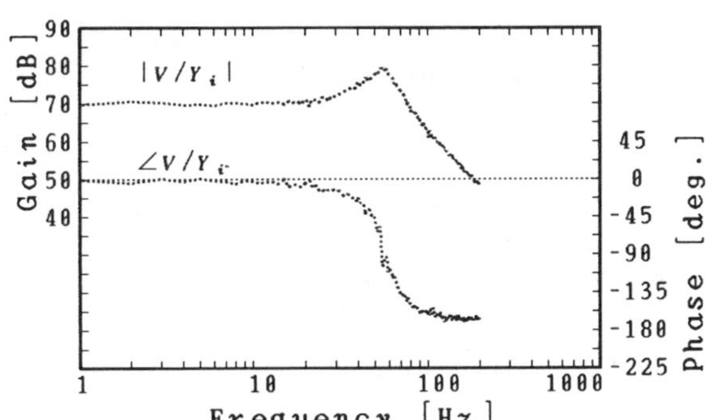


Figure a–2

$$\begin{pmatrix} y_0=2 & i=-0.6 \\ C_p=8 & T_p=1[\text{s}] \end{pmatrix}$$

**Question:** M Sethson
University of Linköping

Have you used different numbers of shift registers? If so, how are the results affected?

**Answer:**

Figure b-1 shows the effects of the number of the shift registers, $C_p$. From the figure, we can see the data obtained for $C_p$=6 and 7 show significant scatter. $C_p$=8 is more suitable for this particular case.

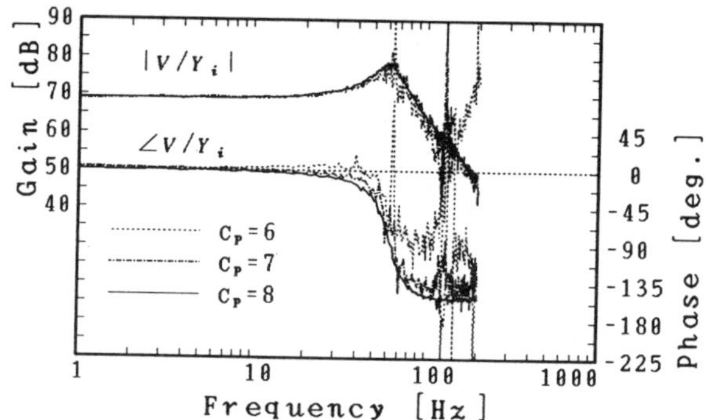


Figure b–1

$$\begin{pmatrix} y_0=2 \quad i=-0.6 \\ T_p=1[\mathrm{s}] \end{pmatrix}$$

# 16. An Experimental Comparative Study of Linear and Nonlinear Adaptive Pressure Regulation

A Bouhal, E Richard *and* S Scavarda


## ABSTRACT

Linear adaptive methods have been successfully used in several applications and particularly in pneumatic and hydraulic equipments [1][2]. However these methods have some limitations in presence of non negligible nonlinearity and can lead in non "ideal" situations to instability problems due for example to large reference input variations or disturbances [3].

An application of a linear and a nonlinear adaptive method to improve a pressure regulation system is proposed and some experimental results are given. The design procedure of the nonlinear adaptive method is systematic and its stability proof uses simple analytical tools. We have pointed out in the experimental results the advantage and the drawback of both methods in order to analyze objectively and to conclude about the improvement acquired by each of these methods.


## 1 INTRODUCTION

The purpose of this paper is to suggest a comparative application of a linear and a nonlinear adaptive method to improve control pressure evolution in the pneumatic process. The linear method is based on a classical pole assignment control and a recursive least squares estimation algorithm. Some additional tools such as deadzone, filtering or data normalization have also been used to improve the robustness of the estimator and the stability of the adaptive scheme. The design procedure of the linear adaptive controller only requires definition of the order and the structure of the sampled linearized model. The nonlinear method was initially introduced by Sastry and Isidori [4] and this method is very helpful in the design of controllers for systems containing both unknown parameters and known nonlinearities.

It consists of an adaptive input-output feedback linearization followed by a tracking control law for the obtained linearized system. The tracking error is then used as a performance index by a parameter estimation algorithm to update the process parameters in order to efficiently cancel the model nonlinear terms by the input-output feedback.

The class of nonlinear system is described (in the single-input, single-output case) by

$$\dot{\mathbf{X}} = f(\mathbf{X}) + g(\mathbf{X})\,u$$
$$y = h(\mathbf{X})$$

with the state $\mathbf{X} \in \mathbf{R}^n$, the input u, the output y, f, g smooth vectors fields and h a smooth nonlinear function (smooth means that we can be cavalier in differentiating as often as we like without worrying about the existence of the derivatives).

The paper is organized as follows. First the description and modeling of the process are given followed by the sampled linearized model and the adaptive pole assignment design procedure. Then the model is reformulated and parametrized in order to design the adaptive nonlinear controller. Finally some practical implementation remarks and experimental results are given and commented on.

## 2 MODELING OF THE EXPERIMENTAL SYSTEM

The experimental system given in figure1 is a pressure regulation process. It consists of a tank, a pressure sensor and a servovalve with negligible dynamics The servoamplifier is considered as a proportional element. The pressure sensor provides the pressure P to the microcomputer via a 12-bit A/D converter with a sensitivity of 1 bit per 3.47mbar.

In order to compare the improvement acquired by both the linear and the nonlinear adaptive methods in different operating conditions some disturbances were introduced, such as increase or decrease of the tank volume V, variation of the supply pressure $P_s$ or the orifice area $A_u$ (ie variation of the output flow).

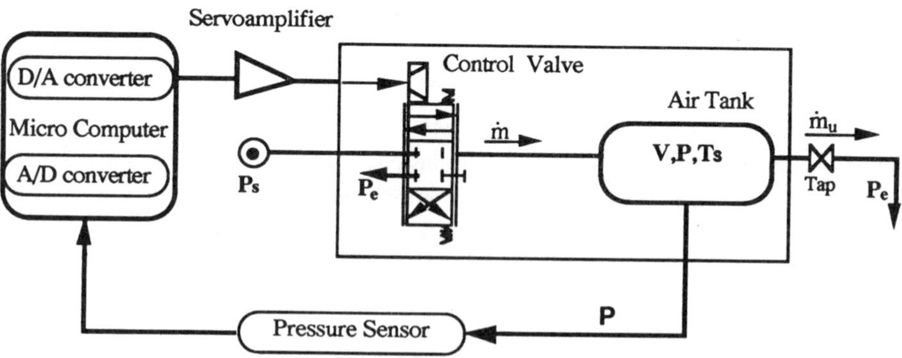


Fig1: The experimental system

In this paper, a simplified nonlinear model has been used [5][6][7] which is sufficiently related to take into account the global behavior of the chamber without being too complex. This model is obtained by assuming that :

- pressure and temperature within the tank are homogeneous,
- the air flow remains turbulent,
- air is a perfect gas with a constant r,
- the process is polytropic with a constant K,
- temperature changes are neglected with regard to the mean temperature $T_S$.

The nonlinear model of the system is:

$$\frac{dP}{dt} = \frac{K\, r\, T_S}{V} \left[ \dot{m}(i,P) - \dot{m}_u(P) \right] \tag{1}$$

$$\dot{m}(i,P) = A_s(i)\, D(P_s,P) - A_e(i)\, D(P,P_e) \tag{2a}$$

$$\dot{m}_u(P) = A_u\, D(P,P_e) \tag{2b}$$

where:

- $\dot{m}$ and $\dot{m}_u$ are respectively the input and the output mass flow rates,
- V and P are respectively the volume and the pressure within the chamber,
- $A_S(i)$ and $A_e(i)$ are the servovalve orifice areas modulated by the control current i,
- $A_u$ is the tap orifice area ,
- $D(P_S,P)$ and $D(P,P_e)$ are the normalized mass flow rate functions.

The functions $D(P_S,P)$ and $D(P,P_e)$ can be obtained analytically by introducing a corrective coefficient which depends on the shape of the orifices or more accurately by using the experimental characteristics given in figure 2 .

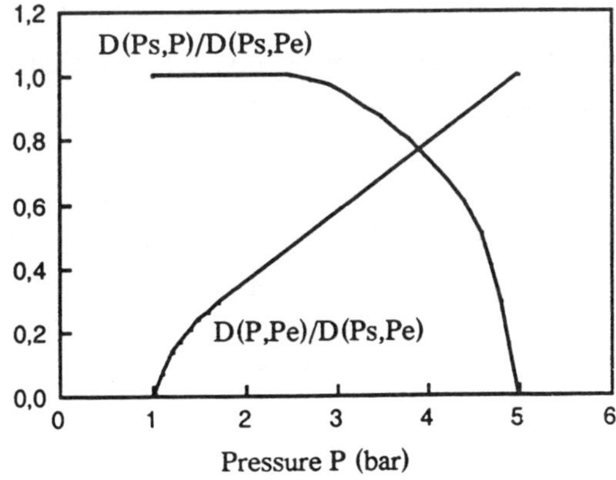


Fig 2: Normalized mass flow rate characteristics

## 3 ADAPTIVE POLE ASSIGNMENT CONTROLLER

In order to design the linear adaptive controller, the system model (1) is linearized around the equilibrium point ($i_0, P_0, P_s, A_u$) by using a Taylor's development limited to the first order and by taking into account the delay introduced by the control computation.

The obtained linearized input-output model is expressed by the following transfer function :

$$P^*(s) = \frac{G\, e^{-TS}}{1+\tau s}\, i^*(s) + \frac{1}{1+\tau s}\left[\, G_{Ps}\, P_s^*(s) - G_{Au}\, A_u^*(s) \,\right] \qquad (3)$$

where $G = \dfrac{G_i}{C_p+\lambda}$, $G_{Ps} = \dfrac{C_s}{C_p+\lambda}$, $G_{Au} = \dfrac{G_u}{C_p+\lambda}$ and $\tau = \dfrac{V}{k\, r\, T_s(C_p+\lambda)}$

with $C_p = -\left.\dfrac{\partial \dot{m}}{\partial P}\right|_0$ $\lambda = \left.\dfrac{\partial \dot{m}_u}{\partial P}\right|_0$ $G_i = \left.\dfrac{\partial \dot{m}}{\partial i}\right|_0$ $C_s = \left.\dfrac{\partial \dot{m}}{\partial P_s}\right|_0$ $G_u = \left.\dfrac{\partial \dot{m}_u}{\partial A_u}\right|_0$

T corresponds to the time delay needed to compute the control input i(t) at time t.

In the following, only step variations of $P_s$ and $A_u$ are taken into account. This assumption means that the introduction of an integral factor in the control loop and the finite differential operator in the identification loop enable us to drop the terms $A_u^*$ and $P_s^*$ from the model.

Transforming equation (3) into the discrete-time form, we obtain:

$$A(q^{-1})\, y^*(k) = q^{-1}B(q^{-1})\, u^*(k) \quad \text{where } A(q^{-1}) = 1 + a_1 q^{-1} \text{ and } B(q^{-1}) = b_1 + b_2 q^{-1} \qquad (4)$$

where k is an integer meaning the time kTe,

$q^{-1}$ is a backward shift operator $q^{-1}\, y^*(k+1) = y^*(k)$.

The algorithm structure of the adaptive pole assignment controller can be summarized in the following steps:

**Step1: Estimation of the process parameters:**

The system input-output $u^*$ and $y^*$ are used in a recursive least squares algorithm to update the process parameters.

The discrete-time model (4) with adjustable parameter vector $\hat{\theta}(k)$ and the observation vector $\phi(k)$ can be written in the form:

$$\hat{y}^*(k) = \phi(k)^T\, \hat{\theta}(k) \qquad (5)$$

where: $\phi(k)^T = [\, -y^*(k-1)\;\; u^*(k-1)\;\; u^*(k-2)]$ and $\hat{\theta}(k)^T = [\hat{a}_1(k)\;\; \hat{b}_1(k)\;\; \hat{b}_2(k)]$

The Parameter Adaptation Algorithm (P.A.A) is given by:

$$\hat{\theta}(k) = \hat{\theta}(k-1) + P(k)\, \Delta\phi_f(k)\, (\Delta y_f(k) - \Delta\phi_f(k)^T\, \hat{\theta}(k-1)) \qquad (6a)$$

$$P(k) = \frac{1}{\lambda}\left[ P(k-1) - \frac{P(k-1)\, \Delta\phi_f(k)\, \Delta\phi_f(k)^T\, P(k-1)}{1 + \Delta\phi_f(k)^T\, P(k-1)\, \Delta\phi_f(k)} \right] \qquad (6b)$$

where P(k) is a positive definite matrix which represents the covariance of the error in the estimates. $\Delta u_f(k)$ and $\Delta y_f(k)$ are constructed by using a finite differential operator and a numerical filter in order to cope with the uncertainty about the equilibrium point and to eliminate the continuous component in the input-output information :

$$\Delta y_f(k) = \frac{1 - q^{-1}}{1 + f_1 q^{-1}} y^*(k) \qquad \Delta u_f(k) = \frac{1 - q^{-1}}{1 + f_1 q^{-1}} u^*(k) \quad \text{with } |f_1| < 1 \tag{7a}$$

$$\Delta\phi_f(k)^T = \left[ -\Delta y_f(k-1) \;\; \Delta u_f(k-1) \;\; \Delta u_f(k-2) \right] \tag{7b}$$

In order to track the time-varying parameters [2], the forgetting factor λ is self adjusted to keep the trace of the matrix P(k) constant and to avoid that the eigenvalues of P(k) vanishes.

$$\text{trace}\,[P(k)] = \text{trace}[P(k-1)] \quad \text{when trace}\,[P(k)] \leq Cst \tag{8}$$

$$\left\{ \begin{array}{ll} \text{if trace}\,[P(k)] \leq Cst & \lambda = 1 - \dfrac{\text{trace}\left[ P(k-1)\, \Delta\phi_f(k)\, \Delta\phi_f(k)^T\, P(k-1) \right]}{\text{trace}[P(k)]\, (1 + \Delta\phi_f(k)^T\, P(k-1)\, \Delta\phi_f(k))} \\ \text{if trace}\,[P(k)] > Cst & \lambda = 1 \end{array} \right\} \tag{9}$$

To improve the stability robustness of this identification scheme with respect to the unmodeled dynamics, the prediction error $e(k) = ( \Delta y_f(k) - \Delta\phi_f(k)^T \hat{\theta}(k-1) )$ is substituted by its deadzone counterpart v(k) defined as:

$$v(k) = \left\{ \begin{array}{lll} e(k) - d, & \text{if} & e(k) > d \\ 0 \quad , & \text{if} & |e(k)| < d \\ e(k) + d, & \text{if} & e(k) < -d \end{array} \right\} \tag{10}$$

d is a constant threshold below which the prediction error e(k) is ignored and adaptation mechanism stopped.

**Step2 : determination of the controller parameters and the control law**

The control objectives can be decomposed into the regulation and the tracking objectives.

To achieve the regulation requirements, the polynomials $R(q^{-1}, k)$ and $S(q^{-1}, k)$ (Fig 3) are designed at each sampling period to assign the closed-loop specified by a desired Hurwitz polynomial $P(q^{-1})$ namely :

$$A(q^{-1}, k)\, (1 - q^{-1}) S(q^{-1}, k) + q^{-1} B(q^{-1}, k)\, R(q^{-1}, k) = P(q^{-1}) \tag{11}$$

The integral factor $(1 - q^{-1})$ is introduced in order to ensure a null steady state error.

The resolvability condition of equation (11) requires that the polynomials $A(q^{-1}, k)$ and $B(q^{-1},k)$ have to be relatively prime at each sampling period.

However if one singularity point occurs ( ie if for a given time k, $A(q^{-1}, k)$ and $B(q^{-1}, k)$ are not relatively prime) we use the last admissible solution obtained.

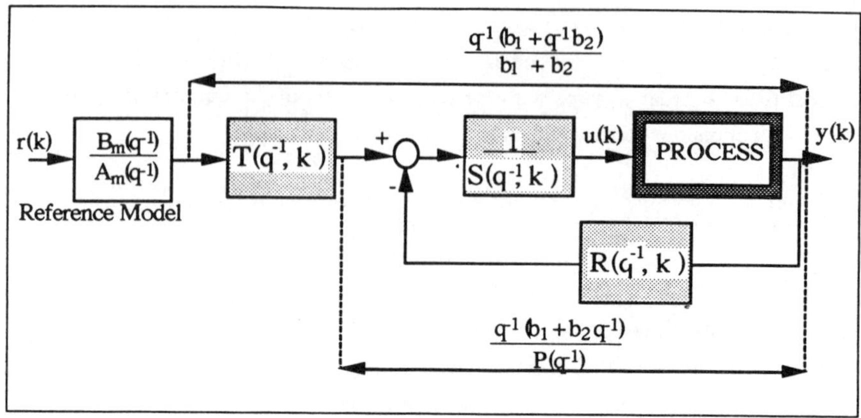


Fig 3: Structure of the pole assignment controller

To achieve the tracking requirements, the reference model $B_m(q^{-1})/A_m(q^{-1})$ is used to specify the tracking dynamic.

The polynomial $T(q^{-1},k)$ is used to cancel the regulation dynamic and to ensure a unite steady state gain between the process output $y(k)$ and the reference output $y_m(k)$:

$$T(q^{-1},k) = P(q^{-1}) \left\{ \frac{R(q^{-1},k)}{P(q^{-1})} \right\}_{q=1} \tag{12}$$

The robust stability proof of this algorithm is treated in [9].

## 4 NONLINEAR ADAPTIVE CONTROL SCHEME

The proposed adaptive nonlinear control scheme is derived for systems which are linearly parametrized and where the driving input appears linearly. Since the process model (1) is nonlinear with respect to the driving input i, then before evolving the controller design procedure the process model (1) needs firstly to be linear with respect to the driving input and secondly to be linearly parametrized with respect to the unknown parameters.

**First step: reformulation of the driving input**

To obtain a linear dependence vis a vis to the driving input, we introduce a new input $A^* = \psi(i)$ [8], where $\psi$ is a continuous bijective function (Fig.4).

This function is obtained by splitting the two servovalve orifice areas into a leakage area $A_l$, supposed to be constant, and a control area $A^*$.

$$\text{if } i \geq i_0 \quad \left\{ \begin{array}{l} A_s(i) = A^*(i) + A_l \\ A_e(i) = A_l \end{array} \right\} \qquad \text{if } i < i_0 \quad \left\{ \begin{array}{l} A_s(i) = A_l \\ A_e(i) = -A^*(i) + A_l \end{array} \right\} \tag{13}$$

where $i_0$ and $A_l$ are given by: $A_s(i_o) = A_e(i_o) = A_l$

By introducing the expression (13) into the equation (2) and (1) we obtain :

$$\frac{dP}{dt} = \frac{K\,r\,T_s}{V}\,[\,A_l\,D(P_s,P) - (A_l + A_u)D(P,P_e) + A^*\,\phi(P,\mathrm{sgn}\,A^*)\,] \tag{14}$$

with :

$$\phi(P,\mathrm{sgn}\,A^*) = \begin{cases} D(P_s,P) & \text{if } A^* \geq 0 \\ D(P,P_e) & \text{if } A^* < 0 \end{cases}$$

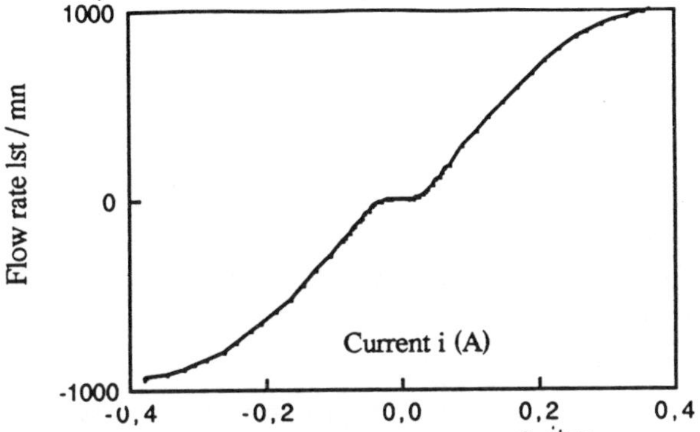

Fig 4: characteristic of the input $A^*=\Psi(i)$

Now the model (14) is linear with respect to the new driving input $A^*$

**Second step : parametrization of the model**

The design procedure of the adaptive linearizing control law requires the equation (14) to be written in the following linearly parametrized form :

$$\dot{P} = f(P) + g(P)\,i \tag{15}$$

$$y = h(P) = P$$

with $f(P) = \sum_{i=1}^{n_1} \theta_i\,f_i(P)$ and $g(P) = \sum_{j=1}^{n_2} \theta_j'\,g_j(P)$

where $\theta_i$ , $i=1\ldots,n_1$ ; $\theta_j'$, $j=1\ldots,n_2$ are unknown parameters, $f_i(P)$ and $g_j(P)$ are smooth known functions. The terms of (5) can be classified as :

(i) - Terms which characterize the process : $\frac{K\,r\,T_s}{V}$ and $A_u$ (unknown parameters)

(ii) - Terms which characterize the servovalve: $A^*=\Psi(i)$ and $A_l$ (unknown parameters)

(iii) - Flow rate functions : $D(P_s,P)$ and $D(P,P_e)$ (known nonlinear terms).

To rewrite the equation (14) in the parametrized form (15), we suppose that: $\psi(i) = G\,i$ , where G is the flow gain evaluated on the linear part of the mass flow rate characteristic $\psi(i)$.

The parameters $\theta_i$ and $\theta'_j$ can be chosen to include the terms of (i) and (ii) :

$$\theta_1 = \frac{K\,r\,T_s}{V} D(P_s,P_e) A_l \quad \theta_2 = -\frac{K\,r\,T_s}{V} D(P_s,P_e) ( A_l + A_u) \quad \theta'_1 = \frac{K\,r\,T_s}{V} G\, D(P_s,P_e) \quad (16a)$$

The functions $f_1(P)$, $f_2(P)$ and $g_1(P)$ (fig 2) are then defined by:

$$f_1(P) = \frac{D(P_s,P)}{D(P_s,P_e)}, \quad f_2(P) = \frac{D(P,P_e)}{D(P_s,P_e)} \quad g_1(P,\text{sgn } i) = \begin{cases} f_1(P) & \text{if } i \geq 0 \\ f_2(P) & \text{if } i < 0 \end{cases} \quad (16b)$$

## 4.1 Exact input-output linearization techniques

### 4.1.1 Linearization

It is well known that a large class of nonlinear systems can be made to have linear input-output behavior through a choice of nonlinear state feedback control laws [10][11].

To review the theory, let us consider the single-input, single-output nonlinear system

$$\dot{\mathbf{X}} = f(\mathbf{X}) + g(\mathbf{X})\,u$$
$$y = h(\mathbf{X}) \quad (17)$$

with $\mathbf{X} \in \mathbf{R}^n$, $u \in \mathbf{R}$, f, g smooth vector fields and h smooth nonlinear function.

Differentiating y with respect to time, one obtains

$$\dot{y} = L_f h + L_g h\, u$$

where:

$$L_f h = \frac{\partial h}{\partial \mathbf{X}} f(\mathbf{X}) \quad \text{and} \quad L_g h = \frac{\partial h}{\partial \mathbf{X}} g(\mathbf{X})$$

stands for the Lie derivatives of h with respect to f,g respectively

If $(L_g h)(\mathbf{X}) \neq 0 \ \forall\ \mathbf{X} \in \mathbf{R}^n$, then the control law: $u = \frac{1}{L_g h}(-L_f h + v)$

yields the new linear system from the input v to y : $\dot{y} = v$

More generally if $\varrho$ is the smallest integer such that $L_g L_f^i h \equiv 0$ for $i=0,..,\varrho-2$ and

$L_g L_f^{\varrho-1} h \neq 0 \ \ \forall\ \mathbf{X} \in \mathbf{R}^n$, then the control law given by

$$u = \frac{1}{L_g L_f^{\varrho-1} h}(-L_f^{\varrho} h + v) \quad \text{yields} \quad y^{(\varrho)} = v$$

and the system (17) is said to have strict relative degree $\varrho$.

Differentiating now the output of (15) with respect to time, one obtains

$$\dot{P} = f_1(P)\theta_1 + f_2(P)\theta_2 + g_1(P,\text{sgn } i)\theta'_1\, i \quad (18)$$

The function $g_1(P,\text{sign } i)$ presents two points of singularity; ($P=P_e$ , $i<0$) and ($P=P_s$ , $i>0$). Notice that those two points are exterior to the "normal" operating conditions of the process, and since $g_1$ depends on the sign of the driving input i, to overpass these singularities points the idea consists of posing i=0 and hence the system (15) has a relative degree $\varrho = 1$.

The control law

$$i = \frac{1}{g_1(P,\text{sgn } i)\theta_1'}\left\{ -\left[f_1(P)\theta_1 + f_2(P)\theta_2\right] + v\right\} \tag{19}$$

yields the linear system $\dot{P} = v$

### 4.1.2 Tracking

We now apply this linear form to the tracking problem. We desire to have P track a given $P_m$ by choosing v as

$$v = \dot{P}_m + \alpha\,( P_m - P) \quad \alpha>0 \tag{20}$$

It can be shown that this control results in asymptotic tracking and bounded state P provided that $P_m$ is bounded.

## 4.2 Adaptive input-output linearization techniques

The chief drawback of the exact input-output linearization techniques presented above (in practical implementation) is that they are based on exact cancellation of the nonlinear terms $[f_1(P)\theta_1 + f_2(P)\theta_2]$ and $[g_1(P,\text{sgn } i)\theta_1']$. If there is any uncertainty in the knowledge of those terms, the cancellation is not exact and the resulting input-output equation is not linear. To deal with those uncertainties, we suggest the use of parameter estimation algorithm to get asymptotically exact cancellation and then asymptotically linearize the system.

The estimates at time t : $\hat{\theta}(t) = \left[\hat{\theta}_1(t) \quad \hat{\theta}_2(t) \quad \hat{\theta}_1'(t)\right]$, are used to compute the control law $\hat{i}$

$$\hat{i} = \frac{1}{g_1(P,\text{sgn } i)\hat{\theta}_1'}\left\{ -\left[f_1(P)\hat{\theta}_1 + f_2(P)\hat{\theta}_2\right] + v\right\} \tag{21}$$

Then substituting (21) in (18) yields after some calculation to $\dot{P} = v + \phi^T\,\mathbf{W(t)}$

where $\phi = \left[\theta - \hat{\theta}(t)\right]$ is the parameter error vector and $\mathbf{W(t)} = \left[f_1(P),\ f_2(P),\ g_1(P,\text{sgn}\hat{i})\hat{i}\right]^T$ (22) is the regressor vector.

Using equation (20) and (21) this yields the following error equation relating the tracking error $e=P-P_m$ to the parameter error $\phi$:

$$\dot{e} + \alpha\, e = \left[\theta \quad - \hat{\theta}\,(\mathbf{t})\right]^T \begin{bmatrix} f_1(P) \\ f_2(P) \\ g_1(P,\text{sgn } \hat{i})\,\hat{i} \end{bmatrix} = \phi^T \mathbf{W(t)} \tag{23}$$

The adaptive tracking theorem for minimum phase systems and relative degree one concludes that lim e(t)=0 when t-->∞ by using the following parameter update law :

$$\dot{\phi} = -\,\mathbf{D}\; e(t)\; \mathbf{W(t)} \tag{24}$$

where **D** is a diagonal positive define matrix

The proof technique use the following Lyapunov function : $V(e,\phi) = \frac{1}{2}\left[\lambda\ e^2 + \phi^T D^{-1}\ \phi\right]$.

To improve the robustness of this control scheme in regard to unmodeled dynamics, a deadzone similar to (10) is used in the estimation algorithm which allows to stop the adaptation when the tracking error falls within the deadzone limits.

## 5 REMARKS CONCERNING THE IMPLEMENTATION

In the P.A.A presented in (6), rounding errors may cause the matrix P(k) to become negative definite and the algorithm unstable. To avoid this problem a solution was proposed in which P(k) is factorized as : $P(k)=U(k)^T D(k) U(k)$

where U(k) is an upper triangular matrix and D(k) is a diagonal matrix. U(k) and D(k) are then update at each iteration by the P.A.A so that the resulting matrix P(k) is guaranteed to be positive definite.

In order to implement on the microcomputer the continuous adaptive linearizing control law (21) and the parameter update law (24), a very short sampling period ($T_e$=2ms) has been used in comparison with the system dynamic. In this way the inner linearized feedback loop may be considered as a continuous loop and the outer feedback loop used for tracking (20) is implemented with a sampling period of $T_e'$=20ms. This sampling technique is used to avoid the design of the linearizing control law in discrete time which is far to be obvious and not necessary for this single-input single-output system.

## 6 EXPERIMENTAL RESULTS

The presented results concern both the linear and the nonlinear control scheme. In the following figures the mixed line represents the input step pressure Pc, the dotted line the reference pressure $P_m$ and the continuous line the process output P.

In the linear case:

To demonstrate the self adjustment ability of the controlled process we present results obtained under different starting conditions of the process; by changing the volume V and the supply pressure $P_s$.

Figures (5a) and (5b) show the process response (pressure P) to a square wave input from 2bar to 2.5bar in the conditions V=10 $cm^3$ - $P_s$ = 5bar and V=2l - $P_s$ =3bar respectively. We remark a good tracking behavior and a zero steady state error.

To show the effect of parametric variations and disturbances sudden changes in the supply pressure $P_s$ and the volume V have been provoked. Figures (6a) and (6b) show respectively the effect provoked by varying V from 2l to 10$cm^3$ and the effect of varying Ps from 5bar to 3bar. We note that after a short adaptation phase the adaptive controller adjusts its parameters to these variations and hence maintains good tracking dynamic behavior.

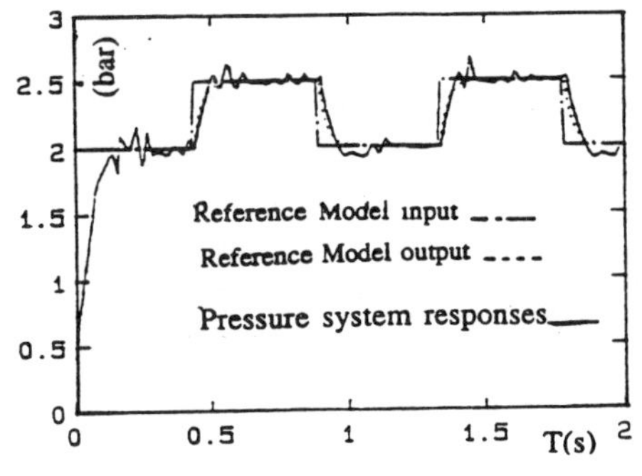


Fig 5a : V= 10 $cm^3$ - $P_s$ = 5bar

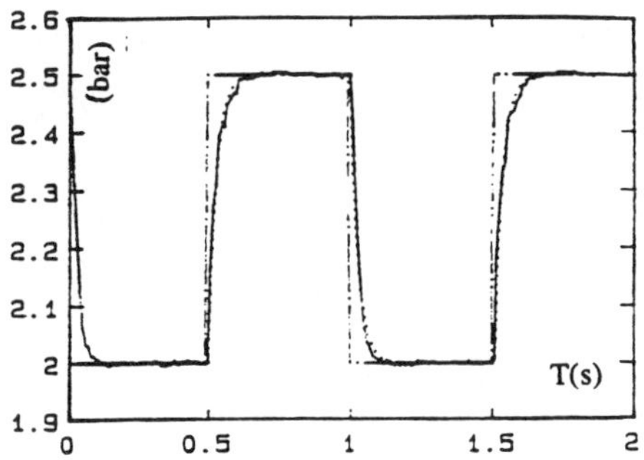


Fig 5b: V = 2 l - $P_s$ = 3bar

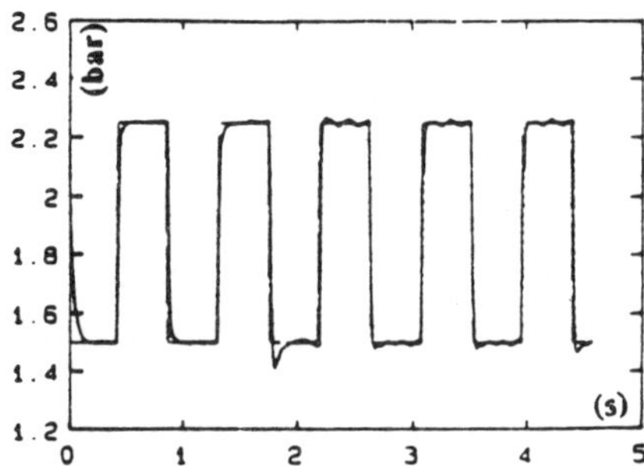


Fig 6a : Variation of V from 2 l to 10 $cm^3$

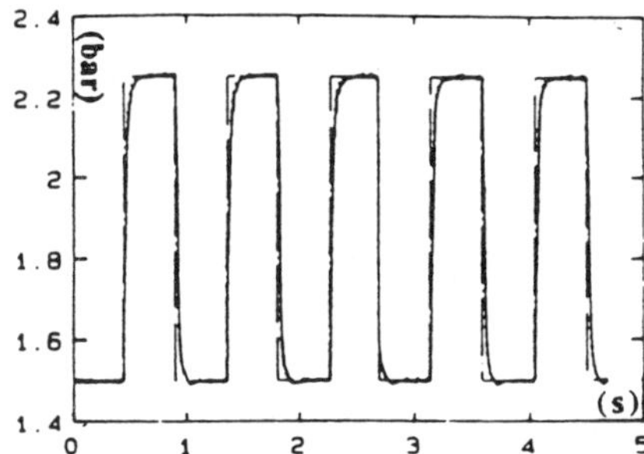


Fig 6b: Variation of Ps from 5bar to 3bar

In the nonlinear case:

To show the self adjustment of the nonlinear adaptive controller, a range of experiments was carried out such as: starting the process without any prior knowledge of the parameters ($\hat{\theta}_1(0)$=0 bar/s, $\hat{\theta}_2(0)$=0 bar/s) and with uncertainty of the servovalve gain G ($\hat{\theta}'_1(0)$=5bar/sA instead of its approximate theoretical value $\theta'_1$=30bar/sA). Figures (7a) (7b) and (7c) show the process response to a square wave input from 2bar to 3bar with the adaptation gain ($D_1$=1, $D_2$=1,$D_3$=10), ($D_1$=10, $D_2$=10, $D_3$=$10^3$) and ($D_1$=$10^2$, $D_2$=$10^2$, $D_3$=$10^5$) respectively (Di are the diagonal elements of the gain matrix **D**).

We show that the evolution of the adaptation rate depends on the gain $D_i$ which should not be too large since the algorithm will be very sensitive to measurement noises and less robust w.r.t unmodeled dynamics. Notice that the process response after a perturbed initial phase follows closely the reference output $P_m$ with however a small steady state error about 7mb and where the control input is given in figure 8.

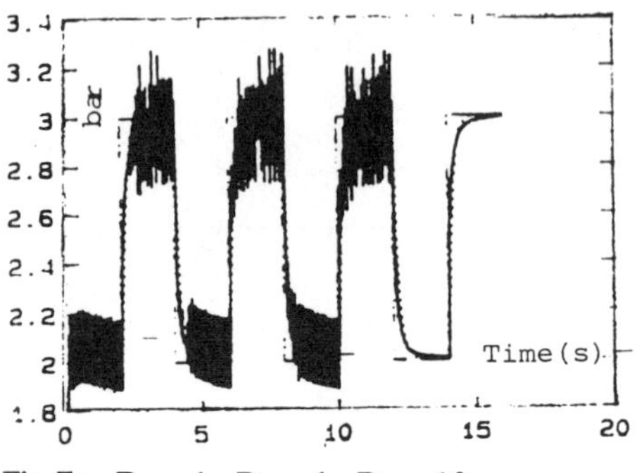

Fig 7a : $D_1 = 1$ , $D_2 = 1$ , $D_3 = 10$

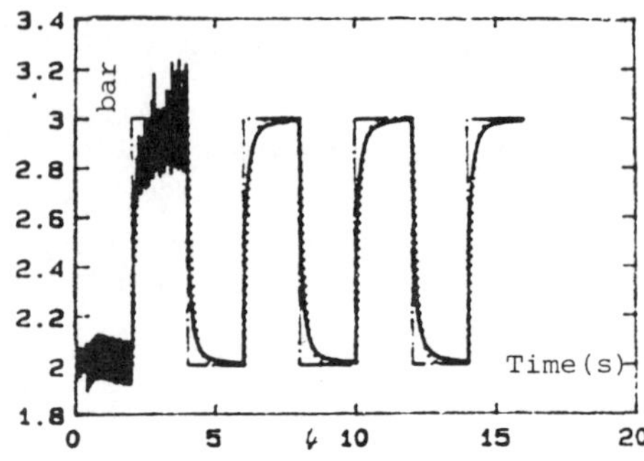

Fig 7b: $D_1 = 10$ , $D_2 = 10$ , $D_3 = 10^3$

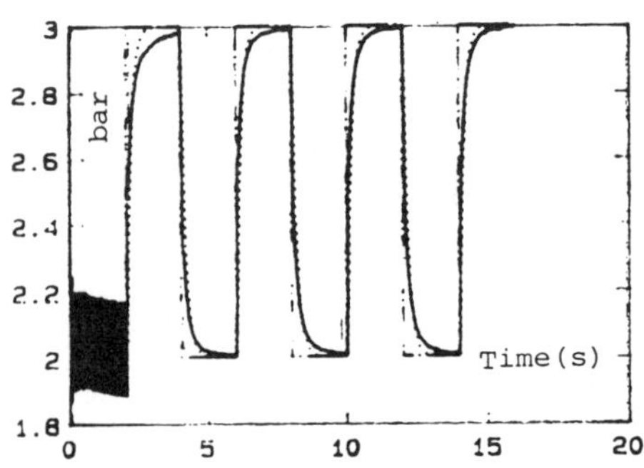

Fig 7c : $D_1 = 10^2$ , $D_2 = 10^2$ , $D_3 = 10^5$

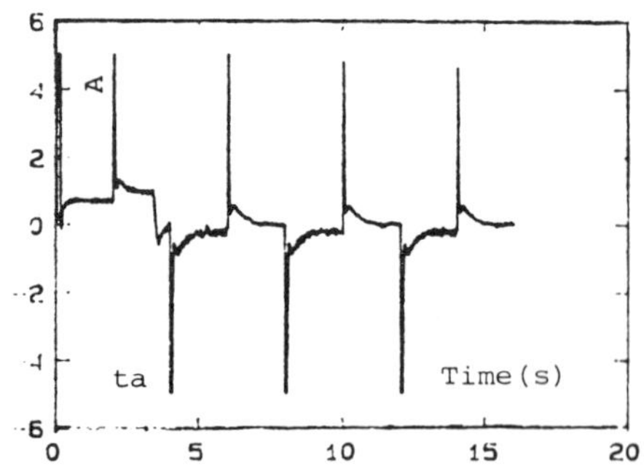

Fig 8: Control input

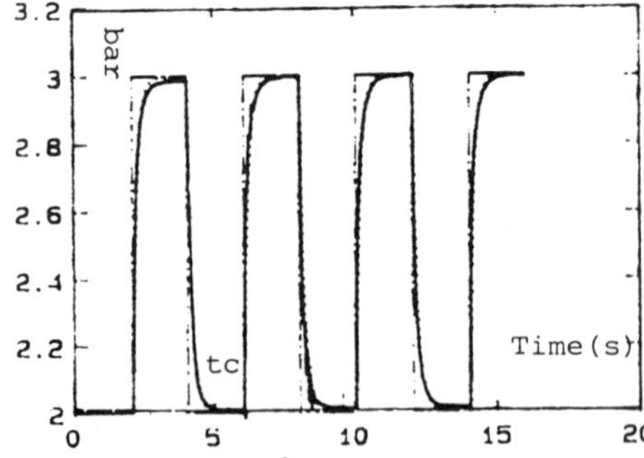

Fig 9a : Variation of $\dot{V}$ from 3.04l to 1l

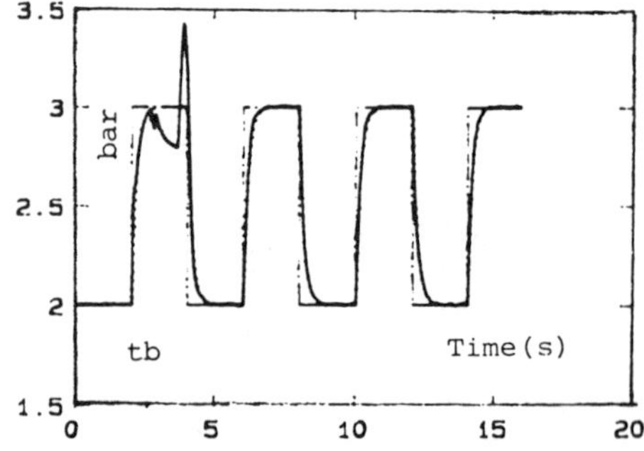

Fig 9b: Sudden closing of the orifice area $A_u$

Figures (9a) and (9b) correspond respectively to sudden variation of the volume V from 3.041 to 1l and to sudden closing of the tape orifice area Au, in both cases the adaptive nonlinear controller behave well and the controlled process is not very affected.

Linear and Nonlinear case:

The effect of a large input variation (2bar) on both linear (Fig 10a) and nonlinear (Fig 10b) adaptive cases is showed. We note that the linear controller is saturated (Fig 11a) compared to the nonlinear controller where the control input is not saturated (Fig 11b) and where the process output follow well the reference model.

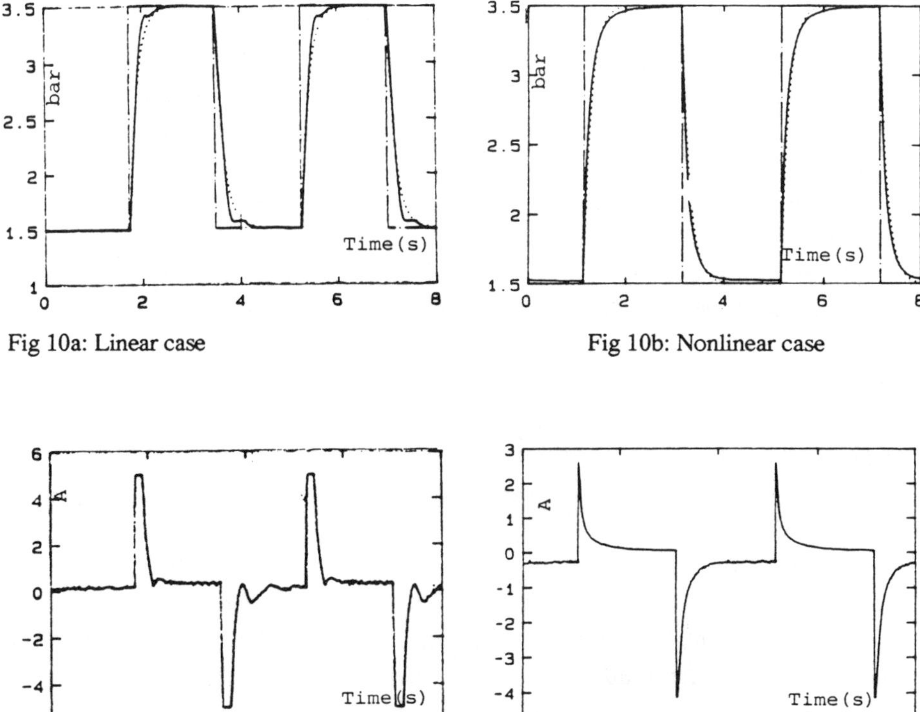


Fig 10a: Linear case

Fig 10b: Nonlinear case

Fig 11a : Linear control input:

Fig 11b: Nonlinear control input

## 7 CONCLUSION

Linear and nonlinear adaptive control designs procedure have been evolved and applied to a pressure regulation problem. The stability of both schemes is not complicated to prove and the design specifications are easily achieved.

The experimental results show that the linear and the nonlinear adaptive controller present approximately the same improvement in the case where the input variations are not very large. However for large input variations, the hypotheses concerning the linear method are violated and lead to control saturation and deterioration of the tracking and regulation performances. The integral action in the linear controller allows a null steady state error but deteriorates the performances (integral windup phenomena).

The tracking performances (dynamic, static behavior and stability) of the nonlinear adaptive approach may resolve the problem of the linear adaptive controller for large input variations.

## REFERENCES

[1] Bouhal, A, Richard, E, Scavarda, S, "Adaptive air pressure control" 10.Aachener Fluidtechnishes Kolloquim, vol 2, pp 393-406, 1992.

[2] Bouhal, A, Richard, E, Scavarda, S, Thomasset, D. "An Application of Adaptive Generalized Predictive Controller to pressure regulation " Fifth Bath International Fluid Power Workshop 1992.

[3] Kokotovic, P.V. "Foundation of Adaptive Control ", Lecture Notes in Control and Information Sciences .Springer-Verlag ,1991

[4] Sastry, S.S and Isidori, A. "Adaptive Control of linearizeable Systems," IEEE Trans.Automat.Contr.,Vol.34,no.11,pp,23-31,Nov 1989

[5] Jebar, H.S. "Design of pneumatic Actuator Systems".Ph.D:Thesis of Univ.of Nottingham,1977,426p.

[6] Shearer, J.L. "Study of Pneumatic Processes in the Continuous Control of Motion with Compressed Air," Part I and II. Trans.Am.Soc.Mech.Eng . 1956,Vol.78,P.233-249.

[7] Burrows, C.R. "Nonlinear pneumatic servomechanisms," Ph.D: Thesis, University of London, 1968, 298p.

[8] Richard, E, Scavarda, S. "Nonlinear control of a pneumatic servo drive" ,Second Bath International Fluid Power Workshop on Components and systems, Bath U.K 1989

[9] Giri , F, M'Saad, M, Dugard, L and Dion, J.M. "Robust Adaptive Regulation with minimal prior knowledge, IEEE Trans on Automatic control Vol 37 No 3, March 1992

[10] Claude, D, Fliess, M, Isidori, A. "Immersion directe et par bouclage, d'un système non linéaire dans un linéaire," C.R. Acad. Sc. Paris, 1983, t.296, Série 1,pp.237-240.

[11] Isidori, A. "Nonlinear control systems:An introduction". New York:Springer-Verlag,1985.

**WRITTEN DISCUSSION**

**An experimental comparative study of linear and nonlinear adaptive pressure regulation**

**A Bouhal, E Richard & S Scavarda (INSA de Lyon, France)**

**Question:** **KMR Sethson**
**University of Linköping, Sweden**

Have you used your analysis for estimating sensitivity to transducer errors/noise?

**Answer:**

In this particular case the pressure sensor sensitivity is about 3mbar/LSB and the information provided is not very noisy.

**Question:** **P Krus**
**University of Linköping, Sweden**

In using the constant trace algorithm for identification there may be some danger of instability in the parameter updates if a too large trace is used together with large excitation signals; $\lambda$ could even become negative. Are there simple ways of avoiding this problem?

**Answer:**

When using the least squares identification algorithm we have to avoid the gain matrix P(k) going to zero or to infinity:

$$P_1 I \leq P(k) \leq P_2 I \text{ with } 0 < P_i < \infty \quad i = 1, 2$$

We have used in our application the constant trace algorithm to assure the upper bound. The lower bound assures that the eigen values of the matrix P(k) are always positive and hence it is easy to see from the constant trace algorithm that the forgetting factor $\lambda$ will always be positive even with a large input excitation.

# 17. Digital Control with Feedforward of Hydraulic Force Actuators

F Conrad, L H Hansen, H S Larsen *and* M Zhang

## ABSTRACT

This paper presents and discusses simulation and experimental results obtained from an investigation of control schemes with feed forward for a hydraulic force actuator. In order to analyse the advantages of applying feed forward of the velocity and the acceleration in the control system first a continuously model was formulated. Based on the very promising simulation results obtained a digital controller was developed, implemented and evaluated. As expected the experimental results confirm that the control performance is improved significant by adding a feed forward of the piston velocity.

## 1. INTRODUCTION

Electro-hydraulic actuators are often applied successfully on both stationary and mobile machines for control of position and speed of heavy loaded machine parts.

A practical application example is the problem of controlling the acting force on a scanning wheel designed for the measurement of the height of sugarbeet relative to the surface of the earth. The scanning wheel has to be in contact with the surface and the top of the sugarbeet by a constant controlled force. The position signal from the scanning wheel controls the reference input to a servo controlled hydraulic cutter system of a developed prototype sugarbeet top cutting machine. Another example is the problem of controlling a force acting between a tool and a workpiece.

A promising solution for both of the above mentioned examples is to use digital controlled hydraulic actuators applying a control scheme with force feed back and feed forward of the piston velocity. It is shown in the paper that this control scheme offers a significant improvement in control performance compared to a traditional PI controller.

## 2. NOTATION

| | | | | |
|---|---|---|---|---|
| A | = | 2.22 E-4 | $m^2$ | Effective piston area |
| $C_L$ | = | 1.019 E-12 | $m^5/(s*N)$ | Leakage coefficient |
| Cl | = | $A/(C_L+K_{qp})$ | = 1.050 E8 $s*N/m^3$ | Gain factor |
| $K_{qo}$ | = | 7.2 E-2 | $m^3/(s*A)$ | Flow gain |
| $K_b$ | = | 62.77 E-5 | V/N | Force transducer gain |
| $K_v$ | = | 1.339 | V/(m/s) | Velocity transducer gain |
| $K_t$ | = | 4.249 | V/Hz | Tacho gain |
| $K_q$ | = | $K_{q0}/R$ | = 3.6E-4 $m^3/(s*V)$ | Servovalve gain |
| $K_{qp}$ | = | 1.096 E-12 | $m^5/(s*N)$ | Flow-pressure gain |
| L | = | 2.52 | H | Inductans in torque motor |
| M | = | 1.3 | kg | Moving mass |
| R | = | 200 | Ω | Coil resistance in torque motor |
| $V_t$ | = | 1.00 E-4 | $m^3$ | Volume both cylinder chambers |
| β | = | 6.865 E8 | $N/m^2$ | Bulk modulus of the oil |
| amp | = | 0.05 | m | Amplitude of the position disturbance |
| ω | = | 2π | rad/s | Frequency |
| $\tau_v$ | = | 6.4 E-3 | s/rad | Time constant for servovalve |
| $\tau_c$ | = | $V_o/(4*\beta*(C_L+K_{qp}))$ = 1.723 E-2 s/rad | | Time constant |
| $\tau_s$ | = | L/R = 1.26 E-3 s/rad | | Time constant |
| $\tau_v$ | = | 6.4 E-3 | s/rad | Time constant for servovalve |

## 3. THE HYDRAULIC FORCE ACTUATOR SYSTEM

In Figure 1 is shown a schematical diagram of the designed hydraulic force actuator system and in Figure 2 is shown the hydraulic force actuator system linked to another hydraulic servo system with a flywheel which is disturbing the force actuator with an position variation generated by the eccentric mechanism. Here the position disturbance is approximated by a sine function which generates the following velocity v and acceleration a:

$$v = \omega * amp * \sin(\omega * t)$$
$$a = \omega^2 * amp * \cos(\omega * t)$$

where t is the time, ω is the frequency, and amp is the position amplitude.

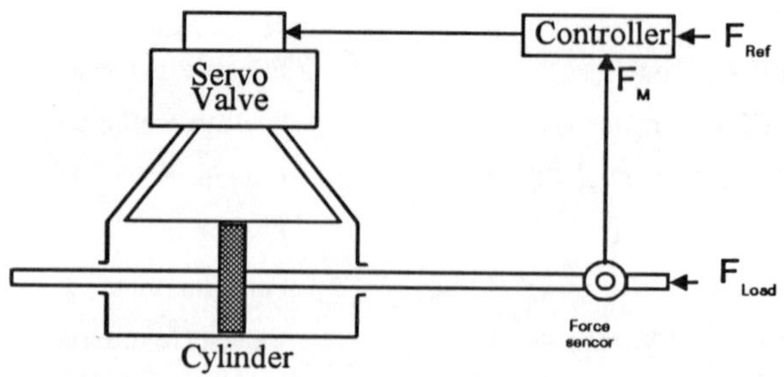


Figure 1. The hydraulic force actuator system.

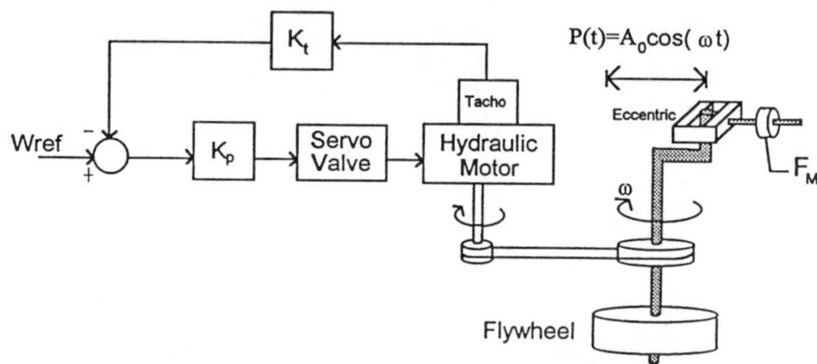


Figure 2. The force actuator connected to the disturbance system.

The implemented experimental set-up is shown on the pictures in Figure 3. The disturbance servo system with the flywheel is seen at the front end, and the hydraulic force actuator system is mounted on the machine table. The other picture is taken from the side.

Figure 3a. Top view of the experimental set-up.

Figure 3b. Side view of the experimental set-up.

## 4. DESIGN OF FORCE CONTROLLERS

In Figure 4 is shown a block diagram of the hydraulic force actuator. In this set-up only the force is fed back. The variation in piston position generated by the disturbance servo system results in a velocity disturbance and an acceleration disturbance.

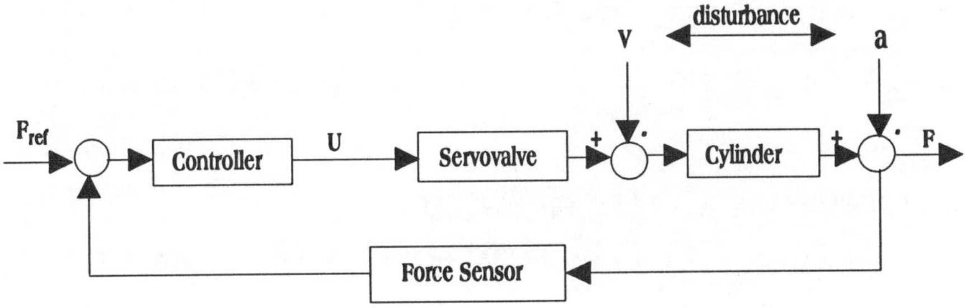


Figure 4. The hydraulic force actuator with force feedback.

For the controller design an approximated linear model of the hydraulic force actuator was developed as shown in Figure 5.

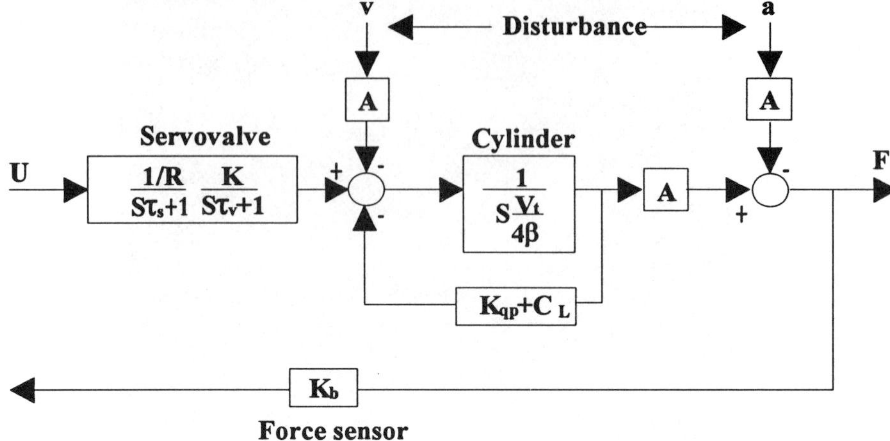


Figure 5. Approximated linear model of the hydraulic force actuator.

Based on the linear model shown in Figure 5 the following transfer function is obtained

$$G(S) = K_b \frac{\frac{1}{R}}{(S\tau_s + 1)} \frac{K}{(S\tau_v + 1)} \frac{\frac{A}{K_{qp}+C_L}}{\left(S\frac{V_t}{4\beta(K_{qp}+C_L)} + 1\right)}$$

$$= \frac{K_b K_q C_l}{(S\tau_s + 1)(S\tau_v + 1)(S\tau_c + 1)} \qquad (1)$$

where Kq=K/R, $C_l = \frac{A}{K_{qp}+C_L}$ and $\tau_c = \frac{V_t}{4\beta(K_{qp}+C_L)}$.

Defining

$$D(S) = K$$

$$C(S) = (S\tau_s + 1)(S\tau_v + 1)(S\tau_c + 1) - KK_b K_q C_l$$

$$= \tau_s \tau_v \tau_c S^3 + (\tau_s \tau_v + \tau_c(\tau_v + \tau_s))S^2 + (\tau_s + \tau_v + \tau_c)S + (1 - KK_b K_q C_l) \qquad (2)$$

and then applying the Routh-Hurwitz stability criteria it is found that stability is obtained for $0<K<0.406$ (assuming K is a positive gain).

**PI - double LEAD controller.**

Since the linear model is a 3rd order model we design the following PI - double LEAD controller.

$$D(S) = K\frac{(S\tau_i + 1)}{S\tau_i}\frac{(S\tau_{D1} + 1)}{(S\alpha\tau_{D1} + 1)}\frac{(S\tau_{D2} + 1)}{(S\alpha\tau_{D2} + 1)} \tag{3}$$

where $\alpha \in [0;1]$.

By doing so we can obtain a higher bandwidth of the closed loop system. In order to optimize the bandwidth we choose: $\tau_i = \tau_C$, $\tau_{D1} = \tau_V$, and $\tau_{D2} = \tau_S$. Again applying the Routh-Hurwitz stability criteria the system is stable for $0 < K < 0.855$.

Referring to Figure 6 the controller is designed to obtain a damping ratio $\xi \geq 0.5$ which corresponds to the gain factor $K = 0.148$.

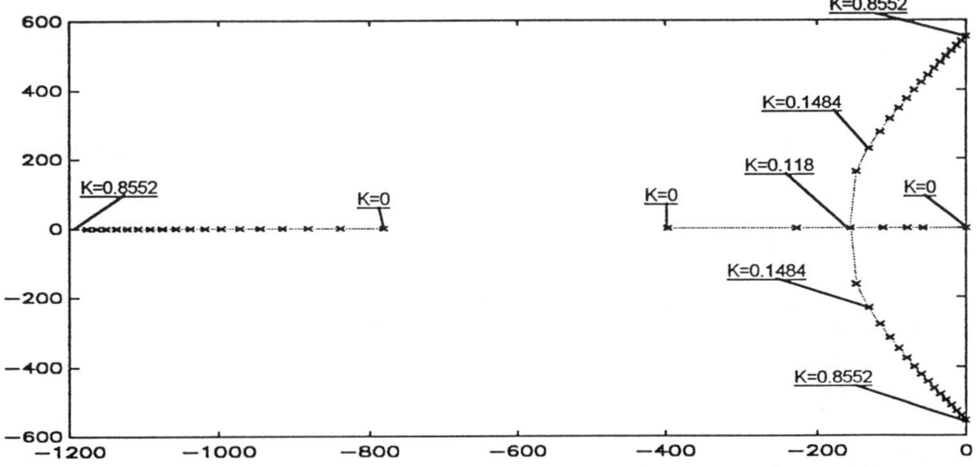


Figure 6. Root locus for the PI - double LEAD controlled force actuator

The controller is then defined by

$$D(S) = 0.1484\frac{(S\tau_c + 1)}{S\tau_c}\frac{(S\tau_v + 1)}{(S\alpha\tau_v + 1)}\frac{(S\tau_s + 1)}{(S\alpha\tau_s + 1)} \tag{4}$$

with $\alpha=0.2$ This value is selected as a compromise taking into account that we want to implement the controller on a PC and obtain a fast computation of the control algorithm.

In Figure 7 is shown an example of the obtained simulation results with the designed PI - double - LEAD controller. It is seen that this controller gives significant control performance improvement compared to the PI controller.

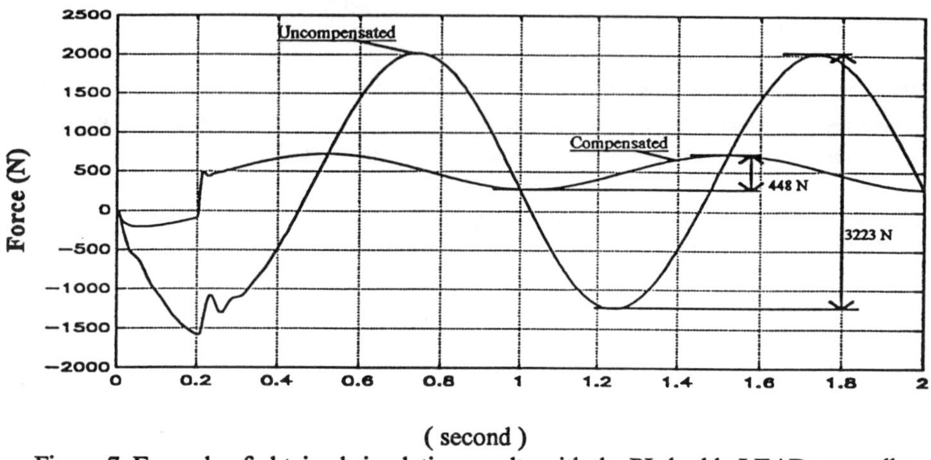


Figure 7. Example of obtained simulation results with the PI-double LEAD controller compared with the P controller.

**PI-double LEAD controller with feed forward.**

The control performance can be improved significantly byadding a feed forward as illustrated in Figure 8. In principle the best result is obtained by using the complete transfer function $F_v(s)$ and $F_a(s)$. For practical reasons of implementation we have chosen the approximated constant values

$$F_v(S) = AR(S\tau_v + 1)\,{}^{1}\!/_{K}(S\tau_s + 1)$$

$$F_a(S) = MR(S\tau_v + 1)\,{}^{1}\!/_{K}(S\tau_s + 1)\frac{C_L + K_{qp}}{A}\left(S\frac{V_t}{4\beta(C_L + K_{qp})} + 1\right) \qquad (5)$$

$$F_v = \lim_{s\to 0} F_v(S) \approx AR\,{}^{1}\!/_{K} = \underline{0.617}\ \frac{Vs}{m}$$

$$F_a = \lim_{s\to 0} F_a(S) \approx MR\,{}^{1}\!/_{K}\frac{C_L + K_{qp}}{A} = \underline{3.44 \cdot 10^{-5}}\ \frac{Vs^2}{m} \qquad (6)$$

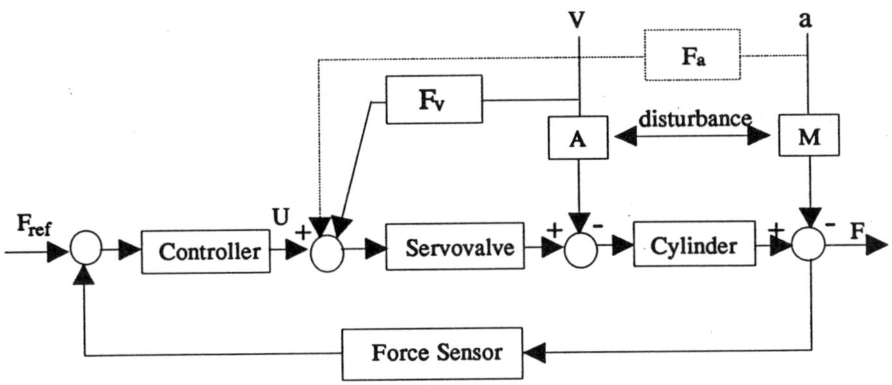


Figure 8. Hydraulic force actuator with feed forward.

Feed forward of both the velocity, v, and the acceleration, a, requires measurement of v and a.

In order to save the cost of an accelerometer only the feed forward of the velocity is implemented. That is reasonable comparing $F_V$=0.617V/m/s and $F_a$=3.44*$10^{-5}$ V/m/$s^2$ as long as the disturbance frequency is low.

In Figure 9 is shown an example of the obtained simulation results with the addition of the velocity feed forward to the PI-double LEAD controller. The effect is that the control performance is improved significantly.

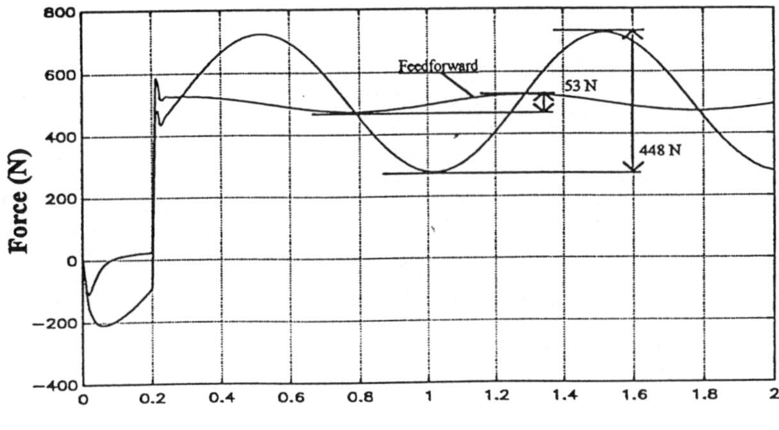


( second )

Figure 9. Simulation of the PI-double LEAD controller with velocity feed forward.

**Digital controller with feed forward.**

Based on the promising simulation results obtained with the analog PI-double-LEAD controller with velocity feed forward a similar digital controller was designed.

Applying the bi-linear transformation we find the digital controller

$$D(Z) = K\frac{\left(1+\frac{T}{2\tau_c}\right)-\left(1-\frac{T}{2\tau_c}\right)Z^{-1}}{1-Z^{-1}}\cdot\frac{\left(1+\frac{T}{2\tau_v}\right)-\left(1-\frac{T}{2\tau_v}\right)Z^{-1}}{\left(\alpha+\frac{T}{2\tau_v}\right)-\left(\alpha-\frac{T}{2\tau_v}\right)Z^{-1}}\cdot\frac{\left(1+\frac{T}{2\tau_s}\right)-\left(1-\frac{T}{2\tau_s}\right)Z^{-1}}{\left(\alpha+\frac{T}{2\tau_s}\right)-\left(\alpha-\frac{T}{2\tau_s}\right)Z^{-1}} \tag{7}$$

To simplify the notation we introduce the following substitutions:

$T_1=1+T/2\tau_C$ $\quad T_2=-1+T/2\tau_C$ $\quad T_3=1+T/2\tau_V$ $\quad T_4=-1+T/2\tau_V$
$T_5=1+T/2\tau_S$ $\quad T_6=-1+T/2\tau_S$
$N_3=\alpha+T/2\tau_V$ $\quad N_4=-\alpha+T/2\tau_V$ $\quad N_5=\alpha+T/2\tau_S$ $\quad N_6=-\alpha+T/2\tau_S$ (8)

and thereby we get:

$$D(Z) = K\frac{T_1+T_2Z^{-1}}{1-Z^{-1}}\cdot\frac{T_3+T_4Z^{-1}}{N_3+N_4Z^{-1}}\cdot\frac{T_5+T_6Z^{-1}}{N_5+N_6Z^{-1}} \tag{9}$$

We have chosen $\alpha=0.2$ and the sampling time T=1mS in order to take into account that the digital controller should be implemented (as an interrupt routine) on a PC. To get a short calculation time, the digital controller is transformed to:

$$\begin{aligned}
D(Z) &= K\frac{T_1+T_2Z^{-1}}{1-Z^{-1}}\cdot\frac{T_3+T_4Z^{-1}}{N_3+N_4Z^{-1}}\cdot\frac{T_5+T_6Z^{-1}}{N_5+N_6Z^{-1}}\\
&= K\frac{T_1T_3+(T_1T_4+T_2T_3)Z^{-1}+T_2T_4Z^{-2}}{N_3+(N_4-N_3)Z^{-1}-N_4Z^{-2}}\cdot\frac{T_5+T_6Z^{-1}}{N_5+N_6Z^{-1}}\\
&= K\frac{T_1T_3T_5+\left[(T_1T_4+T_2T_3)T_5+T_1T_3T_6\right]Z^{-1}+\left[T_2T_4T_5+(T_1T_4+T_2T_3)T_6\right]Z^{-2}+T_2T_4T_6Z^{-3}}{N_3N_5+\left[(N_4-N_3)N_5+N_3N_6\right]Z^{-1}+\left[-N_4N_5+(N_4-N_3)N_6\right]Z^{-2}-N_4N_6Z^{-3}}\\
&= \frac{K\frac{T_1T_3T_5}{N_3N_5}+K\frac{\left[(T_1T_4+T_2T_3)T_5+T_1T_3T_6\right]}{N_3N_5}Z^{-1}+K\frac{\left[T_2T_4T_5+(T_1T_4+T_2T_3)T_6\right]}{N_3N_5}Z^{-2}+K\frac{T_2T_4T_6}{N_3N_5}Z^{-3}}{1+\frac{\left[(N_4-N_3)N_5+N_3N_6\right]}{N_3N_5}Z^{-1}+\frac{\left[-N_4N_5+(N_4-N_3)N_6\right]}{N_3N_5}Z^{-2}-\frac{N_4N_6}{N_3N_5}Z^{-3}}\\
&\equiv \frac{b_0+b_1Z^{-1}+b_2Z^{-2}+b_3Z^{-3}}{1+a_1Z^{-1}+a_2Z^{-2}+a_3Z^{-3}}
\end{aligned} \tag{10}$$

where the constants $a_i$ and $b_i$ are calculated before the regulation starts.

In Figure 10 is shown schematically how the digital force controller is implemented

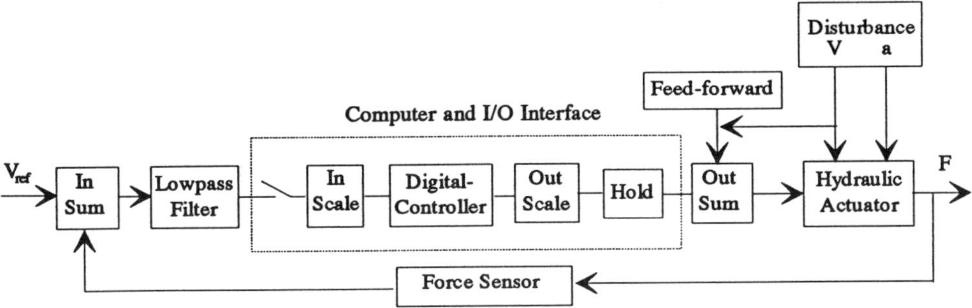


Figure 10. Schematical diagram of the implemented digital force controller.

In Figure 11 is shown an example of the simulation results obtained with the digital force controller with velocity feed forward.

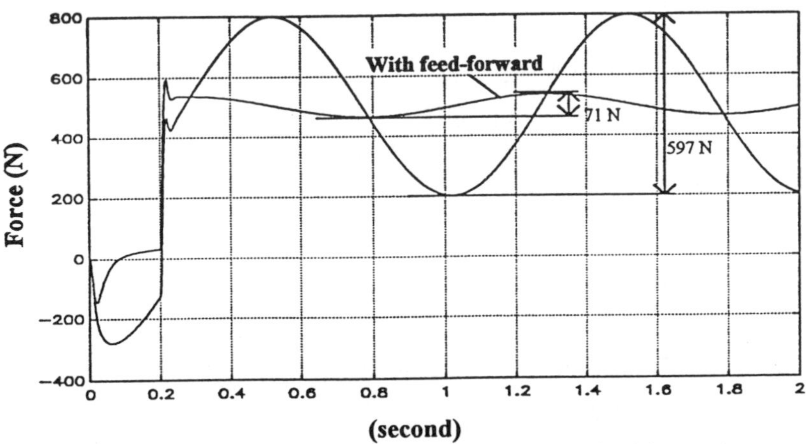


Figure 11. Digital force controller with velocity feed forward.

As expected it can be seen that the control performance is improved significantly by adding the velocity feed forward.

## 5. EXPERIMENTAL RESULTS

The digital force controller with velocity feed forward described above was evaluated by experiments using the experimental set-up (see section 3). Figures 12-14 show examples of the experimental results obtained.

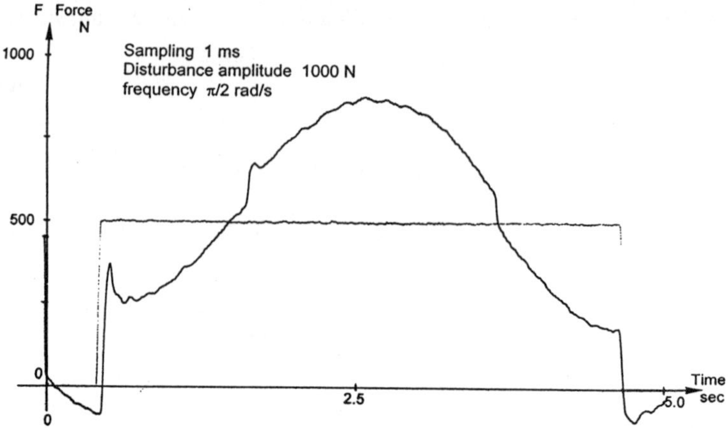

Figure 12. Experimental result obtained with digital P controller.

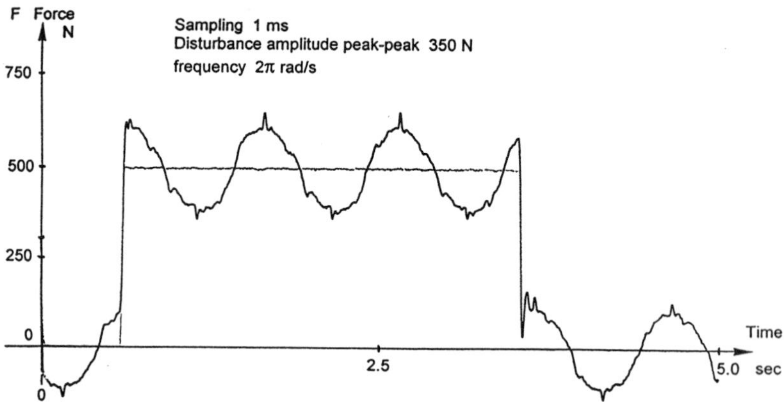

Figure 13. Experimental result. Digital PI controller with double LEAD.

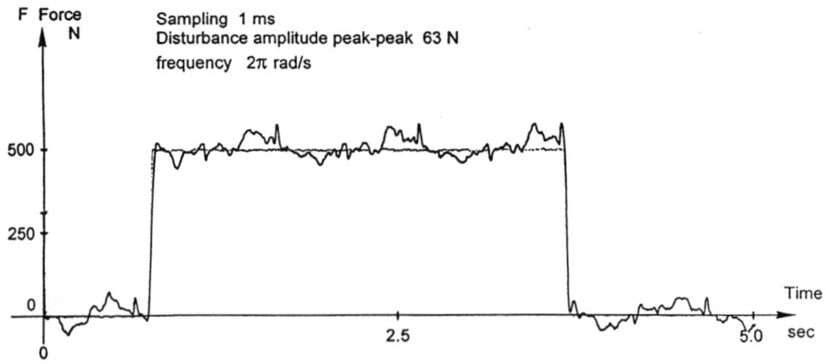

Figure 14. Digital PI controller with double LEAD and velocity feedforward.

As expected from the simulation results it is seen that the control performance is significantly improved with the PI-double LEAD controller compared to the P controller and that it is further improved by adding the velocity feedforward loop.

## 6. CONCLUSION

It is shown through experimental results that the controller can be improved significantly by using a digital PI double LEAD controller compared to a digital P controller with force feedback.

Furthermore, it is shown that the control performance can be improved further by adding a velocity feed forward loop.

**WRITTEN DISCUSSION**
**Digital control with feedforward of hydraulic force actuators**
**F Conrad, LH Hansen, HS Larsen & M Zhang (Technical University of Denmark)**

**Question:** **KA Edge**
**Fluid Power Centre, Bath, UK**

What factors dictated your choice of 1 ms for the sampling period? Do you think you could have achieved similar improvements in performance with a longer sampler period?

**Answer:**

Following the rule of thumb, the sampling period of 1 ms was chosen as 10 times the bandwidth of the open-loop system ($f_{-180°} \approx 100$ Hz), where the bandwidth is defined at the crossover frequency of -180 degrees.

For small sampling periods (KJ Åström and B Wittenmark, 1990: Computer-controlled Systems, Prentice-Hall International Editions), the transfer function of the hold circuit can be approximated as

$$\frac{1 - e^{-sh}}{sh} \approx 1 - \frac{sh}{2} + \cdots \text{ where h is sampling period}$$

The first two terms correspond to the series expansion of exp(-sh/2). That is, for small h, the hold can be approximated by a time delay of half a sampling interval. If a longer sampling period is to be chosen, the phase margin of the system would be decreased. For stability reasons, the open-loop gain would have to be decreased, but this implies a reduction in the error reduction factor, which is undesireable.

For the actual actuator system controlled by using an AT-PC we have obtained approximately the same improvement for a sampling period of 1.43 ms due to the calculation time-delay being a little less than 1 ms, which is relatively large compared to the delay in the hold circuit.

On the other hand, selecting a smaller sampling period on a faster computer could cause problems, since the three poles of the system could result in a non-minimum phase system, if the sampling period is selected too small.

Selection of an appropriate sample period is therefore a compromise.

# 18. On the Application of Neural Networks to the Monitoring of a Simulated Hydraulic Rotary Drive System

S Daley *and* H Wang

## Abstract

Algorithms for fault detection and isolation (FDI) based on the use of mathematical models are becoming a powerful addition to many control systems. Many useful methods have been proposed but the great majority of these are based on the use of linear models. Due to the fact that most practical systems display some nonlinear behaviour, methods that make use of nonlinear representations are being sought. In this paper a method for FDI which makes use of the nonlinear representation capabilites of neural netrworks is proposed. The method proposed is demonstrated through its application to a simulated electro-hydraulic rotary drive system.

## 1 Introduction

A high degree of fault tolerance in complex plants and their control systems has long been recognized as being an essential part of the overall system design. This requirement stems from the fact that failures can lead to substantial material damage involving unacceptable economic loss and can be hazardous to plant personnel. Fault tolerance can be achieved not only by increasing the reliability of individual plant components but also by detecting, locating and diagnosing faults at an early stage in their development.

Traditionally, fault tolerance is achieved through the use of physical redundancy in which several hardware elements are located in parallel in order to provide protection against localised failures. Because of the cost of the extra hardware and due to advances in computer technology, new and more sophisticated approaches based on the use of mathematical models have been developed in recent years. These methods make use of analytical redundancy whereby the model is used to formulate a function of the temporal history of the measured variables which has the property of being small only when the system is operating normally.

Many of the analytical redundancy methods are based on state estimation or system identification approaches which require linear models for successful operation [1]. Because most practical systems display a degree of nonlinearity these methods generally only work well in small regions of the operating space. Recently, several authors have proposed robust methods whereby the linearisation errors are handled as unknown disturbances ([2,3], for example), however, an inevitable by-product is that sensitivity to faults is often also reduced.

The practical limitations of the linear methods are apparent when considering condition monitoring in fluid power systems, particularly those employing electrohydraulic servo-mechanisms. This is due to such factors as the nonlinear relation between flow rate and pressure, variations in oil viscosity and load-dependent characteristics which all make the established analytical redundancy methods difficult to apply. In this paper the use of neural networks to overcome the problems associated with nonlinearity in fluid power systems is explored.

## 2 Process Model and Problem Statement

The method developed is to be applied to the monitoring of a hydraulic test rig, shown schematically in figure 1.

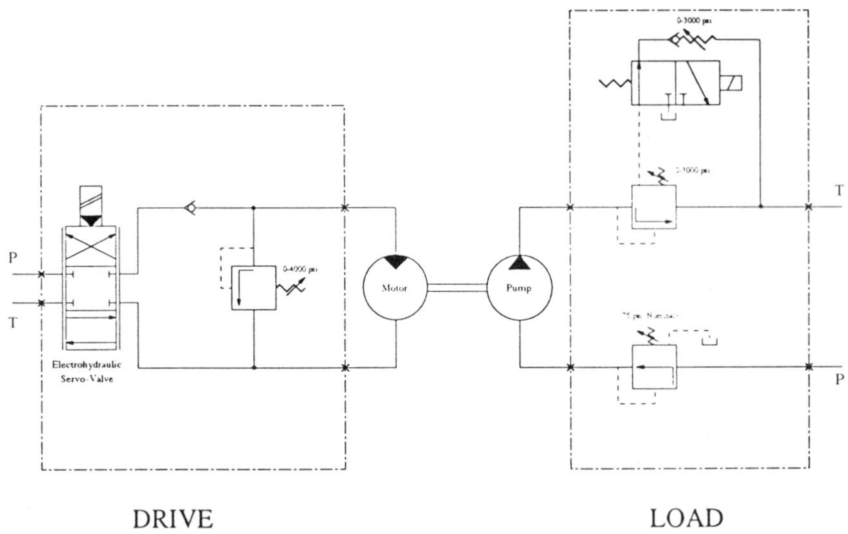


Figure 1. *Schematic diagram of test rig*

The test rig was purpose built as a vehicle for testing advanced control and monitoring strategies [4,5] and consists of a stiff shaft which is driven by a hydraulic motor and loaded with a hydraulic pump. The oil flow to the motor is controlled by an electrohydraulic servo-valve and the pressure differential across the pump can be changed to increase or decrease the loading on the shaft. The rig is intrumented so that the input to the servo-valve, the shaft speed and the pressure differential across the motor can all be monitored. As an example for the neural network methodology, a method is to be developed to monitor the health of the motor. In particular it is required that the motor efficiency and leakage can be determined from the on-line measurements.

Although not essential, a structural model of the test rig is useful in determining the required neural network structure and is given below. It is assumed that the parameters of this model are not known.

The load flow, $Q_v$, (ie. average flow in motor lines) can be approximated by [6]:

$$Q_v = K_\theta X_s (P_s - P_m)^{0.5} \quad (1)$$

where $X_s$ is the spool valve displacement, $P_s$ is the supply pressure, $P_m$ the pressure differential across the motor and $K_\theta$ is the valve flow coefficient. For continuity of flow

$$Q_v = C_r \omega + \left( \frac{V_t}{2B} \right) \dot{P}_m + K_l P_m \quad (2)$$

where $\omega$ is the shaft angular velocity, $C_r$ is the motor displacement, $V_t$ is the total trapped volume, $B$ is the oil bulk modulus and $K_l$ is a leakage coefficient. The motor torque is

$$T_m = P_m C_r \eta_m \quad (3)$$

where $\eta_m$ is the efficiency of the motor. Neglecting static and coulomb friction,

$$T_m = I\dot{\omega} + D\omega + T_p \quad (4)$$

where $I$ is the total inertia of the pump, motor and shaft, $D$ is the viscous friction coefficient and

$$T_p = P_p C_r \eta_p^{-1} \quad (5)$$

where $P_p$ is the pressure differential across the pump and $\eta_p$ is the efficiency of the pump. If it assumed that the dynamics of the servo-valve are much faster than the dynamics of the load, the servo and torque motor can be approximated by a pure gain term, that is

$$X_s \approx K_s u \quad (6)$$

The system is then described by a set of nonlinear, dynamic equations with unknown parameters. Since only two first order differential equations, (2) and (4), are involved, the dynamic order of the system is equal to 2. The system is to be monitored using sampled data and therefore a discrete representation is required. As the shaft speed ($y = \omega$) and pressure difference ($P_m$) are measured outputs and the voltage input to the torque motor ($u$) is a measured input, the discrete model can be represented by

$$y(k) = f(y(k-1), y(k-2), P_m(k-1), P_m(k-2), u(k)) \tag{7}$$

$$P_m(k) = g(y(k-1), y(k-2), P_m(k-1), P_m(k-2), u(k)) \tag{8}$$

$y(k)$, $P_m(k)$ and $u(k)$ are the sampled values of the shaft speed, pressure difference and voltage input, respectively, at sampling time t = kT. T is a pre-specified sampling period and k = 1, 2, ...., f(...) and g(....) are unknown nonlinear functions. The sampled data sequences described by equations (7) and (8) are to be used to provide a monitor of the motor health by estimating the parameters $K_l$ and $\eta_m$.

## 3 Fault Detection and Diagnosis using Neural Networks

In recent years there has been an enormous resurgence of interest in the use of neural networks for the identification and control of complex processes (see [7] for a recent survey). An artificial neural network consists of many simple processing elements each having a number of inputs and a single output. The output of each element is a nonlinear function (termed the activation unit, node or neuron) of a weighted sum of the inputs. The basic elements are interconnected using variable strength links, or weights, which can be modified in order to achieve a desirable input-output relationship. The relevant features of such networks that make them particularly attractive for modelling and control are

(i) any continuous function (linear or nonlinear) can be approximated to arbitrary accuracy using a network consisting of an input layer and output layer and one hidden layer [8,9],

(ii) on-line and off-line adaption for uncertain and non-stationary systems,

(iii) parallel and distributed structure allows fast processing for large-scale systems,

(iv) structure incorporates some redundancy and therefore leads to fault tolerent implementations.

It is features (i) and (ii) that make the neural network ideal for the solution of the problem stated in the previous section.

If it is assumed that training sequences $y(k)$, $P_m(k)$ and $u(k)$ for $0<k<K_0$ are available for the healthy system then these can be used to build a nonlinear neural model to specified accuracy. Once this has been done, the output of the true system can be compared

with the output of the model and any significant deviation used to declare a fault. The modelling error can then be used to locate and determine the magnitude of the fault. These basic principles are expanded in the following sub-sections.

### 3.1 Neural Network Model of Healthy System

Over the years many different types of neural network have been proposed, but the vast majority of applications make use of the multi-layer feedforward network in conjunction with the back-propagation algorithm (or variant) for training. This now established approach is also adopted here. Two models are required, one for the speed represented by equation (7) and one for the pressure represented by equation (8). The strucure of the network for both (although drawn for speed only) is as shown in figure 2.

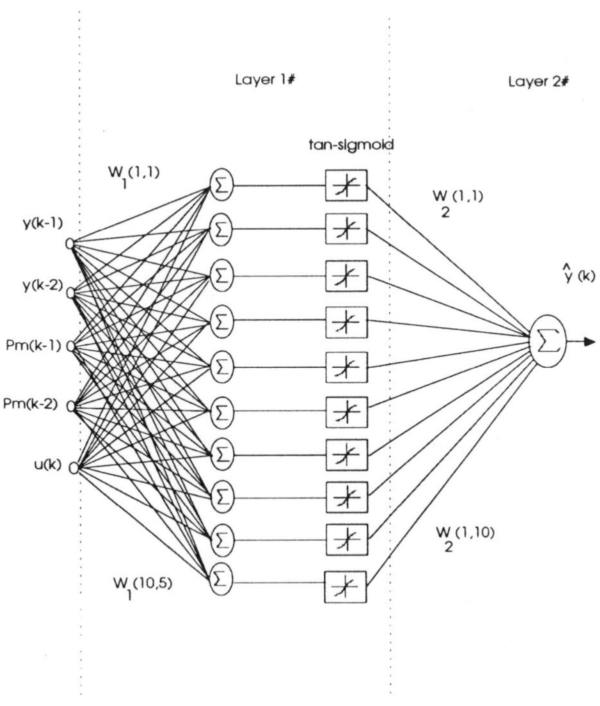


Figure 2 *Neural Network Structure*

This is a so called 1 and 1/2 layer network with a first layer consisting of $n$ sigmoid neurons and a second layer of linear (pure gain) neurons. The output of the ith neuron in layer 1, $N_o(i)$, is given by

$$s_x = [W_1(i,1) \quad W_1(i,2) \quad W_1(i,3) \quad W_1(i,4) \quad W_1(i,5)] \begin{bmatrix} y(k-1) \\ y(k-2) \\ P_m(k-1) \\ P_m(k-2) \\ u(k) \end{bmatrix} + B_1(i) \tag{9}$$

$$N_o(i) = \tanh(s_x) \tag{10}$$

where $W_1(i,j)$ is the weight for the jth input and $B_1(i)$ is a bias term. The neural model output (for speed) is then given by

$$\hat{y}(k) = \sum_{i=1}^{n} W_2(1,i) N_o(i) + B_2 \tag{11}$$

The weights and bias terms are to be determined so that the difference between the training data (collected for $0<k<K_0$) and the network output is minimised. The number of required neurons, $n$, can also be determined by increasing the number until the improvement in modelling error is not significant. Repeated iterations are then made through the training data until the weights converge and the model is a good approximation of the input/output data for the healthy system.

**3.2 Fault Detection**

Denote the neural network model for sub-system (7) as

$$\hat{y}(k) = \hat{f}(y(k-1), y(k-2), P_m(k-1), P_m(k-2), u(k)) \tag{12}$$

and the neural network model for sub-system (8) as

$$\hat{P}_m(k) = \hat{g}(y(k-1), y(k-2), P_m(k-1), P_m(k-2), u(k)) \tag{13}$$

The following residual signal can then be defined

$$\varepsilon(k) = \begin{pmatrix} y(k) - \hat{y}(k) \\ P_m(k) - \hat{P}_m(k) \end{pmatrix} \tag{14}$$

where the first component is the modelling error for sub-system (7) and the second component is the modelling error for sub-system (8). If the real system is not faulty, then both these modelling errors will be zero or very close to zero. This will not be the case in the presence of *significant* faults. As a result of this, $\varepsilon(k)$ can be used as a detection signal to generate an alarm following a fault occurrence. For this purpose, define

$$d(k) = \| \varepsilon(k) \| \tag{15}$$

and then fault detection can be realized according to

$$\begin{cases} d(k) < \delta; & \textit{System is healthy} \\ d(k) \geq \delta; & \textit{Fault has occurred} \end{cases}$$

Where $\delta$ is a small pre-specified threshold dependant upon the modelling error for the nominal healthy system.

### 3.3 Fault Diagnosis

The purpose of the diagnosis is to determine both the location and size of any faults following their detection. Only faults that lead to changes in the leakage and efficiency of the motor and manifest themselves as unexpected abrupt changes in the parameters, $K_l$ and $\eta_m$ are considered. Define

$$K_l = \overline{K}_l(1 + \Delta f_1) \tag{16}$$

$$\eta_m = \overline{\eta}_m(1 + \Delta f_2) \tag{17}$$

where $\overline{K}_l$ and $\overline{\eta}_m$ are the nominal values of the leakage coefficient and the efficiency of the motor, $\Delta f_1$ and $\Delta f_2$ are the abrupt changes. The problem of fault diagnosis for the rig is then transferred to the problem of estimating $\Delta f_1$ and $\Delta f_2$ via the collected input/output data.

Let

$$\Delta F = \begin{pmatrix} \Delta\hat{f}_1 \\ \Delta\hat{f}_2 \end{pmatrix} \tag{18}$$

then the diagnostic matrix,

$$C = \frac{\partial \varepsilon(k)}{\partial (f_1, f_2)} \tag{19}$$

can be constructed with respect to $\Delta f_1 = \Delta f_2 = 0$. This is readily determined by replacing the healthy motor with one possessing known faults and noting the resulting residual change. This can be regarded as a training data set for faults. However, it should be noted that only one data set is required to construct the diagnostic matrix from which all faults in the class can be diagnosed. $C$ is a Jacobian matrix between vector $\varepsilon(k)$ and the vector $\Delta F$ and therefore the following relationship holds

$$\varepsilon(k) = C\Delta F \tag{20}$$

If C is non-singular, the faults are distinguishable and on-line fault diagnosis can be carried out by evaluating

$$\begin{pmatrix} \Delta\hat{f}_1 \\ \Delta\hat{f}_2 \end{pmatrix} = C^{-1}\varepsilon(k) \tag{21}$$

## 4 Simulation Study

To demonstrate the method outlined in the previous section, the system expressed by equations (1) - (6) is simulated using parameters that have previously been verified against the real system. The response of the healthy system to a step change in the voltage input to the servo-valve is simulated and $K_o = 170$ samples taken with $T = 0.5$ mS. This forms the training data set for development of the neural models and is normalised to give better numerical properties.

A backpropagation algorithm with an adaptive learning rate from the Matlab Neural Network Toolbox [10,11] is used to train the weights. 10 neurons were found to be adequate for good representation and 8000 iterations were required for weight convergence. The output of the neural model and the first 40 samples of the training data are plotted for speed and motor pressure in figures 3 and 4 respectively. It can be seen that the neural models provide an excellent description of the real process.

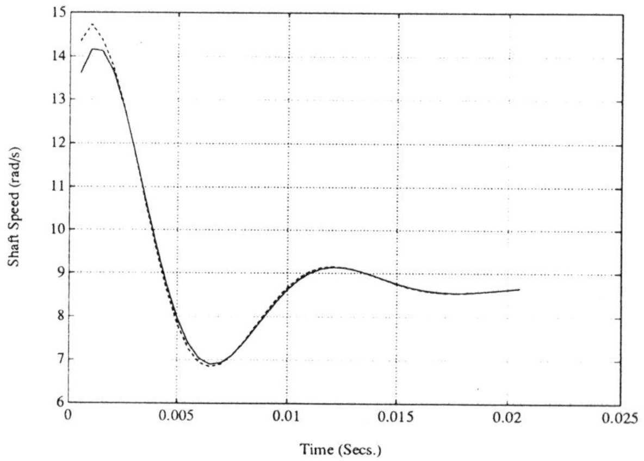


Figure 3 *Network Output and Training Data for Speed*

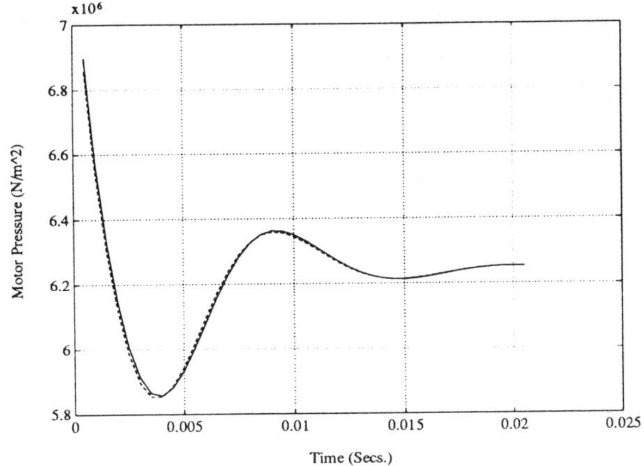


Figure 4 *Network Output and Training Data for Pressure*

Faults are created by setting

$$K_l = 1.15\overline{K}_l$$

$$\eta_m = 0.9\overline{\eta}_m$$

at $k = 500$ ($t = 0.25s$) which is equivalent to making

$$\Delta f_1 = 0.15$$

$$\Delta f_2 = -0.1$$

The detection signal corresponding to equation 15 is shown in figure 5.

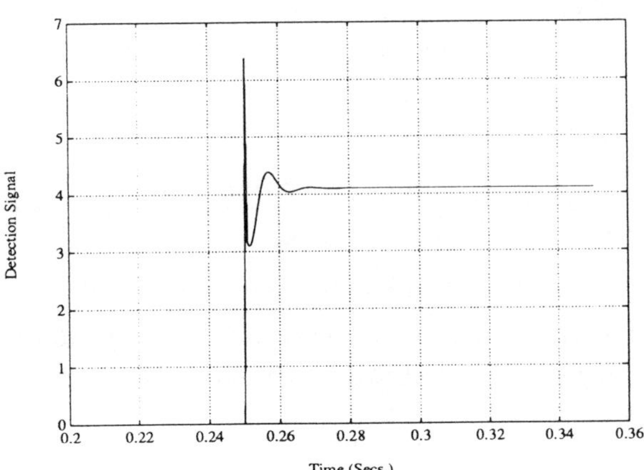


Figure 5 *Detection Signal*

The signal rapidly moves away from the nominal zero value at $t = 0.25s$ giving an almost instantaneous detection of the fault. The corresponding modelling errors following the fault can be seen in figures 6 and 7.

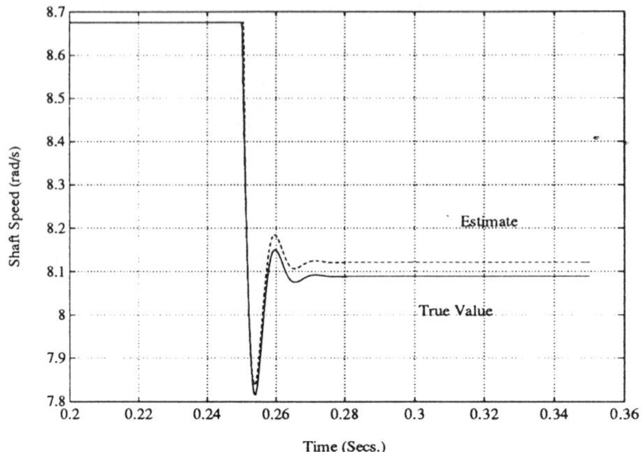


Figure 6 *Simulated Speed and Speed Network Output*

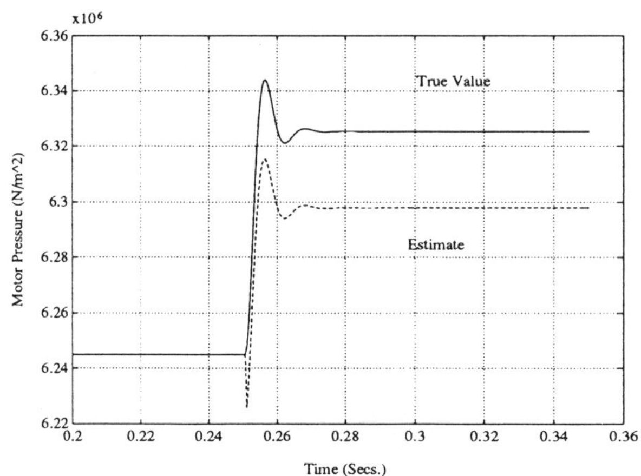


Figure 7 *Simulated Pressure and Pressure Network Output*

Since a parametric model is used for the rig simulation, the diagnostic matrix, $C$, can be determined using deviation techniques. The diagnostic vector corresponding to equation 21 is shown in figure 8.

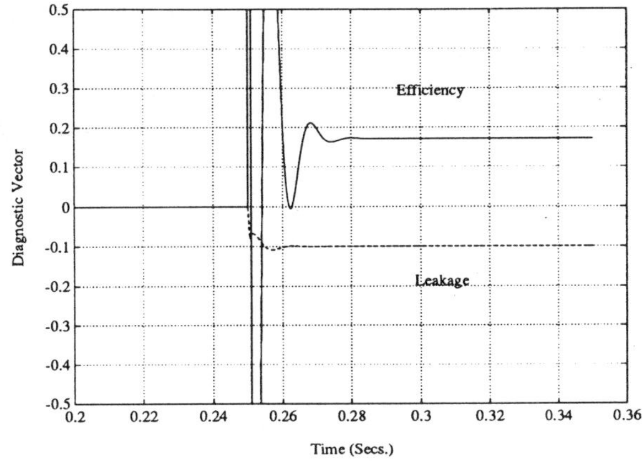


Figure 8 *Behaviour of Diagnostic Vector*

Following some oscillation (particularly in the efficiency estimate) the vector values settle to steady state values of $\Delta\hat{f}_1 = 0.17$ and $\Delta\hat{f}_2 = -0.1$. This represents a very good diagnosis with the small error being due to the linearisation involved in calculating the diagnostic matrix.

## 5 Conclusions

The representation capabilities of neural networks has been exploited to provide a simple method for detecting and diagnosing faults in nonlinear systems. The network is first trained on a healthy system to provide a detection signal that has the property of being small only when the system is operating normally. Using knowledge about the effects of known faults on this signal a diagnostic vector can be constructed to determine the location and size of all similar faults. The method has been successfully demonstrated on a simulated rotary hydraulic system.

## References

[1] R. Isermann, "Process fault detection based on modelling and estimation methods - a survey", Automatica, Vol. 20, 1984, pp387-404.

[2] S. Daley and H. Wang, "On the generation of an optimally robust residual signal for systems with structured model uncertainty", Proceedings of the 1992 American Control Conference, Chicago, pp2104-2108.

[3] R. J. Patton and J. Chen, "A robust parity space approach to fault diagnosis based on optimal eigenstructure assignment", Proc. of IEE Control 91 Conference, Vol. 2, 1991, pp1056-1061.

[4] S. Daley, "Application of a fast self-tuning control algorithm to a hydraulic test rig", Proc. IMechE, Pt C, Vol. 201, 1987, pp285-295.

[5] D. A. Newton, "Design and implementation of an intelligent electro-hydraulic rotary drive system", Machine Actuators and Controls, IMechE Seminar, London, 31st March 1993.

[6] H. E. Merritt, "Hydraulic Control Systems", John Wiley, 1967, p194.

[7] K. J. Hunt, D. Sbarbaro, R. Zbikowski, "Neural networks for control systems - a survey", Automatica, Vol. 28, 1992, pp1083-1112.

[8] G. Cybenko, "Approximations by superpositions of a sigmoidal function", Mathematics of Control, Signals and Systems, Vol. 2, 1989, pp303-314.

[9] K. Funahashi, "On the approximate realization of continuous mappings by neural networks", Neural Networks, Vol. 2, 1989, pp183-192.

[10] T.P.Vogl, Mangis, J.K., Rigler, A.K., Zink, W.T., and Alkon,D.L., "Accelerating the convergence of the backpropagation method", Biological Cybernetics, Vol. 59, 1988, pp.257-264.

[11] H Demuth and Beale, M., "Neural Network Toolbox", The Math Works Inc, 1992.

**WRITTEN DISCUSSION**
**On the application of neural networks to the monitoring of s simulated hydraulic rotary drive system**
**S Daley (European Gas Turbines Ltd, UK) & H Wang (UMIST, UK)**

**Question:** **F Conrad**
**Technical University of Denmark**

Do you have any strategy to guide the network for establishing fault conditions?

**Answer:**

The algorithms are constructed to look for symptoms of specific faults. As a result, the effects that faults have on physical parameters must be known. The approach described in the paper was developed for the ESPRIT II Topmuss project. In this project a software toolbox has been constructed where the designer has access to a range of tools including a database of faults and symptoms.

**Question:** **ND Vaughan**
**Fluid Power Centre, Bath, UK**

1. Why do you use two different networks for speed and motor pressure? Have you tried a single network with two outputs?

2. It appears easier for the net to detect that a fault exists than to identify the fault explicitly. What work is needed to identify the fault?

**Answer:**

1. We tried a single network with 5 inputs and 2 outputs but found that the number of neurons in the first layer had to be very large in order to get the same accuracy as that obtained when using two separate networks. The independent networks for speed and motor pressure were therefore used to reduce the computational load.

2. In the method presented here faults are identified by using a linearised expression for the relationship between physical parameter changes and the modelling error vector. We have also done work where this nonlinear relationship is identified using another neural network thereby improving the reliability of the diagnosis. It is possible to take this one step further by using an additional network to map estimated parameter changes to specific faults.

**Question:** **JB Gamble**
**Vickers Systems Division, Trinova Ltd, UK**

Do you have to train the neural network over the entire *expected* operating range of the System? Also what would happen if the system were operated (correctly) outside the range of operation used to train the network?

**Answer:**

The neural network must be able to represent all normal system dynamics that will be encountered and therefore must be trained over the entire operating range. If the system was operated outside this range and the dynamic properties were significantly different, then a false alarm would be generated; however, this situation could be avoided by allowing the network to retrain.

# 19. Design of a Nonlinear Compensator for Fluid Power Control

S LeQuoc, Y F Xiong *and* R M H Cheng

## Abstract

Adaptive controllers are being employed more and more in fluid control systems because of their nonlinear and time-varying characteristics. A review of adaptive control for hydraulic servos indicates that state-of-the-art designs use as a starting point a linear model of an appropriate low order. However, this paper demonstrates a departure from the conventional approach, by using a nonlinear model for an electrohydraulic position servo using proportional valve control. It also shows the necessity of designing a nonlinear compensator for a deadband. To facilitate the design tasks, an identification method is proposed to estimate the deadband nonlinearity, the time delay and the order of the model. By introducing a piecewise-linear preload element to invert the deadband, a single-input single-output nonlinear model is converted into a double-input single-output model. The parameters of the model are estimated by a linear identification method, so that the piecewise-linear preload element may be used as a compensator to neutralize the deadband in the design of the controller. This reduces the closed-loop behaviour of the nonlinear hydraulic system to that of its linear portion only. A simulation example is included to show the effectiveness of the proposed method.

## 1 Introduction

Electrohydraulic servos are widely used in industrial applications, such as in rolling mills, paper mills, "actuators" in aircrafts, and many kinds of automation and mechanization . This is principally because of the greater power capacity that they can offer, as compared to their d.c. and a.c. counterparts, with good dynamic response and system resolution. However, as a result of the ever-demanding complexity with these applications, the electrohydraulic industry has undergone many changes and made a lot of advances in the last 20 years to take advantage of the rapid development of electronic techniques. The micro-computer together with the many feedback devices that can be used for hydraulic drives has made it possible to develop new control algorithms to achieve better steady-state and dynamic performance in fluid control systems. Most notably, a great deal of research has been published on the application of adaptive control to electrohydraulic servos.

Adaptive control is particularly relevant for systems with unknown or time-varying parameters. An electrohydraulic servo is inherently nonlinear, containing such static nonlinearities as deadband, saturation, friction, hysteresis etc. The characteristics of these nonlinearities are not usually exactly known. Thus the full potential of an electrohydraulic system will not be realized, unless these nonlinearities are properly taken care of in the system design.

This paper attempts to shed some light on the effectiveness of adaptive control algorithms in systems which are characterized with static nonlinearities that are not negligible, and on how to design a powerful controller for these systems. A brief review of research in adaptive control of hydraulic servos is first presented. The review will demonstrate the necessity of designing a nonlinear compensator for fluid power control in order to achieve sufficient accuracy and good control performance.

Different from conventional linear approaches, a nonlinear model of a position servo controlled by an electrohydraulic proportional valve is established.

Based on a recent contribution by LeQuoc et al. [1], a method of designing a nonlinear compensator is developed without the prior knowledge of the model structure. In this method, a nonlinear model of the system is used, contrary to most conventional linear approaches. By using this nonlinear compensator, one can achieve a tight control of the dynamic and steady-state performances, in much the same way as for a linear system. A simulation example is included to show the efficiency of the method.

## 2 Nomenclature

| | |
|---|---|
| $A$ | effective area of piston $m^2$ |
| $B_t$ | viscous damping coefficient of piston and load Ns/m |
| $C_t$ | total leakage coefficient of piston $m^5/Ns$ |
| $d$ | time delay of discrete-time system |
| $F$ | external load force on piston N |
| G | load spring stiffness N/m |
| $I$ | input current of valve from amplifier A |
| $I_v$ | input current of valve from deadband A |
| $K_a$ | gain of servo amplifier A/V |
| $K_c$ | flow-pressure coefficient of valve $m^5/Ns$ |
| $K_f$ | gian of feedback transducer V/m |
| $K_q$ | flow gain of servo valve $m^2/s$ |
| $K_v$ | gain of valve m/A |
| $m$ | total mass of piston and load kg |
| $m_1$ | width of deadband m |
| $P_L$ | pressure drop across the actuator $N/m^2$ |
| $q^{-1}$ | unit delay backward shift operator |
| $Q_L$ | flow through the actuator $m^3/s$ |
| $s$ | Laplace operator |
| $u$ | input voltage to servo amplifier V |
| $V_t$ | total volume of fluid under compression in both chambers $m^3$ |
| $x_v$ | valve spool displacement from neutral m |
| $y$ | displacement of piston m |
| $\beta_e$ | effective bulk modulus of system $N/m^2$ |
| $\delta_v$ | damping ratio of valve |
| $\omega_v$ | natural frequency of valve rad/s |

## 3 Review of adaptive control for fluid power systems

The model of a given hydraulic servo-system may be obtained by linearizing its dynamics about an operating point of interest. Traditionally, a constant-gain feedback controller is designed based on this linear model. However, this controller cannot assure the desired performance when the operating condition varies, since the values of most valve characteristics vary somewhat with the operating point. These constant-gain controllers are rendered even more inadequate in cases when nonlinearities and time-varying parameters simply cannot be ignored. They also react poorly in the presence of many factors, such as load variation, external disturbances, changes in the dynamics of components, and their dimensional tolerances, varying properties of the hydraulic fluid, etc. As a result, designers of such systems turn more and more to adaptive controllers.

Finney et al. [2] and Takahashi et al. [3] are among the first to apply adaptive control to fluid power systems, being motivated by the uncertain and time-varying system characteristics. Very often a system is characterized by a high order and fast dynamic response. For a 5-th order system, the hydraulic drive proposed in [2] needs a closed-loop bandwidth of 19 Hz. Accompanying this requirement is the computational difficulty in using an adaptive control algorithm. The researchers resolve this difficulty by reducing the model to 3-rd order, and by separating the controller synthesis and parameter estimation into two or more distinct sections. The penalty is reduced precision in the closed-loop performance.

Unbehauen et al [4-5] develop in detail the model reference adaptive control (MRAC) of a hydraulic position system. They conclude that MRAC assures zero steady state error for non-negligible loads, and show robust behaviour and satisfactory performance for tracking and regulation as compared to conventional control schemes. The MRAC is based on a linear 3-rd order model. A similar application of MRAC based on a 3-rd order pneumatic servo is reported by Araki and Yomamoto [6].

In order to maintain the fast dynamics of electro-hydraulic systems, some efforts have been made. Hori et al. [7] propose a simplified MRAC which assumes a linear plant, resulting in the algorithm being four times faster than their early method. Xiong et al. [8] develop a very simple self-tuning controller for an electrohydraulic synchronizing position system which optimizes the feedback gain on-line. On the other hand, Koeckemann et al [9] simply reduce the hydraulic positioning servo model to a first order, and design an adaptive control accordingly.

Some special topics have also been considered for adaptive control of hydraulic position systems. Yun and Cho [10] apply an adaptive model following control to a hydraulic position system subjected to unknown, time-varying external load disturbances with good performance in comparison to the conventional constant-gain controller. But the dynamic response is found to be very slow. Hori et al. [11] use the $\delta$-operator approach to replace the conventional z-operator in an attempt to achieve robustness when the sampling rate is very small.

In summary, the afore-mentioned approaches are based on known linear model structures for the hydraulic position systems. These approaches are successful, particularly in cases where the system dynamics are dominated by linear characteristics, or by continuously time-varying nonlinear characteristics.

In most cases, the linear model being derived from first principles gives rise to a high order. Order reduction is subsequently carried out more or less successfully, depending on the experience of the individual. Naturally, the question still remains as to how to select a suitable low-order linear model structure that includes a time delay to represent the real plant adequately before applying adaptive control.

On the other hand, it is salutary to remind oneself that hydraulic systems are inherently characterized by static non-continous nonlinearities. An important example is the deadband. The presence of a deadband is caused by several common factors, such as spool overlap in the servovalve, coulomb friction and so on. Several publications have dealt with compensating for a deadband nonlinearity by employing adaptive control. A model-reference adaptive control has been proposed by Du [12], for a deadband nonlinearity with known parameters. A stochastic adaptive control scheme has been developed by Xiong and Unbehauen [13], while LeQuoc et al [14] propose an adaptive control for a cascade of a preload nonlinearity with deadband followed by a linear portion of the system. Although References [13] and [14] deal with nonlinearities which may not be completely known, there is still a reduced precision in the performance of the control system in comparision with the control of a linear system. An important improvement was made by means of nonlinear model identification [1], but the method assumes that the model structure is known *a priori*.

Thus the task at hand is how to model a nonlinear fluid power system and how to design a nonlinear compensator without knowing the nonlinear model structure *a priori*.

## 4 Modelling of a nonlinear electrohydraulic position system

As an example, an electrohydraulic proportional valve controlled position servo-system as depicted in Figure 1 is used.

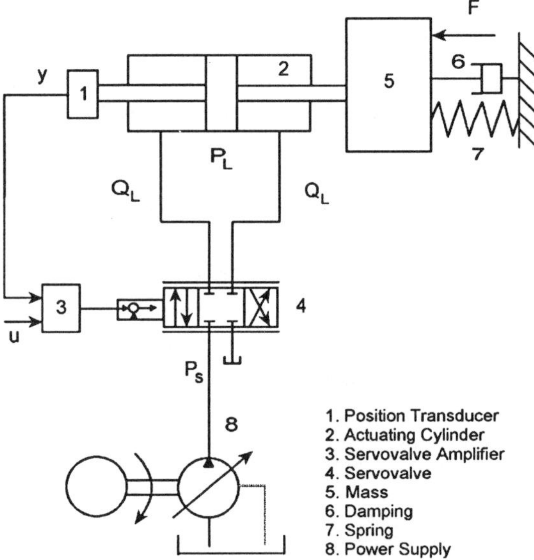


*Figure 1. Configuration of a proportional valve controlled position system*

Taking nonlinearities into account, an accurate model for position control may be derived as follows, which makes use of a Vickers proportional throttle valve KTG4V-5 for position control. The flow rates of this valve at stated pressure drops are as shown in Figure 2. Evidently, the valve exhibits a significant deadband property.

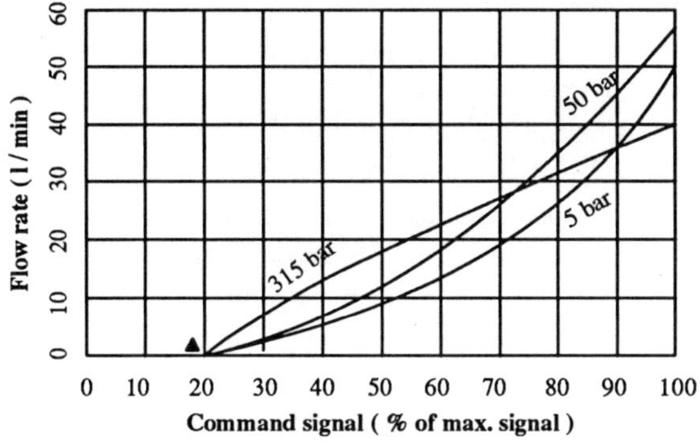


*Figure 2. Flow gain of Vickers proportional throttle valve KTG4V-5*

The dynamics of the spool-displacement of the valve is summarized by the following second-order differential equation:

$$\frac{d^2x_v}{d^2t} + 2\delta_v\omega_v\frac{dx_v}{dt} + \omega_v^2 x_v = \omega_v^2 K_v I_v \tag{1}$$

while the deadband nonlinearity may be described as

$$I_v = \begin{cases} I - m_1 & , \quad if \ \ I > m_1 \\ 0 & , \quad if \ \ |I| \leq m_1 \\ I + m_1 & , \quad if \ \ I < -m_1 \end{cases} \tag{2}$$

The general pressure-flow function of valve has the form:

$$Q_L = Q_L(x_v, P_L) \tag{3}$$

Eq. (3) can be linearized and expressed as a Taylor's series about a particular operating point $Q_{L0}$. For $Q_{L0} = 0$,

$$Q_L = \left.\frac{\partial Q_L}{\partial x_v}\right|_0 x_v + \left.\frac{\partial Q_L}{\partial P_L}\right|_0 P_L \tag{4}$$

Define flow gain $K_q$ and flow-pressure coefficient $K_c$ as

$$K_q = \frac{\partial Q_L}{\partial x_v} \tag{5}$$

and

$$K_c = -\frac{\partial Q_L}{\partial P_L} \tag{6}$$

The linearized equation of the pressure-flow function becomes

$$Q_L = K_q x_v - K_c P_L \tag{7}$$

These two coefficients vary in magnitude with the operating point.
With an assumption that the piston is centered, the continuity expression of oil flow is

$$Q_L = A\frac{dy}{dt} + C_t P_L + \frac{V_t}{4\beta_e}\frac{dP_L}{dt} \tag{8}$$

The force balance on the piston is

$$P_L = \frac{1}{A}\left(m\frac{d^2y}{d^2t} + B_t\frac{dy}{dt} + Gy\right) + \frac{F}{A} \tag{9}$$

The five equations (1), (2), (7), (8), and (9) together yield the transfer function of the system, with the block diagram as shown in Figure 3.

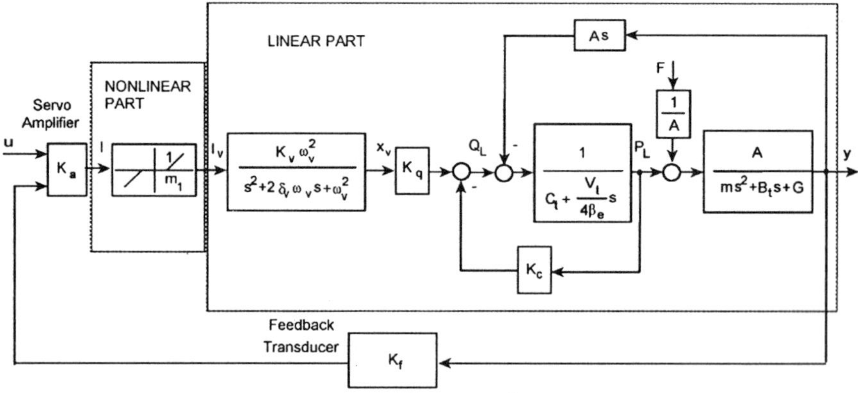


*Figure 3. Block diagram of the hydraulic position system*

This model of a hydraulic position system as shown in Figure 3 includes an important nonlinear feature associated with the servo-valve. Some efforts are previously reported by LeQuoc et al. [15] and by Leung [16]. In those instances, experiments with a rotary electro-hydraulic servo system show that the deadband from the spool-overlap of the servo-valve significantly affects the response of the system, especially during a change of the direction of displacement of the actuator .

## 5 Design of nonlinear compensator

Referring to Figure 3, the width of deadband $m_1$ is not known exactly and signal $I_v$ is not measurable. In order to design a compensator for the deadband, the first task at hand is to utilize the measurable signals $u$ and $y$ to estimate the parameters of the deadband. For this purpose, discretization of the open-loop transfer function of the hydraulic position system as shown in Figure 3 is necessary. Its discrete-time model is shown in Figure 4.

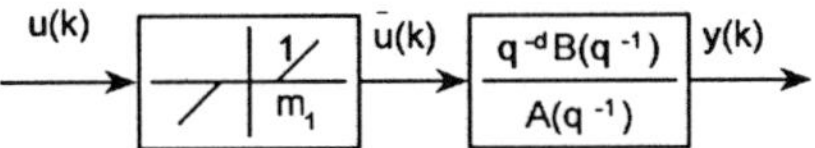


*Figure 4. The discrete-time model with a deadband*

The model is essentially a SISO time-invariant system. The slope of the deadband is normalized to 1 without loss of generality, while the remaining, linear portion of the system may be described by

$$A\, y[k] = q^{-d}\, B\, \bar{u}[k] \tag{10}$$

where $y[k]$ and $\bar{u}[k]$ are respectively the output and input of the linear portion at time $k$, and $A = A(q^{-1})$ and $B = B(q^{-1})$ are polynomials of $q^{-1}$ with unknown orders $n_a$ and $n_b$ respectively, such that

$$A = 1 + a_1 q^{-1} + \cdots + a_{na} q^{-na} \tag{11}$$

$$B = b_0 + b_1 q^{-1} + \cdots + b_{nb} q^{-nb} \tag{12}$$

The deadband portion of the system (Figure 4) is described by

$$\bar{u}[k] = \begin{cases} u[k] - m_1 \,, & if\;\; u[k] > m_1 \\ 0 \,, & if\;\; |u[k]| \leq m_1 \\ u[k] + m_1 \,, & if\;\; u[k] < -m_1 \end{cases} \tag{13}$$

with $m_1 > 0$ being the width of the deadband.

In practice one can find a limiting positive number $m_2$ , such that

$$m_1 \leq m_2 \tag{14}$$

Introduce a preload nonlinearity of a magnitude $m_2$ to precede the deadband, resulting in the equivalent model as in Figure 5.

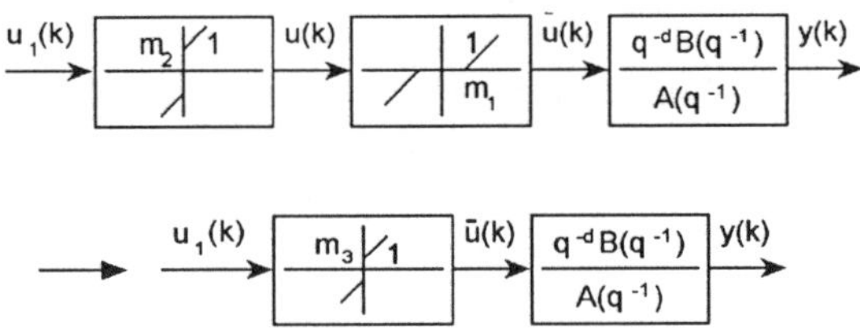


*Figure 5. The equivalent nonlinear model*

The preload inverts the deadband, so that the equivalent nonlinearity, with $m_3 = m_2 - m_1$, can be written as:

$$\bar{u}[k] = \begin{cases} u_1[k]+m_3 & , \quad if\ u_1[k]>0 \\ 0 & , \quad if\ u_1[k]=0 \\ u_1[k]-m_3 & , \quad if\ u_1[k]<0 \end{cases} \tag{15}$$

Define $\beta[k]$ such that

$$\beta[k] = \begin{cases} 1 \ , & if\ u_1[k]>0 \\ 0 \ , & if\ u_1[k]=0 \\ -1 \ , & if\ u_1[k]<0 \end{cases} \tag{16}$$

The nonlinear model can be rewritten as

$$A\ y[k] = q^{-d} B(\ u_1[k] + m_3\ \beta[k]\ ) \tag{17}$$

Eq. (17) can be further arranged as:

$$y[k] = \Phi[k]^T \theta \tag{18}$$

with $$\Phi[k] = [\ y[k-1] \cdots y[k-n_a\ ],\ u_1[k-d] \cdots u_1[k-n_b-d],\ \beta[k-d] \cdots \beta[k-n_b-d]\ ]^T \tag{19}$$

$$\theta = [-a_1 \cdots -a_{na}\ ,\ b_0 \cdots b_{nb},\ b_0^* \cdots b_{nb}^*\ ]^T \tag{20}$$

where $b_i^* = m_3 b_i \ \ , i = 0 \cdots n_b$.
The model expressed in (18) is a linear DISO model (double-input single-output). Since the parameters are time-invariant, a number of linear identification methods are available for parameter estimation. The recursive least square method [17] is employed here.
Denoting the estimate of $\theta$ at time $k$ by $\hat{\theta}[k]$,

$$\hat{\theta}[k] = \hat{\theta}[k-1]+K[k]\ e[k] \tag{21}$$

with $$K[k] = P[k]\ \Phi[k]$$

$$= \frac{P[k-1]\ \Phi[k]}{\lambda + \Phi^T[k]\ P[k-1]\ \Phi[k]} \tag{22}$$

$$P[k] = [\ 1 - K[k]\ \Phi^T[k]\ ]\ P[k-1]\ /\ \lambda \tag{23}$$

$$e[k] = y[k] - \Phi^T[k]\ \hat{\theta}[k-1] \tag{24}$$

where $\lambda$ is the forgetting factor.
The least square method described in eqs. (21) - (24) is available only for parameter estimation. In order to determine the time delay $d$, orders $n_a$ and $n_b$, the model structure selection and model validation method by Ljung [18] is employed to select the structure of model, which is described below.

To determine the time delay $d$, a second order model is selected, which is used to try out every time delay between 1 and $n_d (> 1)$, where $n_d$ is established by experience of the system to be identified. As an alternative, try various values, e.g. $n_d = 10$. The loss function for different models are computed by using the input and output data set of the system. The delay $d$ is selected according to the best fit for the data set. The loss fuction is defined as

$$J_N(d) = \min_{d} \sum_{k=1}^{N} e^2(k) \tag{25}$$

To determine $n_a$ and $n_b$, the fit for combinations of up to $n_{ab}$ $a$-parameters and up to $n_{ab}$ $b$-parameters are checked, for time delay $d$ that has been selected above. $n_a$ and $n_b$ are chosen to minimize the following loss fuction:

$$J_N(n_a, n_b) = \min_{n_a, n_b} \sum_{k=1}^{N} e^2(k) \tag{26}$$

where $n_{ab}$ ( a positive integer number) must be larger than the order of the model and is chosen similarly to $n_d$.
With the time delay $d$, orders $n_a$ and $n_b$ having been determined and $\theta$ identified, the estimates of parameters $\hat{b}_i$ and $\hat{b}_i^*$, $i = 0 \cdots n_b$ are obtained. The value of $m_3$ can be estimated thus:

$$\hat{m}_3 = \frac{\hat{b}_0^* + \cdots + \hat{b}_{nb}^*}{\hat{b}_0 + \cdots + \hat{b}_{nb}} \tag{27}$$

and so,

$$\hat{m}_1 = m_2 - \hat{m}_3 \tag{28}$$

The nonlinear compensator becomes an inversion of the deadband, with the nonlinear system model being converted to a linear one, as depicted in Figure 6.

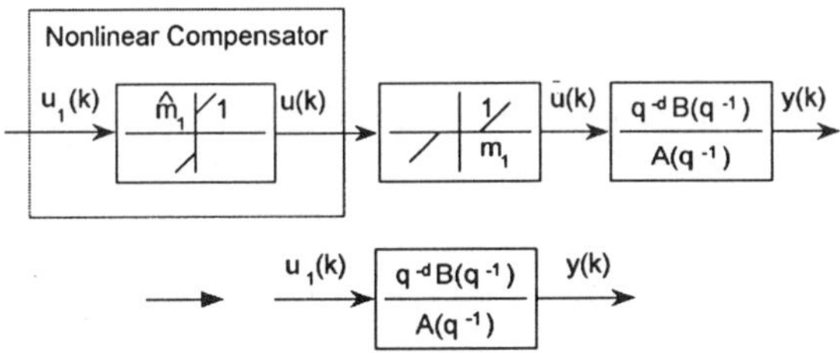


*Figure 6. Eliminating the deadband by an inversion*

At this point, a control method can be employed in combination with the nonlinear compensator. The closed-loop system should exhibit linear properties.

## 6 A design example

The effectiveness of the afore-mentioned identification algorithm may be demonstrated by a simulation example. Because our purpose is to demonstrate how to compensate the deadband in a hydraulic system, a simplified hydraulic position system is obtained from the model presented in Figure 3 by neglecting valve dynamics, piston leakage, compressibility and external load force, as shown in Figure 7.

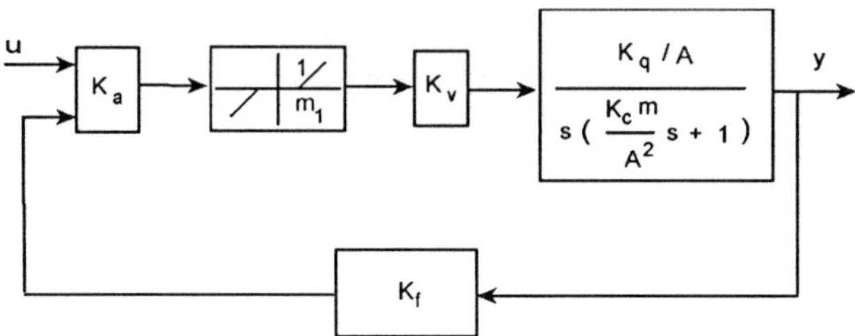


*Figure 7. Simplified model of a hydraulic position system*

The discrete-time counterpart of the model in Figure 7 is shown in Figure 8.

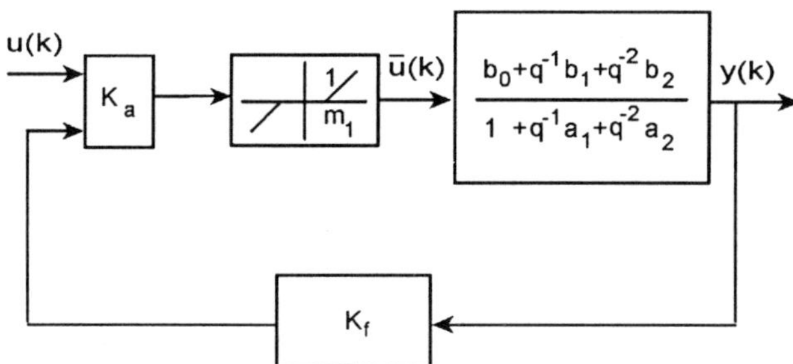


*Figure 8. Discrete-time model of a simplified hydraulic position system*

The open-loop model in Figure 8 is given by

$$y[k] - 1.5y[k-1] + 0.7y[k-2] = \bar{u}[k-1] + 0.5\,\bar{u}[k-2] \qquad (29)$$

$$\bar{u}[k] = \begin{cases} u[k]-m_1 \;, & if\;\; u[k] > m_1 \\ 0 \;, & if\;\; |u[k]| \leq m_1 \\ u[k]+m_1 \;, & if\;\; u[k] < -m_1 \end{cases} \qquad (30)$$

where $a_1 = -1.5$, $a_2 = 0.7$, $b_0 = 0$, $b_1 = 1$, $b_2 = 0.5$, $m_1 = 0.5$ and $K_a = 1$.

The parameter $m_2$ is chosen as 1. Thus, $m_3 = m_2 - m_1 = 0.5$ .
Rewrite (29) as

$$y[k] - 1.5y[k-1] + 0.7y[k-2] = u_1[k-1] + 0.5u_1[k-2] + 0.5\beta[k-1] + 0.25\beta[k-2] \quad (31)$$

with $\beta[k]$ as defined in (16).
The input signal $u_1$ used in this simulation is a zero-mean Gaussian white noise sequence with variance equal to one. The model (31) is identified according to the afore-mentioned method by using a total of 1000 sets of simulated input and output data. The forgetting factor $\lambda$ is selected as 1.
First, determine the time delay $d$. Select $n_d$ equal to 4. The identification result is shown in Table 1.

*Table 1. Identification result of time delay d*

| $J$ | 0.0039 | 1.0135 | 0.9600 | 1.0998 |
|---|---|---|---|---|
| $n_a$ | 2 | 2 | 2 | 2 |
| $n_b$ | 2 | 2 | 2 | 2 |
| $d$ | 1 | 2 | 3 | 4 |

From Table 1, choose $d = 1$ with $J = 0.0039$.
To determine the model orders, choose $n_{ab}$ equal to 4. The result is shown in Table 2.

*Table 2. Identification result of orders $n_a$ and $n_b$*

| $J$ | 9.4749 | 1.3721 | 1.0040 | 1.1245 | 1.0053 |
|---|---|---|---|---|---|
| $n_a$ | 1 | 2 | 2 | 3 | 3 |
| $n_b + d$ | 1 | 1 | 2 | 1 | 2 |
| $d$ | 1 | 1 | 1 | 1 | 1 |
| $J$ | 1.0069 | 1.0932 | 1.0114 | 1.0121 | 1.0116 |
| $n_a$ | 3 | 4 | 4 | 4 | 4 |
| $n_b + d$ | 3 | 1 | 2 | 3 | 4 |
| $d$ | 1 | 1 | 1 | 1 | 1 |

From Table 2, choose $n_a = n_b + d = 2$.

The identification results are $\hat{a}_1 = -1.4925$, $\hat{a}_2 = 0.6937$, $\hat{b}_0 = 0.9979$, $\hat{b}_1 = 0.5326$, $\hat{b}_0^* = 0.5278$ and $\hat{b}_1^* = 0.2688$. The value of $m_3$ is calculated as:

$$\hat{m}_3 = \frac{\hat{b}_0^* + \hat{b}_1^*}{\hat{b}_0 + \hat{b}_1} = 0.5205 \tag{32}$$

and the width of the deadband is

$$\hat{m}_1 = m_2 - \hat{m}_3 = 0.4795 \tag{33}$$

The error of estimation is

$$\Delta = (m_1 - \hat{m}_1) / m_1 = 4.1 \% \tag{34}$$

Thus the afore-mentioned method gives the correct model structure and a satisfactory nonlinear compensator for deadband.

## Conclusion

Experimental and analytical studies of a proportional valve controlled position system show that a deadband nonlinearity significantly influence the performance of a hydraulic system. The authors have attempted to design a nonlinear compensator for the deadband nonlinearity. This attempt is shown to be quite successful by a simulation example.

The key to the success lies in introducing a piecewise-linear preload element to invert the deadband nonlinearity, and in the identification of the model structure. In the process, the nonlinear SISO model is converted to a linear DISO model. Identification is employed to estimate the parameter of the deadband accurately, so as to eliminate its effect completely, leaving behind a linear model in effect. The remaining task is a traditional controller design for any linear system.

An important feature of this method which cannot be over-emphasized is that it does not require *a priori* knowledge of time delay $d$, the orders $n_a$ and $n_b$ of polynomials $A$ and $B$.

## Acknowledgment

This work is supported by the Natural Sciences and Engineering Research Council (NSERC) of Canada, grant OGP0090335.

## References

[1] **LeQuoc, S., Xiong, Y.F. and Cheng, R.H.M.** Identification and Control of Nonlinear Hydraulic System. *SAE International Off-Highway & Powerplant Congress & Exposition*, Paper Series 921622, Milwaukee, Wisconsin, 1992. Also accepted by SAE Transactions.

[2] **Finney, J.M., de Pennington, A., Bloor, M.S. and Gill, G.S.** A Pole-Assignment Controller for an

Ecletrohydraulic Cylinder Drive. *Trans. ASME, Journal of Dynamic Systems, Measurement and Control*, Vol. 107, pp 145-150, 1985

[3] **Takahashi, K., Inoue, M. and Ikeo, S.** Application of the Model Reference Adaptive Technique to an Electrohydraulic Servo System. *Proceedings of the 1st Int. Conference on Fluid Power Transmission and Control*, Hangzhou, China, pp 68-87, 1985

[4] **Unbehauen, H., Du, P. and Keuchel, U.** Application of a Digital Adaptive Controller to a Hydraulic System. *Proc. Int. Conference on Control*, Oxford, pp 598-603, 1988

[5] **Unbehauen, H., Du, P. and Keuchel, U.** Application of Microcomputer-based Model Reference Adaptive Control to a Hydraulic Positioning System. *Proceedings of IFAC Symposium-Adaptive Control and Signal Processing*, Glasgow, 1989

[6] **Araki, K. and Yomamoto A.** Model Reference Adaptive Control of Pneumatic Servo with a Constant Trace Algorithm. *The Journal of Fluid Control*, Vol. 20, No. 4, pp 30-48, 1990

[7] **Hori, N., Ukrainetz, P.R., Nikiforuk, P.N. and Bitner, D.V.** Robust Discrete-time Adaptive Control of an Electrohydraulic Servo Actuator. *Proceedings IEE, Part-D*, Vol.137-2, pp 107-111, 1990

[8] **Xiong, Y.F., LeQuoc, S. and Cheng, R.H.M.** Adaptive Control of a Synchronizing Servo-System. *Preceedings of 1992 SAE International Off-Highway & Powerplant Congress & Exposition*, Paper Series 921621, Milwaukee, Wisconsin, 1992. Also accepted by SAE Transactions.

[9] **Koeckemann, A., Konertz, J. and Pawlik, W.** Adaptive Control Algorithms for Electro-hydraulic Servo Drives. *Preprints of 11th IFAC World Congress*, Tallinn, USSR, Vol. 8, pp 248-253, 1990

[10] **Yun, J.S. and Cho, H.S.** Application of an Adaptive Model Following Control Technique to a Hydraulic Servo System Subjected to Unknown Disturbances. *Trans. ASME, Journal of Dynamic Systems, Measurement and Control*. Vol. 113, pp 479-486, 1991

[11] **Hori, N., Bitner, D.V., Nikiforuk, P.N. and Ukrainetz, P.R.** Robust Adaptive Control of Electrohydraulic Servo Systems Using the Euler Operator. *Proceedings of IEE International Conference on Control*, Edinburgh, pp 671-676, 1991

[12] **Du, P.** *Realisierung eines neuen adaptiven Konzepts für hydraulische Vorschubantriebe mittels Mikroprozessor*. Dissertation, Reihe 8: Mess-, Steuerungs- und Regelungstech-nik, VDI Verlag Düsseldorf, 1990

[13] **Xiong, Y.F. and Unbehauen, H.** Adaptive Control of Systems with Unknown Dead Zone Nonlinearies. In Book *'Mathematics of the Analysis and Design of Process Control'*, eds. P. Borne, S.G. Tzafestas and N.-E. Radhy, North-Holland Press, pp 455-464, 1992

[14] **LeQuoc, S., Xiong, Y.F. and Cheng, R.H.M.** Adaptive Control of Systems with Multiple Amplitude-dependent Nonlinearities. Submitted to *Trans. ASME, Journal of Dynamic Systems, Measurement and Control*, 1992

[15] **LeQuoc, S., Cheng, R.M.H. and Leung, K.H.** Tuning an Electrohydraulic Servo-valve to Obtain High Amplitude Ratio and a Low Resonance Peak. *The Journal of Fluid Control*, Vol. 20, No. 3, pp 30-49, 1990

[16] **Leung, K.H.** *Analytical and Experimental Studies of the Open-loop Frequency Response of a Rotary Electro-hydraulic Servo-system*, M.Eng. Thesis, Mechanical Engineering Department, Concordia University, 1989

[17] **Astrom, K.J. and Wittenmark, B.** *Adaptive Control*, Addison - Wesley Publishing Company, 1989

[18] **Ljung, L.** *System Identification Toolbox - for use with MATLAB*, The MATH WORKS Inc., 1991

# 20. LQG Control of a Flexible Mechanical Structure using Hydraulic Actuators

S Gunnarsson *and* P Krus

## Abstract

This paper presents a control system based on linear quadratic state feedback and state estimation for a hydraulic actuator with a flexible mechanical load. The purpose of the control system is to improve the dynamic properties by reducing the oscillatory behaviour of the load and to eliminate steady state errors in the load position caused by external disturbances.

## 1 Introduction

Systems containing fluid power components have become interesting and important applications of modern automatic control theory. The development of computer power together with new measurement devices has made it possible to apply new control algorithms on fluid power systems. See, for example, [1] for a survey.

An example of such an application is given in this paper, where we study a simplified description of the hydraulic crane shown in Figure 1. This system has been studied previously in [2] and [3].

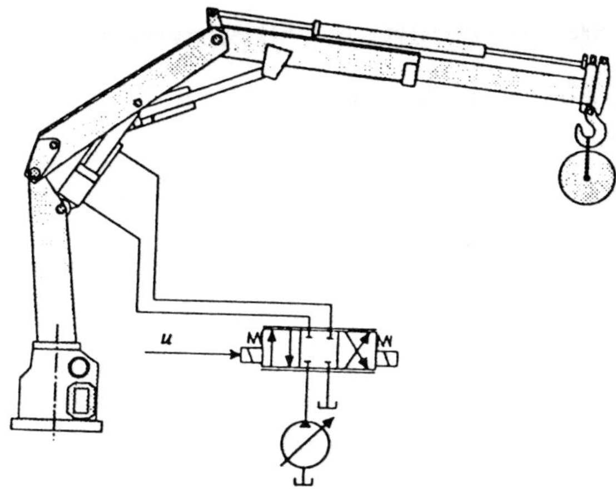


Figure 1: Hydraulic crane

The crane is an interesting control problem also since it represents a flexible mechanical system. Modelling and control of flexible systems has received increasing attention in the recent years, in particular in the area of robotics. Our aim in this paper is to investigate how state feedback and state estimation using the LQG method can be used on this system. This paper presents the main observations while further details are given in [4]. Other applications of state space methods to fluid power systems are given in [5].

## 2 System Description

The crane represents a dynamic system of high complexity and it would be difficult to model the system in detail. We will instead use the simplified description of the system, proposed in [6], which is shown in Figure 2. In using this model we have assumed that the flexibility is restricted to the arm of the crane.

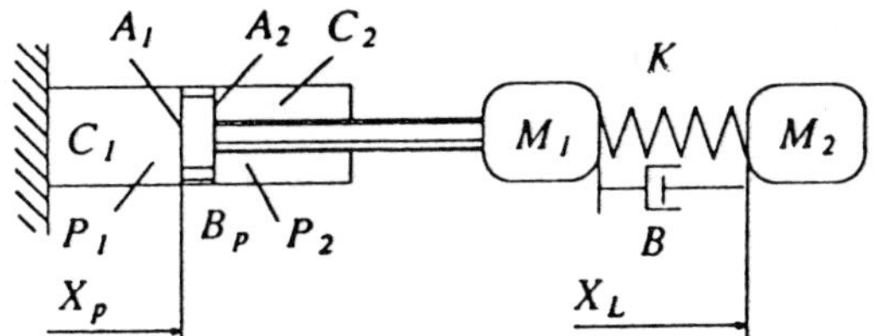


Figure 2: Simplified description.

In Figure 2 the following notations are used:

| | | | |
|---|---|---|---|
| $p_1(t)$ | - pressure in chamber one | $p_2(t)$ | - pressure in chamber two |
| $A_1$ | - piston area in chamber one | $A_2$ | - piston area in chamber two |
| $C_1$ | - capacitance in chamber one | $C_2$ | - capacitance in chamber two |
| $M_1, M_2$ | - masses | $B, B_P$ | - friction coefficients |
| $K$ | - stiffness | $x_C(t)$ | - cylinder position |
| $x_P(t)$ | - piston position | $x_L(t)$ | - load position |
| $f(t)$ | - external force | | |

The problem will be studied using a state space description of the system. Details concerning the derivation of such a description are given in [7]. A natural choice of state variables is position and velocity of piston, load and valve respectively and the pressure in each chamber:

$$\begin{array}{rclrcl} x_1(t) & = & x_P(t), & x_2(t) & = & \dot{x}_P(t), \\ x_3(t) & = & x_L(t), & x_4(t) & = & \dot{x}_L(t), \\ x_5(t) & = & p_1(t), & x_6(t) & = & p_2(t), \\ x_7(t) & = & x_V(t), & x_8(t) & = & \dot{x}_V(t). \end{array} \tag{1}$$

Using these state variables the system can be described by

$$\dot{x}(t) = Ax(t) + B_1 u(t) + B_2 f(t), \tag{2}$$

where

$$A = \begin{pmatrix} 0 & 1 & 0 & 0 & 0 & 0 & 0 & 0 \\ -\frac{K}{M_1} & -\frac{B+B_P}{M_1} & \frac{K}{M_1} & \frac{B}{M_1} & \frac{A_1}{M_1} & -\frac{A_2}{M_1} & 0 & 0 \\ 0 & 0 & 0 & 1 & 0 & 0 & 0 & 0 \\ \frac{K}{M_2} & \frac{B}{M_2} & -\frac{K}{M_2} & -\frac{B}{M_2} & 0 & 0 & 0 & 0 \\ 0 & -\frac{A_1}{C_1} & 0 & 0 & -\frac{K_c}{C_1} & 0 & \frac{K_q}{C_1} & 0 \\ 0 & \frac{A_2}{C_2} & 0 & 0 & 0 & -\frac{K_c}{C_2} & -\frac{K_q}{C_2} & 0 \\ 0 & 0 & 0 & 0 & 0 & 0 & 0 & 1 \\ 0 & 0 & 0 & 0 & 0 & 0 & -\omega_v^2 & -2\delta_v\omega_v \end{pmatrix} \tag{3}$$

$$B_1 = (0 \;\; 0 \;\; 0 \;\; 0 \;\; 0 \;\; 0 \;\; 0 \;\; \omega_v^2)^T \tag{4}$$

and

$$B_2 = (0 \;\; 0 \;\; 0 \;\; -\frac{1}{M_2} \;\; 0 \;\; 0 \;\; 0 \;\; 0)^T. \tag{5}$$

We shall consider the use of two different output signals. These signals are piston position

$$y_1(t) = x_P(t), \tag{6}$$

effective load pressure

$$y_2(t) = p_{Le}(t) = p_1(t) - \frac{A_2}{A_1}p_2(t) = x_5(t) - \frac{A_2}{A_1}x_6(t) \tag{7}$$

Collecting the output signals in a vector $y(t)$ gives

$$y(t) = Cx(t) \tag{8}$$

where

$$C = \begin{pmatrix} 1 & 0 & 0 & 0 & 0 & 0 & 0 & 0 \\ 0 & 0 & 0 & 0 & 1 & -\frac{A_2}{A_1} & 0 & 0 \end{pmatrix} \tag{9}$$

The parameters $K_q$ and $K_c$ denote the flow gain and flow pressure coefficient respectively of the valve, while $\omega_v$ and $\delta_v$ denote resonance frequency and damping in the dynamic description of the valve. We shall consider the system described above using parameter values for which the model gives a good description of the real crane. The parameter values that will be used are given in Table 1.

| Parameter | Value | Unit |
|---|---|---|
| $l$ - piston length | 0.6 | $[m]$ |
| $d$ - cylinder diameter chamber one | 0.06 | $[m]$ |
| $A_1$ - cross section area chamber one ($\frac{d^2\pi}{4}$) | $2.83 \cdot 10^{-3}$ | $[m^2]$ |
| $A_2$ - cross section area chamber two ($0.482A_1$) | $1.36 \cdot 10^{-3}$ | $[m^2]$ |
| $B_P$ - friction coefficient | $2 \cdot 10^4$ | $[Ns/m]$ |
| $\beta$ - bulk modulus | $1 \cdot 10^9$ | $[N/m^2]$ |
| $C_1$ - capacitance chamber one ($\frac{A_1 l}{2\beta}$) | $0.85 \cdot 10^{-12}$ | $[m^5/N]$ |
| $C_2$ - capacitance chamber two ($\frac{A_2 l}{2\beta}$) | $0.41 \cdot 10^{-12}$ | $[m^5/N]$ |
| $C_{Le}$ - effective capacitance | $0.57 \cdot 10^{-12}$ | $[m^5/N]$ |
| $K_c$ - flow pressure coefficient | $1 \cdot 10^{-11}$ | $[m^5/Ns]$ |
| $K_q$ - flow gain | $1 \cdot 10^{-2}$ | $[m^2/s]$ |
| $M_1$ - mass | $2 \cdot 10^3$ | $[kg]$ |
| $M_2$ - mass | $12 \cdot 10^3$ | $[kg]$ |
| $K$ - stiffness | $1.728 \cdot 10^6$ | $[N/m]$ |
| $B$ - friction | $2.4 \cdot 10^3$ | $[Ns/m]$ |
| $\omega_v$ - resonance frequency | 60 | $[rad/s]$ |
| $\delta_v$ -damping | 0.5 | |

Table 1: Parameter values

# 3 Linear quadratic state feedback control

A systematic method to compute a feedback control law for a system given in state space form is to use linear quadratic (LQ) optimization. See [8] for an introduction. Using this method the input signal $u(t)$ is computed as a linear combination of the states

$$u(t) = -Lx(t) + l_0 x_r(t) \tag{10}$$

where $L$ is a $m \times n$ matrix. Here $n$ is the number of states and $m$ is the number of inputs, while $x_r(t)$ is a reference signal. In our case there is only one input, and this implies that

$$u(t) = -l_1 x_1(t) - \ldots - l_n x_n(t) + l_0 x_r(t) \tag{11}$$

where $l_i$ are scalar constants. The constant $l_0$ is calculated to give the closed loop system suitable static gain. The vector $L$ is chosen such that the criterion

$$J = \int_0^\infty [x^T(t) Q_1 x(t) + u^T(t) Q_2 u(t)] dt \tag{12}$$

is minimized. The variables $Q_1$ and $Q_2$ denote a symmetric positive semidefinite and a symmetric positive definite matrices respectively. The gain vector that minimizes the criterion (12) is given by

$$L = Q_2^{-1} B^T S \tag{13}$$

where the matrix $S$ is obtained from the Ricatti equation

$$0 = A^T S + S A^T + Q_1 - S B Q_2^{-1} B^T S \tag{14}$$

From [8] we have that when the open loop system, defined by $A$ and $B$, is controllable and the system defined by $A$ and the output matrix $Q^{1/2}$ is observable, (14) has a unique positive definite solution $S$ such that $L$ gives an asymptotically stable closed loop system. Using Equation (10) in (2) gives the closed loop system

$$\dot{x}(t) = (A - BL)x(t) + B_1 l_0 x_r(t) + B_2 f(t) \tag{15}$$

The dynamic properties of the closed loop system are in this case entirely determined by the location of the closed loop poles via the gain vector $L$.

We then apply the LQ design method to the model given by (2) and (8), with parameter values given in the appendix. Our design objective is to obtain a closed loop system with well damped responses to changes in the command signal and to load disturbances. Due to limitations in the control signal energy the rise time of the step response can not be made too fast. A rise time of around one second seems to be a good compromise between performance and control signal energy. By putting weights on the load position, in order to stabilize the system, and load speed, in order to increase the damping, using

$$Q_1 = \text{diag}(0, 0, 0.4, 0.04, 0, 0, 0, 0) \qquad Q_2 = 1 \tag{16}$$

the step response of the closed loop system satisfies the specifications. The numerical values of the elements in $L$ and the closed loop poles that are obtained for this choice of design variables are given in [4]. The response to a step in the reference signal is shown in Figure 3. The figure shows the load position and effective load pressure, and we see that the load is moved in a well damped way.

# 4 LQG control

A direct application of the feedback law (10) requires that all states can be measured. This is not the case in our application and it will hence be necessary to use a state estimator. The state estimate, here denoted $\hat{x}(t)$, is then given by the state estimator (observer)

$$\dot{\hat{x}}(t) = A\hat{x}(t) + B_1 u(t) + K(y(t) - C\hat{x}(t)) \tag{17}$$

where $K$ is a $n \times p$ matrix, where $p$ denotes the number of outputs. By describing the system in a statistical framework we can use a systematic way to compute $K$ such that an optimal state estimator (Kalman filter) is obtained. See [8] for further information. In the Kalman filter the gain matrix is given by

$$K = PC^T R_2^{-1} \tag{18}$$

where $P$ is given by

$$0 = AP + PA^T + R_1 - PC^T R_2^{-1} CP \tag{19}$$

The matrices $R_1$ and $R_2$ are positive semidefinite and positive definite respectively, and they represent the covariances of the disturbances acting on the states and on the measurements respectively. If the open loop system, defined by $A$ and $C$, is observable and the system given by $A$ and the input matrix $R_1^{1/2}$ is controllable, (19) has a unique positive definite solution such that the observer is asymptotically stable.

Having obtained estimates of the states the idea then is to replace the state vector in the feedback law in (10) by the state estimate. The resulting regulator that is obtained when

$$u(t) = -L\hat{x}(t) + l_0 x_r(t) \tag{20}$$

where $\hat{x}(t)$ given by (17) is denoted an LQG (linear quadratic gaussian) regulator. Using Equation (20) in (2) gives a closed loop system of order $2n$. Representing the closed loop system using the states $x(t)$ and $\tilde{x}(t) = x(t) - \hat{x}(t)$ gives the following representation of the closed loop system

$$\begin{aligned} \begin{pmatrix} \dot{x}(t) \\ \dot{\tilde{x}}(t) \end{pmatrix} &= \begin{pmatrix} A - B_1 L & B_1 L \\ 0 & A - KC \end{pmatrix} \begin{pmatrix} x(t) \\ \tilde{x}(t) \end{pmatrix} \\ &+ \begin{pmatrix} B_1 l_0 \\ 0 \end{pmatrix} x_r(t) + \begin{pmatrix} B_2 \\ B_2 \end{pmatrix} f(t) \end{aligned} \tag{21}$$

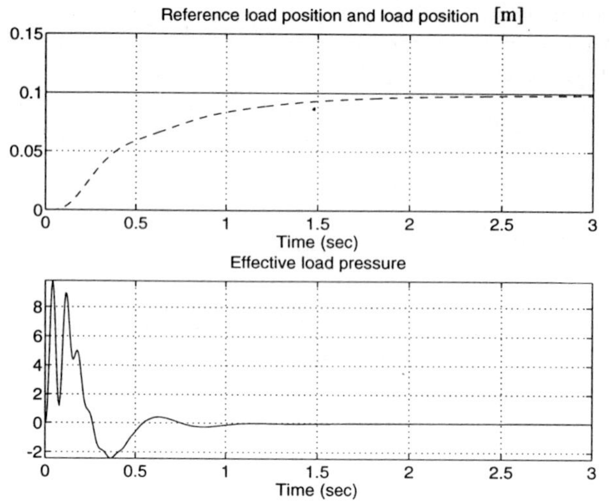


Figure 3: LQ state feedback. Response to step change in reference load position

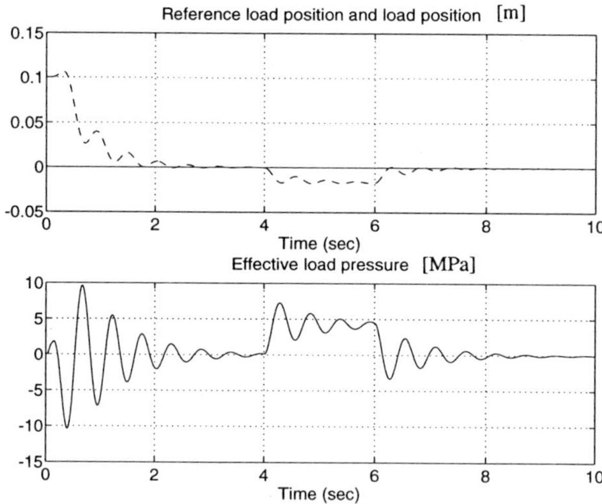


Figure 4: LQG control. Feedback from $x_P(t)$. Response to initial condition and pulse disturbance.

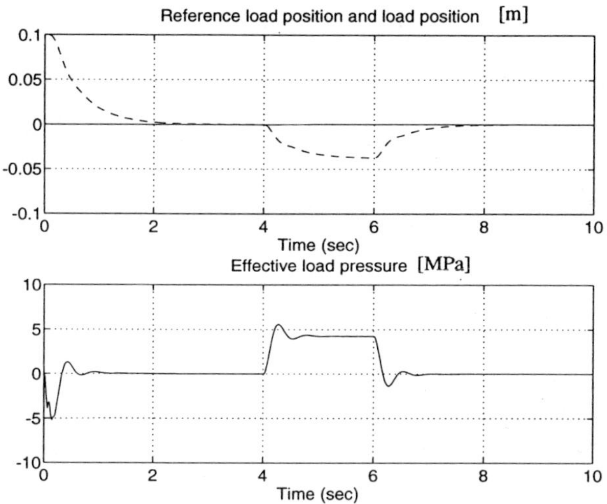


Figure 5: LQG control. Feedback from $x_P(t)$ and $p_{Le}(t)$. Response to initial condition and pulse disturbance

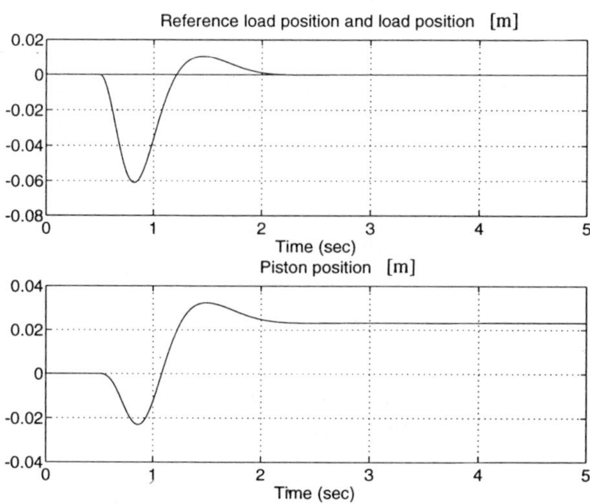


Figure 6: LQG control with integral action, disturbance and compression estimation. Response to step disturbance

When $\tilde{x}(t) = 0$ and we only consider changes in the reference signal, the choice of the gain matrix $K$, i.e. the observer poles, does not affect the closed loop properties. If we on the other hand consider the effects of the disturbance $f(t)$ or an erroneous initial state in the observer the choice of observer dynamics will have substantial influence on the closed loop dynamics. This will be further illustrated below.

In order to be able to compute the gain matrix of the Kalman filter we have to assign suitable values to the elements of the matrices $R_1$ and $R_2$. Experiments have shown that

$$R_1 = \text{diag}(0.01, 0, 0.01, 1, 0, 0, 0, 0) \tag{22}$$

is a good choice in the cases under consideration. The elements $R_1(1,1)$ and $R_1(3,3)$ are due to uncertainties in the initial values of the piston and load position, while the element $R_1(4,4)$ comes from the external disturbance $f(t)$. We will present two experiments with LQG control. First we consider feedback only from the piston position and second feedback from piston position and load pressure. The covariance matrix of the measurement errors is chosen as

$$R_2 = 10^{-4} \tag{23}$$

in the first case, and

$$R_2 = \text{diag}(1 \cdot 10^{-4}, 10) \tag{24}$$

when both position and pressure are used. The gain matrix $K$ and the estimator poles for the two different cases are presented in [4].

In Figures 4 and 5 in the appendix we show the desired and actual load position and effective load pressure for the two feedback configurations. To show the properties of the state estimator we consider the effects of incorrect initial conditions in the estimator and load disturbances. In the simulations shown in Figures 4 - 5 the reference load position $x_r(t)$ is zero while

$$x(0) = (0.1\ 0\ 0.1\ 0\ 0\ 0\ 0\ 0)^T \tag{25}$$

and

$$\hat{x}(0) = (0\ 0\ 0\ 0\ 0\ 0\ 0\ 0)^T. \tag{26}$$

The disturbance that is applied is given by

$$f(t) = \begin{cases} 0 & 0 \le t < 5 \\ 1.2 \cdot 10^4 & 5 \le t < 6 \\ 0 & t \ge 6 \end{cases} \tag{27}$$

Figure 4 shows the result when only the piston position is used, while Figure 5 shows the behaviour when both piston position and effective load pressure feedback are used. The simulations show that the addition of effective load pressure feedback has a substantial effect on the system dynamics. Using only the piston position signal the system dynamics becomes oscillatory while a well damped system is obtained by adding the second signal.

## 5 Load Disturbances

A desirable property of the control system is that it is able to handle load disturbances, i.e. we would like the regulator to be able to position the load properly even though a constant external force is acting on the system. Here we shall discuss a method to handle this problem. The method is based on a combination of integral action and estimation of the load force. We therefore introduce the extended state vector $X(t)$, given by

$$X(t) = (x^T(t) \;\; f(t) \;\; \int_0^t [x_r(s) - x_P(s)]ds)^T \tag{28}$$

i.e. the original state vector is extended by the disturbance $f(t)$ and the integral of the piston position error. Assuming $f(t)$ to be piecewise constant the model can be expressed as

$$\dot{X}(t) = \bar{A}X(t) + \bar{B}_1 u(t) + \bar{B}_3 x_r(t) \tag{29}$$

$$y(t) = \bar{C}X(t) \tag{30}$$

where

$$\bar{A} = \begin{pmatrix} A & B_2 & 0 \\ 0 & 0 & 0 \\ -C_1 & 0 & 0 \end{pmatrix} \qquad \bar{B}_1 = \begin{pmatrix} B_1 \\ 0 \\ 0 \end{pmatrix} \tag{31}$$

$$\bar{B}_3 = \begin{pmatrix} 0 \\ 0 \\ 1 \end{pmatrix} \qquad \bar{C} = (C \;\; 0 \;\; 0) \tag{32}$$

and $C_1$ denotes the first row of $C$. With the control signal

$$u(t) = -\bar{L}\hat{X}(t) + l_0 x_r(t) \tag{33}$$

we achieve that the steady state error in piston position caused by the external force is eliminated. When the piston position is correct we know that the error in the load position is only caused by the compression of the spring, and we then have the following static relationship between the load force and the compression

$$K\Delta_x(t) = K(x_P(t) - x_L(t)) = f(t) \tag{34}$$

where $K$ denotes the spring stiffness. Using the spring stiffness and the estimate of the force we obtain an estimate of the compression $\Delta_x(t)$ from

$$\hat{\Delta}_x(t) = \frac{1}{K}\hat{f}(t) \tag{35}$$

This quantity can then be added to the piston position reference signal, which means that the integrator state is formed according to

$$\hat{x}_{10}(t) = \int_0^t [\frac{1}{K}\hat{x}_9(s) + x_r(s) - x_P(s)]ds \tag{36}$$

With this modification we achieve a closed loop system that is able to control the load position without steady state error. To illustrate the performance of the control system we consider a situation where the reference position is zero, i.e. $x_r(t) = 0$, and a step disturbance is applied after one second. More precisely we apply the force

$$f(t) = \begin{cases} 0 & 0 \leq t < 1 \\ 1.2 \cdot 10^4 & t \geq 1 \end{cases} \tag{37}$$

Applying this idea results in a regulator that is able to control the load position without steady state error. This is shown in Figure 6, which shows the load and piston position respectively. To compensate for the compression of the spring the piston is positioned according to the estimated deformation.

# 6 Transfer Function Formulation

It is sometimes useful to describe the LQG regulator using the corresponding transfer functions. This means that that we express the regulator as

$$U(s) = F_{X_r}(s)X_r(s) - F_{X_P}(s)X_P(s) - F_{P_{Le}}(s)P_{Le}(s) \tag{38}$$

where $F_{X_r}(s), F_{X_P}(s)$ and $F_{P_{Le}}(s)$ denote transfer functions from reference signal, piston position and load pressure to control signal. These transfer functions can be derived as follows. Consider the observer corresponding to the extended state space model (29)

$$\dot{\hat{X}}(t) = \bar{A}\hat{X}(t) + \bar{B}_1 u(t) + \bar{B}_3 x_r(t) + \bar{K}(y(t) - \bar{C}\hat{X}(t)) \tag{39}$$

and the feedback

$$u(t) = -\bar{L}\hat{X}(t) + l_0 x_r(t) \tag{40}$$

Inserting (40) into (39) and taking laplace transform yield that the transfer functions in (38) can be expressed as

$$\begin{aligned} F_{X_r}(s) &= (l_0 - \bar{L}(sI - \bar{A} + \bar{K}\bar{C} + \bar{B}\bar{L})^{-1})(\bar{B}_1 l_0 + \bar{B}_3) \\ F_Y(s) &= \bar{L}(sI - \bar{A} + \bar{K}\bar{C} + \bar{B}\bar{L})^{-1}\bar{K} \end{aligned} \tag{41}$$

$F_Y(s)$ is here a row vector with two transfer functions as elements since we can measure two signals.

The next step is to calculate these transfer functions using the crane model and the gain matrices $L$ and $K$ that were computed above. This results in the Bode diagrams shown in the appendix. Figures 7 - 9 show the amplitude curves of the transfer functions $F_{X_r}(s), F_{X_P}(s)$ and $F_{P_{Le}}(s)$ respectively. The integral action is seen in the first two figures where the gain curve increases as the frequency decreases.

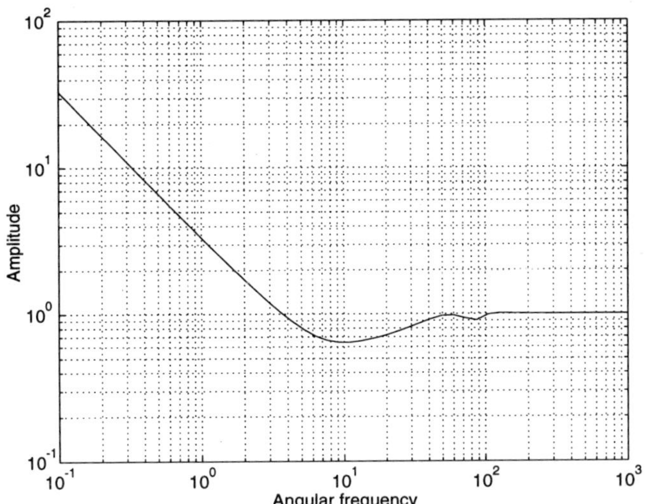


Figure 7: Reference position to control signal.

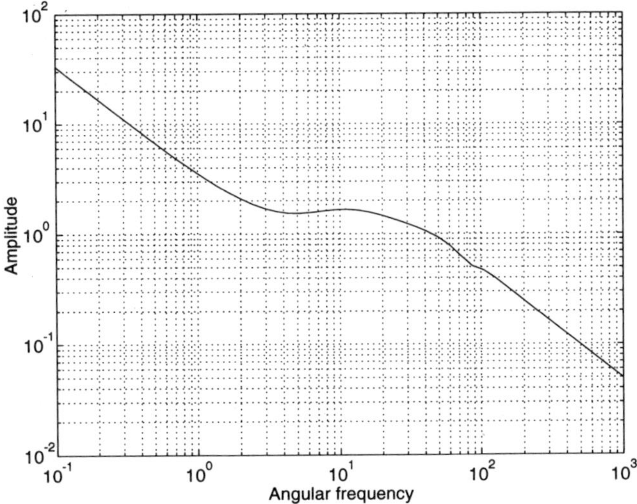


Figure 8: Piston position to control signal.

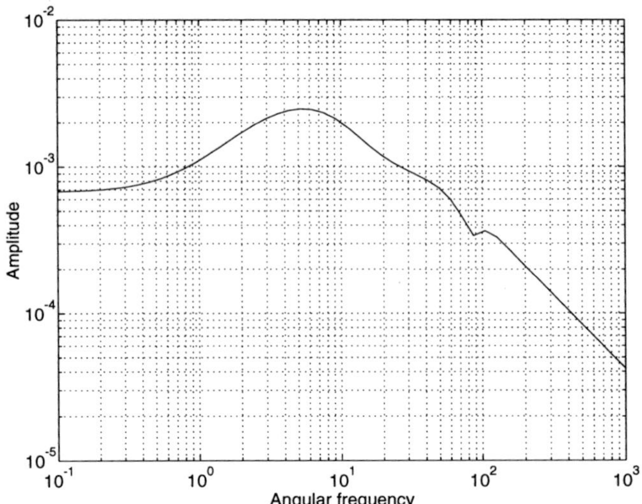


Figure 9: Load pressure to control signal.

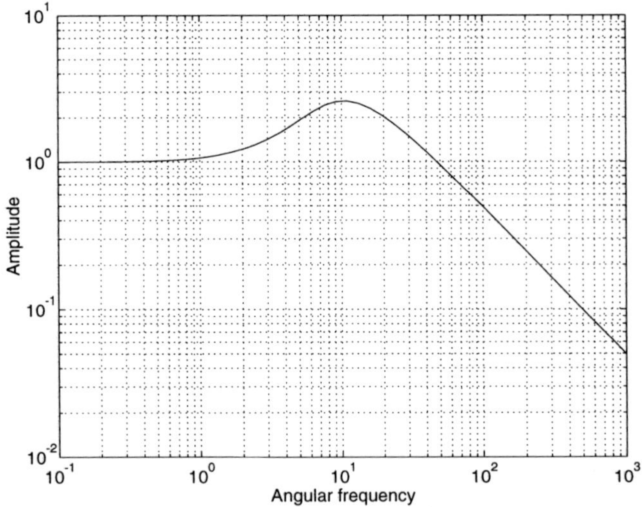


Figure 10: Piston position feedback in modified structure.

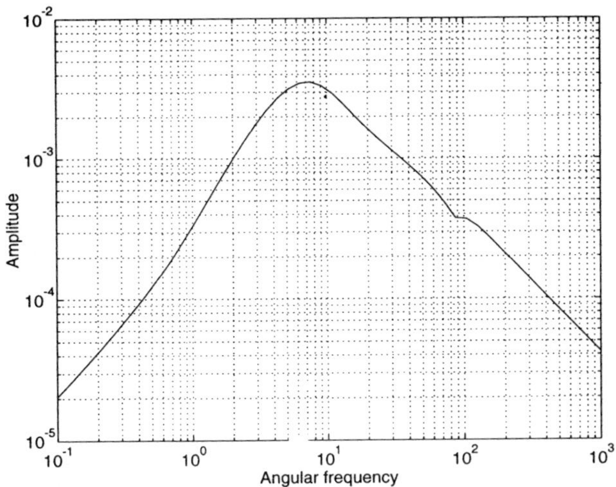


Figure 11: Load pressure feedback in modified structure.

A straightforward implementation of the regulator according to Equation (38) is, however, not possible, since we then have an integrator outside the feedback loop. An alternative implementation is to move the transfer function $F_{X_r}(s)$ into the feedback loop and compute the input signal according to

$$U(s) = F_{X_r}(s)[X_r(s) - \bar{F}_{X_P}(s)X_P(s) - \bar{F}_{P_{Le}}(s)P_{Le}(s)] \tag{42}$$

$$\bar{F}_{X_P}(s) = F_{X_P}(s)/F_{X_r}(s) \qquad \bar{F}_{P_{Le}}(s) = \bar{F}_{P_{Le}}(s)/F_{X_r}(s) \tag{43}$$

The reconfiguration results in the Bode diagrams of $\bar{F}_{X_P}(s)$ and $\bar{F}_{P_{Le}}(s)$ in Figures 10 and 11. The resulting control system has a fairly simple and intuitively appealing structure. The piston position feedback can be seen as a lead compensator in series with a first order low pass filter, while the load pressure feedback is a band pass filter with pass band concentrated at the resonance frequency of the structure. The pressure feedback also has zero static gain, which is a desirable property in order to avoid problems with the DC level in the pressure signal. The two feedback signals are then fed through $F_{X_r}(s)$ which mainly is a PI-regulator.

# 7 Experimental evaluation

Even though we concentrate on experimental evaluation of the LQG regulator it should be mentioned that implementation on the real crane has been carried out. As can be seen in the Bode diagrams in the appendix the transfer functions in the regulator have fairly simple structure despite the relatively large model order. We have therefore chosen to approximate the LQG regulator by a regulator of lower order that has the main properties of the LQG regulator. Results of such an implementation are given in [9] and [10].

# 8 Conclusions

We have in this report studied how linear quadratic state feedback, using estimated states, can be used to control an hydraulic actuator with a flexible mechanical load. The response to changes in the reference signal and disturbance signal shows a well damped behaviour. By introducing the disturbance as an extra state and integral action in the regulator we have also been able to design a control system that is able to eliminate the steady state error in load position when an external disturbance is applied.

# 9 Acknowledgements

This work was sponsored by the Center for Industrial Information Technology (CENIIT) at Linköping University.

# References

[1] W. Backé and P. Anders. "The influence of microelectronics on fluid power control". In *Prepr. 10th IFAC World Congress*, pages 267–277, Munich, German Federal Republic, 1987.

[2] P. Krus. *On Load Sensing Fluid Power Systems - With Special Reference to Dynamic Properties and Control Aspects.* PhD thesis, Linköping University, Linköping, Sweden, Oct. 1988.

[3] P Krus and J.O. Palmberg. "Damping of fluid power systems in machines with high inertia loads". In *Prepr. First JHPS International Symposium on fluid Power*, pages 9–16, Tokyo, Japan, 1989.

[4] S. Gunnarsson and P. Krus. "LQG control of an hydraulic actuator with a flexible mechanical load". Technical report, LiTH-ISY-1209, Department of Electrical Engineering, Linköping University, Linköping, Sweden, 1991.

[5] F. Conrad and C.J.D. Jensen. "Design of hydraulic force control systems with state estimate feedback". In *Proc. 10th IFAC World Congress*, pages 267–277, Munich, German Federal Republic, 1987.

[6] T.J.Viersma. *Analysis, synthesis and design of hydraulic servosystems and pipelines.* Elsevier Scientific Publishing Company, Amsterdam, The Netherlands, 1980.

[7] S. Gunnarsson and P. Krus. "Modelling of a flexible mechanical system containing hydraulic actuators". Technical report, LiTH-ISY-1078, Department of Electrical Engineering, Linköping University, Linköping, Sweden, 1990.

[8] B.Friedland. *Control system design. An introduction to state space methods.* McGraw-Hill, New York, 1986.

[9] S. Gunnarsson and P. Krus. "LQG control of a hydraulic actuator with a flexible mechanical load". In *12th IFAC World Congress*, 1993.

[10] P. Krus and S. Gunnarsson. "Adaptive control of a hydraulic crane using on-line identification". In *Prepr. The Third Scandinavian International Conference on Fluid Power*, volume 2, pages 363–388, Linköping, Sweden, 1993.

**WRITTEN DISCUSSION**
**LQG Control of flexible mechanical structure using hydraulic actuators**
**S Gunnarsson & P Krus (University of Linköping, Sweden**

**Question:** **ND Vaughan**
**Fluid Power Centre, Bath, UK**

1. Does the model shown in Figure 2 represent one axis or the complete crane?

2. Are you using differential pressure or just the load pressure as a feedback signal?

**Answer:**

1. It represents one axis of the crane.

2. The feedback signal used (effective load pressure) is defined as:

   $P_{Le} = P_1 - P_2 A_2 / A_1$

   However, when the algorithm was applied to a real crane it was found that it was sufficient to measure just the load pressure. $P_{Le} = P_1$. This can be justified since the pressure variations on the low pressure side of the piston are small due to the valve geometry.

**Question:** **A Bouhal**
**INSA de Lyon, France**

1. Integral action in the LQG controller decreases its robustness properties. What are your views on this?

2. Does the error positioning depend only on fluid compressibility? What about friction?

**Answer:**

1. We agree that the integral action may decrease the stability robustness, but we have not, however, encountered any substantial difficulties due to this. On the other hand, the integral action improves the performance properties with respect to small modelling errors.

2. The main cause of the error in positioning is deformations in the mechanical structure. Errors due to fluid compressibility and friction become negligible due to the integral action in the controller.

# Index